QP 75073
551
.T6813   Tonomura, Yuji

Muscle proteins, muscle contraction
and cation transport.

MUSCLE PROTEINS, MUSCLE CONTRACTION AND CATION TRANSPORT

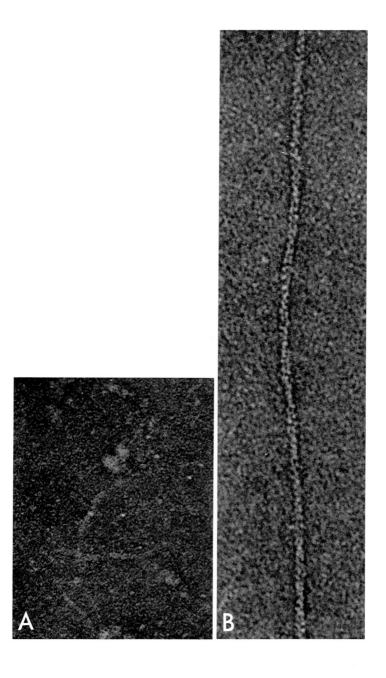

Frontispiece: (A) Electron micrograph of a negatively stained myosin molecule. ×370,000. Note that the myosin molecule has two ellipsoidal heads and a long tail with a hinge (see chapter 2). (B) Electron micrograph (under focus) of negatively stained F-actin. ×370,000. Note that F-actin consists of two mutually intertwined strands crossing at intervals of 350Å (see chapter 5). (Courtesy of Dr. K. Takahashi, Faculty of Agriculture, Hokkaido University.)

MUSCLE PROTEINS, MUSCLE CONTRACTION AND CATION TRANSPORT

by
YUJI TONOMURA
translated by
TSUNEICHI TAKESHITA

UNIVERSITY PARK PRESS
Baltimore • London • Tokyo

UNIVERSITY PARK PRESS
Baltimore·London·Tokyo
Library of Congress Cataloging in Publication Data

Tonomura, Yūji.
  Muscle proteins, muscle contraction, and cation transport.

  1. Muscle proteins. 2. Muscle contraction. 3. Adenosinetriphosphate.
4. Minerals in the body.
I. Title.
QP551.T6813   612'.741   72-10296
ISBN 0-8391-0739-0

© UNIVERSITY OF TOKYO PRESS, 1973
UTP 3045-67757-5149
Printed in Japan.

All rights reserved. No part of this publication may be reproduced or transmitted in any form or by any means, electronic or mechanical, including photocopy, recording, or any information storage and retrieval system, without permission in writing from the publisher.

Originally published by
UNIVERSITY OF TOKYO PRESS

PREFACE

In 1950 Dr. S. Watanabe, Dr. K. Yagi, Dr. H. Shiokawa and I at Hokkaido University began to study the mechanism of the interaction of actomyosin with ATP. Three years later, on the basis of our initial results, we proposed a mechanism of muscle contraction involving (i) a cyclic pathway for the reaction of actomyosin with ATP, (ii) activation of myosin by coupling with the splitting of ATP, leading to superprecipitation of actomyosin, and (iii) regulation by trace Ca^{2+}. Later work has successfully substantiated and elaborated our original theory. Our many co-workers in this endeavor have included Professor S. Kubo, Professor T. Yasui, Dr. S. Kitagawa, Dr. S. Tokura, Dr. F. Morita, and Dr. K. Takahashi of Hokkaido University; Dr. T. Kanazawa, Dr. K. Shibata-Sekiya, Dr. T. Mori, Dr. K. Imamura, Dr. M. Tada, Dr. N. Kinoshita, Dr. H. Nakamura, Dr. Y. Hayashi, and Dr. Y. Kozuki of Osaka University; Dr. J. Yoshimura of Kyoto University; Dr. H. Tokuyama of Kanazawa University; Dr. T. Shimada of Niigata University; and Dr. H. Matsumiya and Dr. T. Tokiwa of Nomura Institute of Technology and Economics.

Because of the complexity of muscle contraction and the gradual development of our ideas, our published work is fragmented among a variety of journals and has never been presented as a unified whole. Unfortunately, our studies have often been cited by other workers with misunderstanding of our interpretation of the experimental results and even of our results themselves. Therefore I have felt strongly that there is need for a comprehensive summary of our work in its entirety, and have been encouraged in this by Professors M.F. Morales and S. Watanabe of the San Francisco Medical Center. This monograph is therefore mainly a review of our own work and is not intended to present an objective survey of the entire field. However, I believe that this imbalance has been compensated by the inclusion of a general description of each topic at the beginning of the appropriate chapter. Furthermore, each chapter is provided with many references by others on the related subjects.

A second objective of this monograph is to compare the mechanisms of hydrolysis of ATP by three different ATPases—myosin-ATPase, the Ca^{2+}-Mg^{2+}-dependent ATPase of the sarcoplasmic reticulum, and the Na^+-K^+-dependent membrane ATPase. A knowledge of the mechanism of muscle contraction is useful for understanding not only the contractile process but also other energy transduction systems, but there has been insufficient biochemical study at the molecular level. Consequently Dr. T. Kanazawa, Dr. M. Saito and Dr. T. Yamamoto at Osaka University, Dr. K. Taniguchi at Hokkaido University, Mr. S. Yamada at Shizuoka University, and others in our laboratory, have carried out investigations to compare transport ATPase with myosin-ATPase. Our knowledge of the intermediates formed in enzyme reaction in general has advanced rapidly over the past few years. However, while this knowledge has been useful to elucidate the action mechanism of particular enzymes, it has not usually been helpful in understanding various physiological phenomena at a molecular level, although kinetic studies are now becoming an important source of such understanding, especially in the field of allosteric regulation. Investigation of the reaction intermediates formed with enzymes such as ATPases which are directly related to physiological functions is relevant to an understanding of these functions, and the comparison of several enzymes catalysing the same reaction is worthwhile in order to clarify the relationship between

PREFACE

enzymatic reaction and physiological function. In this monograph, the similarity between the reaction mechanisms of contractile and transport ATPases is emphasised in chapters 11 and 12, and the reaction mechanisms of G-factor-dependent ribosomal GTPase and ATP-synthesis in biological membranes are also discussed in the last chapter. Basing on our studies on the reaction mechanisms of various nucleoside triphosphatases, the general principles for the biological energy transduction are presented in the last two chapters. Thus the monograph presents a biochemical approach to the understanding of the molecular mechanisms of biological energy transductions in general, especially of muscle contraction and active transport of cation, and I hope that it will be useful to research workers and to advanced students in the field of bioenergetics.

I wrote the first draft of this monograph for lectures in Osaka, Sapporo, Tokyo, and Nagoya. In 1969 I rewrote it together with contributors, and added four new chapters—10, 11, 12 and 14. Finally, I revised the complete manuscript. I am indebted to Dr. Tsuneichi Takeshita of E. I. Du Pont de Nemours and Co., Inc., for the translation into English; to Dr. T. A. Cooper of the same company for correcting the English translation; to Mrs. A. Muraki of the Research Institute for Catalysis, Hokkaido University, for typing the first English draft; to Professor K. Maruyama of the Faculty of Science, Kyoto University, for reading the manuscript and making many useful comments; and to the various contributors whose names are shown at the beginning of the chapters concerned. I express my sincere appreciation to Mr. T. Miyazaki and the other members of the University of Tokyo Press for their skill, patience and unremitting courtesy in translating my manuscript into a form suitable for publication. We are grateful to the National Institutes of Health, U.S.A., the Muscular Dystrophy Associations of America, Inc., the Toyo Rayon Science Foundation and the Ministry of Education of Japan for the financial support that made this work possible. Finally, I thank my wife, Hiroko, for her patient co-operation in the many phases of the preparation of this monograph.

Osaka, May 1972   Yuji Tonomura

CONTENTS

Preface vii
Abbreviations xvii

INTRODUCTION 1

1 THE STRUCTURE AND FUNCTION OF MUSCLE CELLS
   1. The structure of muscle cells 6
   2. Physiology of muscle contraction 11
   3. History of the biochemistry of muscle contraction 18

2 THE STRUCTURE OF THE MYOSIN MOLECULE
   1. The preparation of myosin 28
   2. Methods for determination of molecular weight 29
   3. The molecular weight of myosin 31
   4. The molecular shape of myosin 34
   5. The subunit structure of the myosin molecule 34
   6. The submolecular structure of the myosin molecule ... 47
   7. The formation of myosin filaments 53
   8. Changes in secondary structure at the active sites of myosin 55

3 REACTION MECHANISM OF MYOSIN-ATPase
1. Energy changes accompanying the hydrolysis of ATP .. 69
2. The steady-state reaction of myosin with ATP 72
3. The Michaelis complex formed by ATPase 74
4. The effect of pH on myosin-ATPase 76
5. The pre-steady-state condition of the hydrolysis of ATP by myosin 81
6. The reaction of myosin-ATPase with ATP in the presence of high concentrations of Mg^{2+} 84
7. The properties of the reactive myosin-phosphate-ADP complex 91
8. The reaction of myosin-ATPase with ATP in the presence of low concentrations of Mg^{2+} 103

4 THE CHEMICAL STRUCTURES OF THE ACTIVE SITES OF MYOSIN
1. The chemical structure of the myosin molecule 117
2. The number of active centres in myosin 118
3. Chemical modification of myosin-ATPase 123
4. The effect of chemical modification on the ability of myosin to bind with actin 131
5. The effect of chemical modification on the formation of myosin filaments 132
6. The chemical structure of the active site 132

5 THE STRUCTURE AND FUNCTION OF ACTIN
1. G-Actin 142
2. The binding of G-actin with ATP 144
3. Chemical modification of actin 145
4. F-Actin 151
5. The G⇌F transformation of actin 153

6 THE FORMATION OF ACTOMYOSIN AND ITS DISSOCIATION BY ATP
1. Electron micrographs of actomyosin 164
2. The binding ratio of actin with myosin 164

3. The nature of the binding forces 170
   4. Dissociation of actomyosin by ATP 173
   5. Kinetic studies of the reaction of actomyosin with PP_i .. 177
   6. Kinetic studies of the dissociation of actomyosin with ATP 180
   7. A molecular model for the dissociation of actomyosin, acto-H-meromyosin and acto-S-1 complexes by ATP ... 186

7 THE SUPERPRECIPITATION OF ACTOMYOSIN WITH ATP, AND THE ACTOMYOSIN-ATPase SYSTEM
   1. Actomyosin-ATPase 192
   2. Mechanism of the reaction of F-actin with myosin-ATPase 200
   3. Mechanism of the reaction of acto-H-meromyosin with ATP 205
   4. Superprecipitation and the actomyosin-ATPase activity 208
   5. Role of ADP bound to F-actin in superprecipitation ... 211
   6. Molecular mechanism of superprecipitation of myosin B 213
   7. Effect of various factors on the reaction of actomyosin with ATP 219
   8. Comparative studies on the myosin-actin-ATP system .. 227

8 CONTRACTION OF MUSCLE MODELS BY ATP
   1. Glycerol-treated muscle fibres 238
   2. Contraction of isolated sarcomeres by ATP 242
   3. Thread models 244
   4. Relationship between development of tension, splitting of ATP and sarcomere length 244
   5. Kinetic and viscoelastic properties of cross-bridges 249

9 ATP ANALOGUES
   1. Synthesis of ATP analogues 259
   2. Decrease in light-scattering intensity of myosin B solution 260
   3. Hydrolysis by myosin 266
   4. Contraction of myofibrils 268

10 THE REGULATION OF MUSCLE CONTRACTION BY CALCIUM IONS AND PROTEIN FACTORS

1. The role of Ca^{2+} in the regulation of contraction and relaxation 274
2. Properties of tropomyosin 276
3. The discovery of the relaxing protein factor 278
4. Properties of the relaxing protein 281
5. Mechanism of action of the relaxing protein 284
6. Functions of the components of the relaxing protein ... 291
7. Myosin and the regulation of muscle contraction 301

11 THE Ca^{2+}-Mg^{2+}-DEPENDENT ATPase AND THE UPTAKE OF Ca^{2+} BY THE FRAGMENTED SARCOPLASMIC RETICULUM

1. Morphology and physiological functions of the sarcoplasmic reticulum 306
2. Ca^{2+}-uptake and the ATPase of the sarcoplasmic reticulum 309
3. The reaction mechanism and the phosphorylated intermediate of the Ca^{2+}-Mg^{2+}-dependent ATPase 312
4. The formation and decomposition of the phosphorylated intermediate 316
5. Elementary steps in cation transport 324
6. The change in affinity for calcium and magnesium ions 333
7. Regulation of the Ca^{2+}-Mg^{2+}-dependent ATPase 336
8. Remaining problems 344
9. Other Ca^{2+}-Mg^{2+}-dependent ATPases of membranes and regulation of the Mg^{2+}-dependent ATPase by Ca^{2+} ions and cyclic AMP 347

12 THE Na^+-K^+-DEPENDENT ATPase OF MEMBRANES

1. Active transport of Na^+ and K^+, and the ATPase 357
2. Phosphorylated intermediates of the membrane ATPase 361
3. The reaction mechanism of the membrane ATPase 362
4. The role of phospholipids in the transport ATPase 374
5. The transport ATPase and myosin-ATPase 376

13 THE MOLECULAR MECHANISM OF MUSCLE CONTRACTION AND THE ACTIVE TRANSPORT OF CATIONS

1. Theories of the molecular mechanism of muscle contraction 382
2. Experimental data relevant to the molecular mechanism of contraction 385
3. The molecular mechanism of muscle contraction 389
4. The molecular mechanism of active transport 394
5. Evolution of high-energy phosphate compounds 402

14 CONTRACTION, TRANSLOCATION, TRANSPORT AND PHOSPHORYLATION

1. The G-factor-linked ribosomal GTPase 409
2. The intermediates of the ribosomal GTPase reaction .. 411
3. Phosphorylation of transport ATPase coupled with cation translocation 419
4. Oxidative phosphorylation 428

ABBREVIATIONS

ADP	adenosine diphosphate
AMP	adenosine monophosphate
ATP	adenosine triphosphate
ATPase	adenosine triphosphatase
CM-	carboxyaminomethyl
CP	carrier protein
CrP	phosphocreatine
Cyclic AMP	cyclic 3′,5′-adenosine monophosphate
DDPM	N-(4-dimethyl-amino-3,5-dinitrophenyl) maleimide
DEAE-	diethylaminoethyl
DGM	dithiodiglycolic acid dimethyl ester
DHT	diazonium-1 H-tetrazole
DNP	2,4-dinitrophenol
DOC	deoxycholate
DP	diphosphate
DTT	dithiothreitol
EDTA	ethylenediamine tetraacetate
EGTA	ethyleneglycol bis (β-aminoethylether)-N, N′-tetraacetate
EP	phosphorylated intermediate

EPR	electron paramagnetic resonance
ES	enzyme-substrate complex
G	G-factor
GDP	guanosine diphosphate
GTP	guanosine triphosphate
H-MM	heavy meromyosin
I	ionic strength
IAA	monoiodoacetamide
IAA*	N-2,2,6,6 tetramethyl piperidine nitroxide iodoacetamide
IDP	inosine diphosphate
ITP	inosine triphosphate
K_{ADP}	dissociation constant of enzyme-ADP complex
K_f	Michaelis constant for formation of enzyme-phosphate complex or phosphoryl enzyme
K_m	Michaelis constant
L-MM	light meromyosin
LPC	lysophosphatidyl choline
LPE	lysophosphatidyl ethanolamine
LPS	lysophosphatidyl serine
MAL*	N-2,2,6,6 tetramethyl piperidine nitroxide maleimide
$\bar{M}_n$	number-average molecular weight
$\bar{M}_w$	weight-average molecular weight
MW	molecular weight
NEM	N-ethylmaleimide
NTP	p-nitrothiophenol
NTP-	p-nitrothiophenyl
PC	phosphatidyl choline
PCMB	p-chloromercuribenzoate
PE	phosphatidyl ethanolamine
PEP	phosphoenol pyruvate
PI	phosphatidyl inositol
P_i	inorganic phosphate
PL	phospholipid
PP_i	inorganic pyrophosphate
PS	phosphatidyl serine

ABBREVIATIONS

Pyr	pyruvate
R	ribosome
S-l	subfragment-1 of myosin
SDS	sodium dodecyl sulfate
s. l.	sarcomere length
Sph	sphingomyelin
SR	sarcoplasmic reticulum
TBS	trinitrobenzene sulphonate
TCA	trichloroacetic acid
TL	translocase
TNP-	trinitrophenyl
TP	triphosphate
T-system	transverse system
V_{max}	maximum velocity

INTRODUCTION

Molecular biology is becoming a focal point of modern biology whereby the phenomena of life can be understood in terms of the conformations of, and the information contained in, the molecules composing living organisms. The branch of molecular biology which has achieved the most spectacular advances in recent years is undoubtedly molecular genetics. At the same time molecular physiology, one of the remaining major areas, is making steady progress. However, there has been insufficient advance in our knowledge of molecular mechanisms of energy transformation in living organisms, a central theme of modern molecular physiology.

Characteristic of all animals is their ability to move. The study of the function of muscle cells, which are highly developed to perform this movement, and of the molecular mechanism of 'contraction,' is important in advancing our understanding of biological energy transformation. Muscle contraction is one of the oldest areas of energy transformation studied, since it can be easily observed and the accompanying change is large. For example, the indirect wing muscles of some Diptera and Hymenoptera convert energy at a rate similar to the maximum continuous output of an aircraft piston engine[1]. Equally remarkable is the control exerted over this process, such that the fastest muscle can attain its maximum tension from rest within several milliseconds. During this time, metabolic activity increases several hundred- to several thousand-fold. Thus, muscle is a very interesting subject for the study of energy transformation and its regulation, and consequently a great deal of work has been done, although, in the author's opinion, little significant information about the molecular mechanism had been obtained until 1940. Lately, studies in this area have made tremendous progress because of recent methodological advances such as the isolation and purification of the constituents of living cells, biophysical studies of biopolymers and especially of the molecular properties of proteins, elucidation of the fine and ultrafine structure of the living cell and the use of isotopes. The important discovery that muscle protein, 'myosin,' is active as an ATPase was made by Engelhardt and Ljubimova[2] in 1939. This was followed by Szent-Györgyi's[3] and Weber's [4] detailed studies on the interaction of the structural protein of muscle with ATP, which resulted in the current theory that muscle contraction occurs basically through the interaction of three components—actin, myosin, and ATP. Based on their work on the fine structure of muscle, Hanson and H. E. Huxley[5] and A. F. Huxley and Niedergerke[5a] proposed the important 'sliding theory' of contraction which describes muscle contraction as the sliding of an actin filament and a myosin filament past one another, and this theory has been widely accepted. However, the relationship of the several elementary reactions occurring in the actin-myosin-ATP system to the various macroscopic phases observed in muscle contraction is still uncertain. Recently, the system regulating these basic reactions has attracted attention[6] but even here the elucidation of the molecular mechanism has only just begun.

This monograph summarises our work, carried out since 1950, on the properties of muscle proteins and is directed towards understanding muscle contraction at the molecular level. Our main objective is to present a systematic picture of the role of the various elementary reactions in the phases of contraction. We also hope to demonstrate the significance of our research results for the mechanism of energy transformation in general. The functions of the Ca^{2+}-Mg^{2+}-dependent ATPase[7] of the sarcoplasmic reticulum and of the Na^+-K^+-dependent ATPase[8,9] of membranes are described, and their reaction mechanisms are compared in order to point out common elements in the workings of ATPases in general. Furthermore, the reaction mechanisms of G-factor-dependent ribosomal GTPase and ATP-synthesis in biological membranes are also discussed. Basing on the reaction mechanisms of various nucleoside triphosphatases, the general principles of biological energy transduction are emphasised.

The monograph is not an overall review of the physiology and biochemistry of muscle contraction and active transport of cations but is directed towards the objectives outlined above, with main emphasis on our own work. The range of topics discussed is biased accordingly. It is hoped that the interested reader will refer to Refs. *3–6, 10–13b* for a general description of the physiology and biochemistry of muscle, and to Refs. *7, 8, 14–19* for a general description of properties of transport of molecules.

REFERENCES

1 T. Weiss-Fogh, *Proc. Roy. Soc.*, *B***237**, 1 (1952).
2 W. A. Engelhardt and M. N. Ljubimova, *Nature*, **144**, 668 (1939).
3 A. Szent-Györgyi, "Chemistry of Muscular Contraction," 1st ed. (1947) and 2nd ed. (1951), Academic Press, New York.
4 H. H. Weber and H. Portzehl, *Progr. Biophys. Biophys. Chem.*, **4**, 60 (1954).
5 J. Hanson and H. E. Huxley, *Symp. Soc. Exp. Biol.*, **9**, 228 (1955).
5a A. F. Huxley and R. Niedergerke, *Nature*, **173**, 971 (1954).
6 S. Ebashi and M. Endo, *Progr. Biophys. Mol. Biol.*, **18**, 123 (1968).
7 W. Hasselbach, *Progr. Biophys. Mol. Biol.*, **14**, 167 (1964).
8 R. Whittam, *in* "The Neurosciences," ed. by G. C. Quarton, T. Melnechak and F. O. Schmitt, Rockefeller Univ. Press, New York, p. 313 (1967).
9 J. C. Skou, *Biochim. Biophys. Acta*, **42**, 6 (1960).

10 "Physiology of Volumentary Muscle," *Brit. Med. Bull.*, **12**, No. 3 (1956).
11 H. E. Huxley, *in* "The Cell," ed. by J. Brachet and A. E. Mirskey, Academic Press, New York, Vol. 4, 365 (1960).
12 G. H. Bourne, "The Structure and Function of Muscle," Academic Press, New York, Vols. 1 & 2 (1960).
13 *Symp. Soc. Exp. Biol.*, Cambridge Univ. Press, Cambridge, No. **22** (1968).
13a R. J. Podolsky, "Contractility of Muscle Cells and Related Processes," Prentice-Hall, Engelwood Cliffs, N.J. (1971).
13b D. M. Needham, "Machina Carnis: The Biochemistry of Muscular Contraction in Its Historical Development," Cambridge Univ. Press, Cambridge (1971).
14 A. Kleinzeller and A. Kotyk, "Membrane Transport and Metabolism," Publishing House of the Czechoslovak Academy of Sciences, Praha (1961).
15 H. N. Christensen, "Biological Transport," W. A. Benjamin, New York (1962).
16 W. D. Stein, "The Movement of Molecules across Cell Membranes," Academic Press, New York (1967).
17 E. E. Bittar, "Membranes and Ion Transport," Wiley-Interscience, London, Vols. 1 & 2 (1970).
18 A. Kotyk and K. Janáček, "Cell Membrane Transport, Principles and Techniques," Plenum Press, London (1970).
19 A. B. Hope, "Ion Transport and Membranes, A Biophysical Outline," Butterworths, London (1971).

1
THE STRUCTURE AND FUNCTION OF MUSCLE CELLS*

This introductory chapter is intended to describe the important background work on the structure and function of muscle cells, and particularly on contractile protein. This is prerequisite to the understanding of the research described in this monograph, the main objective of which is to elucidate the mechanism of muscle contraction at a molecular level by analysis of the interactions between muscle proteins, the reactions of these proteins with ATP, and the workings of the associated regulatory system.

There are currently three approaches used to investigate the contraction of muscle cells. The first is the morphological study of muscle and the electron microscopy of its structure; the second is the physiological study of the sequence of muscle excitation, contraction and relaxation, and the third is the biochemical study of energy metabolism in muscle, contractile protein, and the associated regulatory system. One could say that modern

* Contributor: Naokazu Kinoshita

developments in muscle contraction have involved a magnificent synthesis of the results from these three lines of attack.

1. The Structure of Muscle Cells[1]

A specialised tissue called muscle has evolved in all animals with a tissue level of organization, and its contraction is the source of their motility. Depending on its function and location, in vertebrates there is a further specialisation into skeletal, cardiac, and smooth muscle. This review deals only with skeletal muscle, particularly that of rabbit, since this has been most extensively studied and was used in our own research.

A muscle cell—commonly called a muscle fibre because it usually takes the form of a long fibre—contains not only a contractile structure but also a structure to regulate contraction and a structure to supply energy, and it thus forms an ordered unit which carries out a harmonized physiological function. Muscle fibres vary in size from 10 to 100μ in diameter and from 20μ to several centimetres in length. They may contain several hundred nuclei.

The muscle cell is surrounded by a membrane called the sarcolemma, which corresponds to the plasma membrane found in all other cells. The interior of the muscle cell is divided into the contractile structure called myofibrils, and sarcoplasm, which usually contains many nuclei, mitochondria, fat droplets, glycogen and other components. In addition, the muscle cell has a membranous structure, the sarcoplasmic reticulum, surrounding the myofibrils, which was first observed towards the end of the 1800's by Retzius[2,3] using an optical microscope. Sjöstrand[4], Anderson[5], Porter and Palade[6] and Anderson-Cedergren[7] later studied it in more detail by electron microscopy. The structure and physiological function of this membrane are described in chapter 11.

As early as 1840 Bowman[8] described the striated appearance of skeletal muscle. The contractile structure occupies the major portion of the muscle cell, and each muscle fibre enclosed in its sarcolemma is composed of a large number of myofibrils $1-2\mu$ in diameter lying parallel to the fibre axis. Bowman showed that each myofibril has a striated pattern which forms part of the overall striated pattern of the entire fibre. The striation was not caused by differences in light absorption but was suggested to be due to differences in refractive index. Later Dobie[9] described a line of high refractive index (now called the Z line), bisecting a broad band of low refractive index, and Brücke[10] found a broad birefringent band, also of high refractive index. The isotropic band of low refractive index is called the I band and the optically anisotropic region of high refractive index is called the A band. The centre of the A band has a lower refractive index than the edges and is called the H-zone after Hensen[11,12], who first described it, although in a somewhat unsatisfactory manner. The portion between two Z lines, the sarcomere, is the unit of contractile structure of striated muscle. It is about 2.2μ long in resting skeletal muscle of rabbit.

For about 50 years it was not known whether the change on muscle contraction occurred mainly in the A or the I band. However, in the 1950's this change on contraction of living muscle and isolated myofibrils was investigated in detail, using interference and phase-contrast microscopy to observe the differences in refractive indices. A. F. Huxley and Niedergerke[13] studied both extended and contracted frog muscle fibres by interference microscopy and found that only the length of the I band changed, while that of the A band did not. New bands called 'contraction bands' appear around the Z line only when contraction has occurred to such an extent that the adjacent A bands are in contact. Hanson and H. E. Huxley[14,15] investigated isolated myofibrils by phase-contrast microscopy and found that upon addition of ATP a 'contraction band' developed, and that the length of the A band remained constant when the myofibrils were extended or contracted in the presence of ATP. Furthermore the length of the H-zone changed by the same amount as the alteration in the length of the sarcomere. Based on these results, it was predicted that, since the band pattern was due to differences in refractive index, *i.e.* content of protein, there must be two kinds of filaments in the myofibrils, both having a fixed length. Contraction must therefore occur by means of a change in their internal position through sliding of the filaments. H. E. Huxley's comparative study[16], by low-angle X-ray diffraction, of resting muscle in the presence of ATP and of muscle in rigor after removal of all ATP, also suggested that the contractile materials in striated muscle comprise two sets of filaments and showed for the first time that the lattice of the filaments of the A and I bands maintains a constant volume during contraction. As will be described in detail in chapter 13, Elliott *et al.*[17,18] and H. E. Huxley and Brown[19] measured the periodicity of X-ray patterns of two pairs of filaments from frog sartorius muscle at several sarcomere lengths. These period lengths did not change from the active state to the resting state.

The study of the fine structure of muscle by electron microscopy has advanced greatly because of progress in the technique of ultra-thin section-

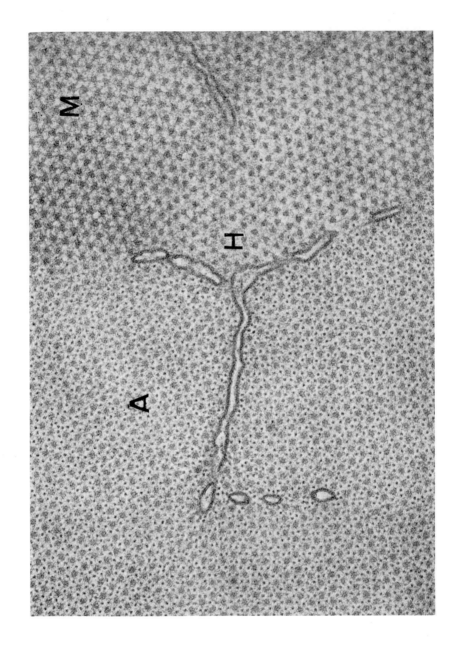

ing. Morgan et al.[20] and Hodge et al.[21] found that the filaments are arranged in a hexagonal manner within the myofibril. H. E. Huxley[22] found from cross-sections taken at different points along the sarcomere that the A band contains thick filaments of about 100Å diameter arranged hexagonally, with six thin filaments of about 50Å diameter surrounding each thick filament. In contrast, in the I band only thin filaments were observed, while the H-zone contained no thin filaments but only thick ones (Fig. 1). Therefore, it was concluded that the thick filaments are found only in the A band while the thin filaments extend from the Z line to the edge of the H-zone, there being a regular hexagonal arrangement of the two types of filament. These discoveries strengthened the hypothesis of Hanson and H. E. Huxley[15], and A. F. Huxley and Niedergerke[13] that contraction occurs by means of the two kinds of filament in the muscle sliding past each other.

Hanson and H. E. Huxley[23,24], and Hasselbach[25] treated muscle with a high ionic strength salt solution containing ATP or pyrophosphate (PP_i) which selectively extracts myosin. This resulted in the disappearance of the high refractive index and the birefringence from the A band. Muscles treated this way contained no thick filaments; only the thin filaments remained and extended from the Z line to the edge of the H-zone. Further treatment of this muscle with KI solution to extract actin removed the remaining thin filaments, leaving only the Z line. Later H. E. Huxley and Hanson[26,27] showed by interference microscopy and biochemical studies that at least four fifths of the myosin is contained in the A band, while Perry and Corsi[28] observed that treatment of myofibrils by a method which dissolves the I band completely did not extract myosin but removed tropomyosin (see chapter 10, section 2) and a modified product of actin. Thus it is almost certain that myosin is present in the thick filaments of the A band, whereas actin is present in the thin filaments of the A and I bands. These results, and the relationship between the sarcomere length and tension to be described in chapter 8, led to the acceptance of the 'sliding theory' of muscle contraction, in which the thick filaments of

←Fig. 1. Cross-section of frog sartorius muscle. The myofibrils are surrounded by the sarcoplasmic reticulum, and each myofibril is sectioned at a slightly different level so that the arrangements of the thin and thick filaments are different. In the A band (A), the thick filaments are arranged hexagonally. Each thin filament is surrounded by a triangle of three thick filaments, while each thick filament is surrounded by six thin filaments. In the H-zone (H), only thick filaments are observed whereas at the M line (M), an M-bridge is observable connecting each thick filament. × 78,000. (Courtesy of Professor K. Hama, Institute of Medical Science, University of Tokyo.)

myosin and the thin filaments of actin alter only their mutual positions but not their lengths. This is one of the most important concepts in the mechanism of muscle contraction.

H. E. Huxley[29,30] presented a model of the thick filaments based on electron microscopy of longitudinally sliced muscle and of the aggregates of myosin at low ionic strength. The model indicates that the thick filaments are a polymer of myosin with a helical axis of symmetry and are composed of a trunk portion with projections. The trunk is formed by the aggregation of the rod-shaped L-meromyosin, and the projections, which protrude from the trunk in a helical arrangement, consist of H-meromyosin (see chapter 2 for the two portions of the myosin molecule called L- and H-meromyosin). Two halves of the filament must have opposite polarity because of the tail-to-tail association at the centre. The projections of H-meromyosin from the thick filaments are thought to form bridges touching the thin filaments, as has been observed in longitudinal sections of muscle.

H. E. Huxley and Brown[19,31] extended their X-ray diffraction studies of live striated vertebrate muscle to elucidate the structure of the thick filaments. The projections on the thick filaments were found to be arranged on a 6/2 helix. At a given level, two bridges projected out directly opposite each other from the backbone of the thick filament. The next two bridges occurred 143Å further along the filament and were rotated relative to the first pair by 122°. The structure as a whole was repeated at intervals of 3×143, or 429Å. A significant feature of the cross-bridge pattern was the fact that, although the pattern was strongly developed at low angles (at spacings greater than about 50Å), the reflections, especially the off-meridional ones, faded out very quickly at high angles. This showed that on any given thick filament there must be a considerable amount of disorder in the helical arrangement of the bridges.

Hanson and Lowy[32] observed an isolated thin filament under the electron microscope and found that the structure could not be distinguished from the fibrous (F)-actin formed as a double-stranded helix by the polymerisation of globular (G)-actin, as described in chapter 5. However, as shown in chapter 10, the thin filament is not just F-actin but also probably contains other proteins, namely, tropomyosin and troponin. H. E. Huxley and Brown[19,31] elucidated the fine structure of the thin filaments by X-ray diffraction. The X-ray reflections from the actin filaments showed that G-actin units are arranged on a nonintegral helix with subunits repeating at intervals of 54.6Å along either of two chains which are staggered relative to each other by half a subunit period (27.3Å), and which twist around each other with crossover points 360 to 370Å apart, so that the pitch of the helix formed by either of the two chains is 720 to 740Å. Again, the actin reflections showed something of the same disorder that characterises the reflections from the myosin filaments. In this case, prominent meridional reflections out to 6 Å or less showed good ordering in a purely axial sense, but off-meridional reflections at higher angles were extremely weak, indicating relatively poor helical ordering of the subunits. This suggested that the helix may be able to twist and untwist to some extent, but that the helical repetition of the subunits remains rather constant.

H. E. Huxley[30] isolated the thin filament bundle attached to the Z line and added to it H-meromyosin, which contains the site at which myosin binds to actin. Electron microscopic studies of this complex filament, together with the results from the electron microscopy of the thick filaments, showed that both filaments have the polarity required from the 'sliding theory.' These studies, done mainly by H. E. Huxley and Hanson have further contributed to the widespread acceptance of the sliding theory of contraction. The structural basis of muscle contraction has been well reviewed recently by H. E. Huxley[32a].

2. Physiology of Muscle Contraction

The contraction of a skeletal muscle can be conceptually divided into three stages: excitation of the muscle plasma membrane, contraction of the myofibril, and relaxation. The sequence is very rapid, taking at most one tenth of a second in the case of one stimulus of a frog skeletal muscle.

The muscle plasma membrane or sarcolemma plays an important role in muscle excitation[33]. In skeletal muscle, each sarcolemma is normally connected to the motor nerve end plates at 1 to 3 points along its length. Transmission of an impulse through the motor nerve is thought to give rise to the secretion of acetylcholine, which reacts with the plasma membrane surface near the nerve end plate to generate an end plate potential, which then causes production of the action potential. This action potential is then propagated very rapidly throughout the muscle plasma membrane by a mechanism similar to that of nerve fibre excitation[34]. However, depolarization of the plasma membrane does not itself directly cause the contraction of the myofibril.

Kamada and Kinoshita[35], Heilbrunn and Wiercinski[36,37] and Sandow[38] had already found in the 1940's that Ca^{2+} ions directly induce con-

traction of the myofibrils. But, as pointed out by Hill[39], if Ca^{2+} ions or some other substances which are combined with the plasma membrane were released on its depolarisation and transported to the myofibrils by diffusion, this would not explain the fact that contraction even of the myofibrils located in the middle of the muscle fibre occurs within 10 msec of depolarisation. Therefore there must be some special mechanism for transmitting the excitation from the membrane to the interior of the contractile protein. Investigation of this mechanism was initiated by the electron microscopy of the detailed membrane structure within the muscle cell by Sjöstrand[4], Anderson[5] and in particular by Porter and his colleagues[6,40,41]. Their results (Fig. 2) showed that there is a tubular structure—the transverse (T) system—which extends from the exterior membrane into the interior of the cell, where it joins with the sarcoplasmic reticulum (SR) which surrounds the myofibrils. The contact of the T-system with the SR forms a structure called the triad. The excitation occurring in the plasma membrane is transmitted to the SR through the T-system, and this is thought to release the Ca^{2+} ions present in the SR, thus causing the contraction of the myofibril. This mechanism is supported by A. F. Huxley and Taylor's[42,43] experiments in which a local contraction of significant magnitude of a muscle fibre occurred only when the plasma membrane surface was locally depolarised by a microelectrode exactly at the position of the opening of the T-system. Fujino and his co-workers[44,45] observed that the return of a muscle to an isotonic Ringer solution, following treatment of the muscle with solutions made hypertonic with glycerol, induced an irreversible loss of the twitch which was not accompanied by any effect on the resting or action potential. Electron microscopic and physiological studies made by Howell[46] revealed that when the muscle fibre is immersed in 0.4M glycerol-Ringer solution, which destroys only the T-system, no contraction is caused by excitation of the plasma membrane. However, contraction does occur when Ca^{2+} ions are released from the sarcoplasmic reticulum by the addition of caffeine (see chapter 11, section 1). Podolsky[47], using a Natori skinned muscle from which the sarcolemma had been removed, also showed that the addition of a minute amount of Ca^{2+} ions undoubtedly caused the contraction of myofibrils.

←Fig 2. Sarcoplasmic reticulum of frog sartorius muscle. The left-hand photograph shows the surface view of the reticulum. The transverse system (T) and particularly the terminal sack (Tc) of the reticulum are clearly observed. The right-hand photograph shows a section of the reticulum. At the position of the Z line (Z), the structure called the triads (Tr) are seen. M indicates the M line. ×36,000. (Courtesy of Professor K. Hama, Institute of Medical Science, University of Tokyo.)

Bozler[48] and Watanabe[49] prepared glycerinated muscle fibres (see chapter 8) by destroying the function of the membrane with glycerol and water and added to them a small amount of EDTA. In the presence of Mg-ATP the fibres relaxed, but the subsequent addition of a small amount of Ca^{2+} caused contraction. Ebashi[50] found a correlation between the ability of various metal-chelating agents to form complexes with Ca^{2+} and the ability to effect relaxation, and he thus concluded that relaxation is induced by removal of small amounts of Ca^{2+}. Weber[51] and Ebashi[52] also showed that a trace amount of Ca^{2+} is necessary for the high ATPase activity of actomyosin, and for superprecipitation (see chapter 7).

In 1951 Marsh[53] isolated from living muscle a component—the Marsh factor—causing relaxation, in the presence of ATP, of actomyosin gel, a kind of muscle model (see chapter 8), but the composition of this component was not then clear. However, in the 1960's the important work of Ebashi and Lipmann[54] and Portzehl[55] suggested that this component originates in the microsomes of muscle homogenates, probably in the sarcoplasmic reticulum. Ebashi[52] and Hasselbach and Makinose[56] demonstrated a strong uptake of Ca^{2+} ions by the sarcoplasmic reticulum in the presence of Mg^{2+} and ATP, the mechanism of which will be described in detail later (see chapter 11). The foregoing experimental results have led to the general acceptance of the Ca^{2+} theory. This postulates that the sarcoplasmic reticulum releases Ca^{2+} ions which react with the contractile protein to cause muscle contraction, while the reverse withdrawal of Ca^{2+} back into the sarcoplasmic reticulum causes relaxation[57].

Recently Ebashi[58] and Szent-Györgyi and Kaminer[59] discovered a protein factor which plays an important role in the regulation of muscle contraction by Ca^{2+}. This factor is called native tropomyosin, or metin, and it is contained in normal specimens of actin. Ebashi et al.[60] showed it to be a complex of another protein, troponin, with tropomyosin (which had been isolated by Bailey[61] as one of the structural proteins of muscle). Addition of native tropomyosin to systems containing actin and myosin but no native tropomyosin showed that it is this protein factor which is the source of the sensitivity of superprecipitation and of ATPase activity to Ca^{2+} ions. Detailed investigation of this substance is now in progress, and chapter 10 describes its regulation of contraction and relaxation in systems containing actin, myosin and ATP.

The dynamic and thermodynamic aspects[62-68] of muscle contraction will now be briefly summarised. These topics do not seem to be directly related to the subjects discussed in the remaining chapters, but biochemical descriptions of muscle must, in the final analysis, be able to explain the results of physiological research. At present, only imprecise analyses of the chemical compounds contained in muscle before and after contraction are possible, and only overall changes are observable. In contrast, the heat liberated by contraction can be measured very precisely and rapidly. Heat generation is, of course, chemically unspecific, but it can provide a framework to determine what sort of chemical changes are taking place. Therefore, the measurement of heat generated by muscle contraction has attracted the interest of many physiologists, and it was first performed on a separated frog leg muscle by Helmholtz[69] in 1848, using a thermoelectric method. These measurements have been continually improved in A. V. Hill's [70-72] laboratory since 1911, with excellent results. He has shown that the heat generated by contraction is of two kinds, the initial heat and the recovery heat. The initial heat is liberated independently of oxygen and is observed simultaneously with contraction. The recovery heat is generated slowly after completion of the contraction and relaxation and is not observed in the absence of oxygen or when the muscle has been treated with monoiodoacetate. Hence the recovery heat is related to the regeneration of ATP and phosphocreatine (CrP), while the initial heat is directly associated with contraction. An important point to note is the absence of heat generation during relaxation[73].

The energy, E, needed to produce a twitch in a loaded muscle is related to the work, W, done by the muscle, and to the initial heat, Q; namely, $E = W + Q$. The energetic efficiency of muscle contraction can be approximately expressed as W/E, and its maximum is 0.45—a much higher efficiency than that obtained from a heat engine[74]. From detailed heat measurements, Hill concluded that Q consists of two components, $A + \alpha x$, where A is the heat of activation, x is the shortening in length of the muscle, and αx is the heat of shortening. The heat of activation, A, is related to the conversion of the muscle from the resting state to the active state and is independent of the degree of shortening. Quite recently, Mommaerts and co-workers[74a] measured the heat of activation A separately from $W + \alpha x$ using frog semitendinosus muscle stretched to a length at which tension was eliminated (cf. chapter 8, section 4). At 0°C, A is elaborated in two phases, a very rapid one and a smaller and slower one. The rapid phase is not temperature-dependent, and is regarded as the enthalpy effect of the liberation of Ca^{2+} ions and their interaction with the regulatory proteins; the slow phase has a high temperature coefficient, and is ascribed to the uptake of Ca^{2+} ions by the sarcoplasmic reticulum.

If the velocity of shortening is v, the rate of energy consumption accompanying the shortening is given by

$$\frac{d(P+\alpha)x}{dt}=(P+\alpha)v,$$

where P is the tension of the muscle. On the other hand, the rate of energy consumption is proportional to (P_0-P), where P_0 is the maximum tension at a given muscle length. Hence,

$$(P+\alpha)v=\beta(P_0-P), \qquad (1)$$

where β is a constant with the dimension of velocity. Further, the relationship between the tension and the velocity of shortening during an isotonic contraction was shown by Hill[75] to be expressed by

$$(P+a)v=b(P_0-P), \qquad (2)$$

where a and b are constants with the dimensions of force and velocity respectively. Comparison of Eq. (1), based on heat measurements, with Eq. (2), derived from dynamic measurements, leads to the identities $\alpha=a$, $\beta=b$. However, a very important correction was made recently by Hill himself[75]. A detailed measurement of the heat of shortening at 0°C revealed that α is not a constant but depends on P as follows:

$$\alpha=0.16P_0+0.18P, \qquad (3)$$

so that α is not identical with a. Despite this, according to Hill the original principle that the rate of energy consumption $(P+\alpha)v$ is proportional to (P_0-P) need not be discarded, since better agreement with the experimental data can be obtained by adding a term which decreases in proportion to the velocity of shortening, as in

$$-\frac{dE}{dt}=(P+\alpha)v=\beta(P_0-P)-\gamma P_0 v, \qquad (4)$$

where P_0 in the last term serves to correct the dimensionality of v, and γ is a pure number. Substitution of Eq. (3) and comparison to Eq. (2) gives $\gamma=0.135$, and hence $\frac{a}{P_0}=0.25$, and $b=\frac{\beta}{1.18}$. The former value is in good agreement with the experimental results obtained by many researchers. Therefore, the idea that a and α are identical has been abandoned, but since a and α are always of the same order of magnitude, this denotes some relationship between them.

It has long been desirable to substantiate by metabolic studies Hill's conclusions about the energy consumption of muscle, which he derived from his dynamic and thermal measurements, as shown above. Recently Carlson et al.[76] and Wilkie[77] measured the decrease in phosphocreatine (CrP), an apparent energy source (see the next section), simultaneously with the work done, W, and the heat produced, Q, during contraction. These studies led to discovery of the relationship $E = K_1 + K_2 W$, and comparison of this equation with that of Hill showed that in the decrease of phosphocreatine, E, there is no term corresponding to the heat of shortening. Mommaerts et al.[78] also obtained similar results. Thus, shortening heat is not derived from CrP (or ATP) splitting, and the nature of shortening heat remains a large question in energetics of muscle contraction. In previous papers by Wilkie[77] and others, it was reported that the relation between enthalpy change $(-(Q+W))$ and ΔCrP was always the same in a variety of different types of tetanus, and corresponded to an *in vivo* molar enthalpy change of hydrolysis of CrP of -11.0 ± 0.2 kcal/mole. However the value of -11.0 kcal/mole is considerably higher than the most probable value of enthalpy change for CrP-splitting *in vivo*, -9 kcal/mole[68], and quite recently Gilbert et al.[78a] have clearly shown that the heat production in very short tetanic contraction cannot be accounted for by splitting of CrP or ATP. They suggested binding of Ca^{2+} ions to proteins and conformational changes in the proteins (see chapter 10, section 5) as the source of this heat.

The effect of stretching during a maintained contraction gives important information about the relationship between energy changes and chemical changes during the active state. Hill and his colleagues[79-82] found that when a muscle is stretched during a maintained contraction the whole of the work done in stretching it may apparently disappear, *i.e.* it is not accounted for either as heat or as elastic potential energy. They stated that the experiments left little doubt that the chemical reactions which normally occur during contraction can be reversed by stretching, as a result of the mechanical work supplied. Since the breakdown of ATP seemed to be the immediate energy source for muscle contraction, Davies et al.[83] repeated some of Hill's experiments with muscle pretreated with fluorodinitrobenzene, which inhibits creatine kinase, to see if ATP was resynthesised during stretch. It was found that ATP was quite definitely not resynthesised under these conditions. Maréchal et al.[83a] also showed that the negative work done on the muscle depresses the rate of breakdown of high energy phosphates, but does not reverse the energy-producing chemical reactions. These experimental results do not negate the idea that splitting of ATP is the immediate energy source for contraction, as concluded by these researchers. One should propose instead a molecular mechanism of contraction in which external work performed in stretching an activated muscle ruptures weak bonds such as those within myosin molecules and between myosin and F-actin.

Many problems remain in relating Hill's studies of the energetics of metabolism to the data from chemical studies of metabolism. Nevertheless, the experimental results obtained in the foregoing dynamic and thermodynamic investigations must be considered when postulating any muscle contraction mechanism. As of 1972 we have no theory completely compatible with both the physiological studies and the biochemical results described in the following sections. Because of the limited scope of the present review we will simply point out that A. F. Huxley[84] proposed a molecular theory of muscle contraction involving 'sliding' which could explain Hill's results (cf. chapter 13).

3. History of the Biochemistry of Muscle Contraction[84a]

Elucidation of the mechanism of muscle contraction requires an understanding of the relationship of the reactions of each muscle component to contraction. Modern chemical and physical techniques have clarified the association between the structure and function of the contractile system. The historical development of the investigation of the source of energy for motion will now be briefly summarised. As stated above, muscle fibre is extremely suitable for the study of the source of energy in living organisms, since the movements and mechanical characteristics as well as the accompanying energetic chemical changes can all be easily recorded. This explains why many biochemical studies of the metabolism of living organisms were carried out using muscle specimens and why the development of the biochemistry of metabolism has for a long time centred around muscle tissues.

According to Muralt[85] there are four 'periods' in the history of the biochemistry of muscle—the pre-lactic acid, lactic acid, phosphorylation, and myosin periods.

About 140 years ago Berzelius performed the first chemical study of muscle and discovered that lactic acid appears in muscle fatigued by vigorous movement. The significance of this work was neglected for a long time despite its confirmation by du Bois-Reymond[86] and Heidenhain[87] around the 1860's. The first search for the source of energy for motion was undoubtedly that of Fick and Wislicenus[88] in 1865. These workers analysed their urine after climbing a high mountain and demonstrated the fallaciousness of Liebig's then current theory which postulated that protein is an energy source. This finding attracted attention to the metabolism of muscle carbohydrate, and in particular to glycogen, and led to the discovery of the decrease of muscle glycogen during motion (Weiss[89]) through hydrolysis to glucose (Panormoff[90]).

The 'lactic acid period' began with the classical paper of Fletcher and Hopkins[91] in 1907. They unambiguously demonstrated the formation of lactic acid during contraction and its absence in resting muscle. This was followed by Meyerhof's[92] study of lactic acid metabolism and by Hill's measurement of the heat generation accompanying contraction. These studies led to the accumulation of an enormous amount of data on the biochemistry and physiology of muscle contraction.

In 1930 Lundsgaard[93] showed clearly that muscle to which monoiodoacetate had been added did not produce any lactic acid during its contraction, and later discovered the consumption of CrP during muscle contraction[94]. This discovery was taken up by Meyerhof, and led to the period of phosphorylation study, which became the 'golden age,' not only of muscle biochemistry, but of biochemistry in general. Meyerhof, Parnas and their colleagues studied many metabolic reactions involving intermediates such as adenosine phosphates (ATP, ADP and AMP), CrP and fructose-2-P, culminating in the discovery of the Parnas-Meyerhof scheme of glycolysis and in the idea of the energetic coupling between biochemical reactions, which was finally unified by Lipmann[95] into the concept of the high energy phosphate bond. As described above, contraction was shown to occur so long as the dephosphorylation of ATP and CrP continued, even when glycolysis was inhibited by monoiodoacetate. In particular, Lohmann[96] discovered, in 1934, a creatine kinase which catalysed

$$CrP + ADP \rightleftharpoons Cr + ATP,$$

suggesting that the chemical reaction most likely to provide the energy for contraction is

$$ATP + H_2O \rightarrow ADP + P_i.$$

At about this time, the Krebs tricarboxylic acid cycle was established, originating from the work of Szent-Györgyi.

Using frog sartorius muscle in which the creatine kinase, glycolysis and respiration were all inhibited to keep the amount of CrP constant before and after a single twitch, Davies and his colleagues[97-99] only recently demonstrated that the direct energy source for muscle contraction is ATP. The decrease in ATP corresponded to the energy consumption of the muscle and to the increase in ADP and P_i. Mommaerts and Wallner[100] have confirmed the findings by Davies et al. and also shown that the ATP breakdown occurs exclusively within the contraction phase, not during relaxation. Using muscle pretreated with dinitrofluorobenzene to inhibit the creatine kinase, Kushmerick and Davies[101] further showed that the overall efficiency was over 60% at a constant contraction rate of 2 cm/sec on the basis of a free energy of hydrolysis for ATP of 10 kcal/mole. Measurement of the ATP consumed during an entire cycle of excitation, contraction and relaxation when little or no work was performed suggested that the heat of shortening cannot represent degraded free energy from ATP[102].

In contrast to this brilliant work on the energetics of metabolism in muscle contraction, there was much delay in the study of muscle structural protein which is an integral part of the contractile system. However, the structural protein of muscle protoplasm has been investigated by a small, but capable, group. Consequently, our knowledge advanced considerably in the 1930's. As early as 1859, Kühne[103] discovered a protein called 'myosin' in muscle. In 1930 Muralt and Edsall[104,105] measured the flow-birefringence of the protein extracted from muscle by a highly concentrated salt solution and concluded that its molecule was very slender and long. In 1934 a simple method was devised by Weber[106] to prepare a 'myosin' thread from 'myosin' solution, and toward the end of the decade studies of this thread showed that 'myosin' is a structural protein of the muscle protoplasm. In 1939, Engelhardt and Ljubimova[107] published a famous paper, 'Myosin and Adenosinetriphosphatase', showing that myosin contains an enzyme which hydrolyses ATP. At about this time, dephosphorylation of ATP, which supplies chemical energy, had been postulated to be the reaction most directly coupled with muscle contraction, as described above, while myosin was believed to be the factor most likely to be responsible for contraction. With the benefit of hindsight one wonders why no one tried the reaction of ATP with myosin until 1939. This discovery, coming on top of such a background, attracted the attention of many biochemists and was confirmed by Szent-Györgyi[108], Needham[109] and others, and as a result many new facts about muscle protein were discovered, helping to establish the 'myosin period'.

Straub[110] and co-workers, having isolated actin from muscle, made the interesting observation that it exists in two forms, globular G-actin in the absence of salt and fibrous F-actin upon addition of salt. G-Actin is bonded to ATP while F-actin is bonded to ADP[111] (see chapter 5). Szent-Györgyi[112] purified a protein now called myosin and discovered that the material previously called 'myosin' was actually a complex of myosin and actin, namely, actomyosin. Furthermore he[113] and Needham and his colleagues[114] showed that there was a drastic decrease in viscosity and flow-birefringence of the actomyosin solution upon addition of ATP. Szent-Györgyi[113] observed that the addition of ATP to a suspension of actomyosin at low ionic strength caused sudden precipitation (superprecipitation) in the presence of Mg^{2+} (see chapter 7), whereas the actomyosin gel shrank upon addition of ATP. Spicer[115] found that the addition of a high concentration of ATP dissolved actomyosin suspensions (the clearing response) at low ionic strength (see chapter 7).

This reaction of ATP with actomyosin is basic to muscle contraction as was shown by the following experiments using various muscle models. Szent-Györgyi[116] found that ATP caused contraction of glycerol-treated muscle, which is mainly composed of actin and myosin and is prepared by treatment with glycerol in order to nullify the function of the membrane. Weber and Portzehl[117,118] observed that when glycerinated muscle was placed in an ATP solution, it developed tensions similar to those obtained with living muscle. Muscle treated with glycerol maintains its contractile

TABLE I. Reactions of ATP with Muscle Proteins and Muscle Models

Muscle protein or muscle model	Reaction
Myosin	Catalyses $ATP + H_2O \longrightarrow ADP + P_i$
Actin	G-Actin-ATP + $H_2O \xrightarrow{salt}$ F-actin-ADP + P_i
Actomyosin	High ionic strength: ATP dissociates actomyosin to F-actin and myosin
	Low ionic strength, low concentration of ATP: superprecipitation, actomyosin-type ATPase
	Low ionic strength, high concentration of ATP: dissolution, myosin-type ATPase
Actomyosin thread	ATP addition facilitates shrinkage or elongation
Oriented actomyosin thread	Addition of ATP causes shortening
Isolated sarcomeres Isolated myofibrils Glycerinated muscle	Addition of ATP shortens the I band with no change in the A band; actomyosin-type ATPase
Living muscle	Excitation causes shortening of the I band only, with no change in the A band; $ATP + H_2O \longrightarrow ADP + P_i$

component structure and X-ray diffraction and electron microscopy show that this is similar to that of living muscle. Portzehl[119] and Hayashi[120] showed that an actomyosin thread specially prepared so that the actomyosin is oriented to a considerable degree in the direction of the longitudinal axis, shortened only in this axial direction with little accompanying change in volume on the addition of ATP, similar to the behaviour of a living muscle. Chapter 8 will describe in detail the reaction between these muscle models and ATP. Table I summarises briefly the muscle proteins, the properties of muscle models, and their reaction with ATP.

The work described above shows that the basic process in muscle contraction involves the interaction of three components—actin, myosin and ATP. Hence it is understandable that the mechanism of interaction of these three compounds is currently the central problem in the investigation of the molecular mechanism of muscle contraction. It is certainly impossible to explain every aspect of such a complex physiological phenomenon by a simple molecular mechanism. Nevertheless, we have studied this system since 1950, with the working hypothesis that the intrinsic nature of the contraction mechanism could be elucidated by determination of the mechanism of reaction of actomyosin and myosin with ATP. This monograph emphasises the investigation of the molecular structure and function of myosin, and in particular the reaction mechanism of the ATPase (chapters 2–4). Chapters 5–9 attempt to show how the complicated interaction of myosin, actin and ATP can be explained on the basis of this reaction mechanism. Chapters 10 and 11 describe the system for the regulation of muscle contraction. In chapter 13 we propose a molecular model of muscle contraction based on the mechanism of interaction of actin, myosin and ATP. The role of ATP in active transport of cation is also discussed, and its hydrolysis by transport ATPase is compared to that by myosin-ATPase (chapters 11 and 12), and a molecular mechanism for active transport based on this is propounded in chapter 13. Finally, in chapter 14 the general principles for the biological energy transduction are presented.

REFERENCES

1 H. S. Bennett, *in* "Biophysical Science—A Study Program," ed. by J. L. Oncley, John Wiley & Sons, New York, p. 394 (1959).
2 G. Retzius, *Biol. Untersuchungen*, **1**, 1 (1881).
3 G. Retzius, *Biol. Untersuchungen, Neue Folge*, **1**, 51 (1890).
4 F. S. Sjöstrand, *Int. Rev. Cytol.*, **5**, 455 (1956).

5 E. Anderson, *in* "Electron Microscopy (Proc. Stockholm Conf.)," Academic Press, New York, p. 208 (1956).
6 K. R. Porter and G. F. Palade, *J. Biophys. Biochem. Cytol.*, **3**, 269 (1957).
7 E. Anderson-Cedergren, *J. Ultrastructure Res., Suppl.,* **1** (1959).
8 W. Bowman, *Phil. Trans.*, **130**, 457 (1940).
9 W. M. Dobie, *Ann. Mag. Nat. Hist., 2nd Ser.*, **3**, 109 (1849).
10 E. Brücke, *Denkschr. Akad. Wiss. Wien*, **15**, 69 (1858).
11 V. Hensen, *Arb. Kieler Physiol. Inst.*, **1868**, 1 (1869).
12 T. W. Englemann, *Pflügers Arch. ges. Physiol.*, **7**, 33 (1873).
13 A. F. Huxley and R. Niedergerke, *Nature*, **173**, 971 (1954).
14 J. Hanson, *Nature*, **169**, 530 (1952).
15 H. E. Huxley and J. Hanson, *Nature*, **173**, 973 (1954).
16 H. E. Huxley, *Proc. Roy. Soc.*, *B* **141**, 59 (1953).
17 G. F. Elliott, J. Lavy and B. M. Millman, *Nature*, **206**, 1357 (1965).
18 G. F. Elliott, *J. Gen. Physiol.*, **50**, 171 (1967).
19 H. E. Huxley and W. Brown, *J. Mol. Biol.*, **30**, 383 (1967).
20 C. Morgan, G. Rosza, A. Szent-Györgyi and R.W.G. Wyckoff, *Science*, **111**, 201 (1950).
21 A. J. Hodge, H. E. Huxley and D. Spiro, *J. Exp. Med.*, **99**, 201 (1954).
22 H. E. Huxley, *Biochim. Biophys. Acta*, **12**, 387 (1953).
23 J. Hanson and H. E. Huxley, *Nature*, **172**, 530 (1953).
24 J. Hanson and H. E. Huxley, *Symp. Soc. Exp. Biol.*, **9**, 228 (1955).
25 W. Hasselbach, *Z. Naturforsch.*, **8b**, 449 (1953).
26 H. E. Huxley and J. Hanson, *Biochim. Biophys. Acta*, **23**, 229 (1957).
27 J. Hanson and H. E. Huxley, *Biochim. Biophys. Acta*, **23**, 250 (1957).
28 S. V. Perry and A. Corsi, *Biochem. J.*, **68**, 5 (1958).
29 H. E. Huxley, *J. Mol. Biol.*, **7**, 281 (1963).
30 H. E. Huxley, *J. Biophys. Biochem. Cytol.*, **3**, 631 (1957).
31 H. E. Huxley, *Science*, **164**, 1356 (1969).
32 J. Hanson and J. Lowy, *J. Mol. Biol.*, **6**, 46 (1963).
32a H. E. Huxley, *Proc. Roy. Soc.*, *B***178**, 131 (1971).
33 B. Katz, *Brit. Med. Bull.*, **12**, 210 (1956).
34 B. Katz, "Nerve, Muscle, and Synapse," McGraw-Hill Book Co., New York (1966).
35 T. Kamada and H. Kinoshita, *Jap. J. Zool.*, **10**, 469 (1943).
36 L. V. Heilbrunn, *Physiol. Zool.*, **13**, 88 (1940).
37 L. V. Heilbrunn and F. J. Wiercinski, *J. Cell. Comp. Physiol.*, **29**, 15 (1947).
38 A. Sandow, *Yale J. Biol. Med.*, **25**, 176 (1952).
39 A. V. Hill, *Proc. Roy. Soc.*, *B***136**, 399 (1949).
40 H. S. Bennett and K. R. Porter, *Amer. J. Anat.*, **93**, 61 (1953).
41 C. Franzini-Armstrong and K. R. Porter, *J. Cell Biol.*, **22**, 675 (1964).
42 A. F. Huxley and R. E. Taylor, *Nature*, **176**, 1068 (1955).

43　A. F. Huxley and R. E. Taylor, *J. Physiol.*, **144**, 426 (1958).
44　M. Fujino, T. Yamaguchi and K. Suzuki, *Nature*, **192**, 1159 (1961).
45　T. Yamaguchi, T. Matsushima, M. Fujino and T. Nagai, *Jap. J. Physiol.*, **12**, 129 (1962).
46　J. N. Howell, *J. Physiol.*, **201**, 515 (1969).
47　R. J. Podolsky, *J. Physiol.*, **170**, 110 (1964).
48　E. Bozler, *J. Gen. Physiol.*, **38**, 149 (1954).
49　S. Watanabe, *Arch. Biochem. Biophys.*, **54**, 559 (1955).
50　S. Ebashi, *J. Biochem.*, **49**, 150 (1960).
51　A. Weber, *J. Biol. Chem.*, **234**, 2764 (1959).
52　S. Ebashi, *J. Biochem.*, **50**, 236 (1961).
53　B. B. Marsh, *Nature*, **167**, 1065 (1951).
54　S. Ebashi and F. Lipmann, *J. Cell Biol.*, **14**, 389 (1962).
55　H. Portzehl, *Biochim. Biophys. Acta*, **26**, 377 (1957).
56　W. Hasselbach and M. Makinose, *Biochem. Z.*, **333**, 518 (1961).
57　A. Sandow, *Pharm. Rev.*, **17**, 265 (1965).
58　S. Ebashi, *Nature*, **200**, 1010 (1963).
59　A. Szent-Györgyi and B. Kaminer, *Proc. Natl. Acad. Sci. U. S.*, **50**, 1033 (1963).
60　S. Ebashi, A. Kodama and F. Ebashi, *J. Biochem.*, **64**, 465 (1968).
61　K. Bailey, *Nature*, **157**, 368 (1946).
62　A. V. Hill, *Proc. Roy. Soc.*, *B***159**, 319 (1963).
63　A. Sandow, in "Biophysics of Physiological and Pharmacological Actions," ed. by A. M. Shanes, American Association for the Advancement of Science, Washington, D. C., p. 413 (1961).
64　A. V. Hill, "Trails and Trials in Physiology," Arnold, London (1965).
65　D. R. Wilkie, *Annu. Rev. Physiol.*, **28**, 17 (1966).
66　F. F. Jöbsis, *Curr. Topics Bioenergetics*, **3**, 279 (1969).
67　W. F. H. M. Mommaerts, *Physiol. Rev.*, **49**, 427 (1969).
68　R. C. Woledge, *Progr. Biophys. Mol. Biol.*, **22**, 39 (1971).
69　H. Helmholtz, *Arch. Anat. Physiol.*, **144** (1849).
70　A. V. Hill, *J. Physiol.*, **42**, 1 (1911).
71　A. V. Hill, *J. Physiol.*, **43**, 35 (1912).
72　A. V. Hill and W. Hartree, *J. Physiol.*, **54**, 84 (1920).
73　A. V. Hill, *Proc. Roy. Soc.*, *B***136**, 220 (1949).
74　A. V. Hill, *Proc. Roy. Soc.*, *B***126**, 136 (1938).
74a　E. Homsher, W. F. H. M. Mommaerts, N. V. Richiuti and A. Wallner, *J. Physiol.*, **220**, 601 (1972).
75　A. V. Hill, *Proc. Roy. Soc.*, *B***159**, 297 (1964).
76　F. D. Carlson, D. J. Hardy and D. S. Wilkie, *J. Gen. Physiol.*, **46**, 851 (1963).
77　D. R. Wilkie, *J. Physiol.*, **195**, 157 (1968).
78　W. F. H. M. Mommaerts, K. Seraydarian and G. Marechal, *Biochim. Biophys. Acta*, **57**, 1 (1962).

78a C. Gilbert, K. M. Kretzschmar, D. R. Wilkie and R. C. Woledge, *J. Physiol.*, **218**, 163 (1971).
79 B. C. Abbott, X. M. Aubert and A. V. Hill, *Proc. Roy. Soc.*, **B139**, 86 (1951).
80 B. C. Abbott and X. M. Aubert, *Proc. Roy. Soc.*, **B139**, 104 (1951).
81 A. V. Hill and J. V. Howarth, *Proc. Roy. Soc.*, **B151**, 169 (1959).
82 A. V. Hill, *Science*, **131**, 897 (1960).
83 A. A. Infante, D. Klanpiks and R. E. Davies, *Science*, **144**, 1577 (1964).
83a G. Maréchal, W. F. H. M. Mommaerts and K. Seraydarian, *J. Physiol.*, **214**, 40P (1971).
84 A. F. Huxley, *Progr. Biophys. Biophys. Chem.*, **7**, 225 (1957).
84a D. M. Needham, "Machina Carnis: The Biochemistry of Muscular Contraction in its Historical Development," Cambridge University Press, Cambridge (1971).
85 A. von Muralt, *Biochim. Biophys. Acta*, **4**, 126 (1950).
86 E. du Bois-Reymond, *Monatsber. Berl. Akad.*, 288 (1859).
87 R. Heidenhain, "Mechan. Leitung bei der Muskeltätigkeit," Leipzig (1864).
88 A. Fick and J. Wislicenus, *Vierteiljahresschr. Naturforsch. Ges. Zürich*, **10**, 317 (1865).
89 S. Weiss, *Sitzher. Akad. Wiss. Wien*, **64**, 1 (1871).
90 C. Panormoff, *Z. Physiol. Chem.*, **17**, 596 (1893).
91 W. M. Fletcher and F. G. Hopkins, *J. Physiol.*, **35**, 247 (1907).
92 O. Meyerhof, "Die Chemischen Vorgänge in Muskel," Berlin (1930).
93 E. Lundsgaard, *Biochem. Z.*, **217**, 162 (1930).
94 E. Lundsgaard, *Biochem. Z.*, **269**, 308 (1934).
95 F. Lipmann, *Advan. Enzymol.*, **1**, 99 (1941).
96 K. Lohmann, *Biochem. Z.*, **271**, 264 (1934).
97 D. F. Cain and R. E. Davies, *Biochem. Biophys. Res. Commun.*, **8**, 361 (1962).
98 A. A. Infante and R. E. Davies, *Biochem. Biophys. Res. Commun.*, **9**, 410 (1962).
99 D. F. Cain, A. A. Infante and R. E. Davies, *Nature*, **196**, 214 (1962).
100 W. F. H. M. Mommaerts and A. Wallner, *J. Physiol.*, **193**, 343 (1967).
101 M. J. Kushmerick and R. E. Davies, *Proc. Roy. Soc.*, **B174**, 315 (1969).
102 M. J. Kushmerick, R. E. Larson and R. E. Davies, *Proc. Roy. Soc.*, **B174**, 293 (1969).
103 W. Kühne, *Arch. Anat. Physiol.*, 748 (1859).
104 A. L. von Muralt and J. T. Edsall, *J. Biol. Chem.*, **89**, 315 (1930).
105 A. L. von Muralt and J. T. Edsall, *J. Biol. Chem.*, **89**, 351 (1930).
106 H. H. Weber, *Pflügers Arch. ges. Physiol.*, **235**, 205 (1934).
107 W. A. Engelhardt and M. N. Ljubimova, *Nature*, **144**, 668 (1939).
108 A. Szent-Györgyi and I. Banga, *Science*, **93**, 158 (1941).
109 D. M. Needham, *Biochem. J.*, **36**, 113 (1942).
110 F. B. Straub, *Stud. Inst. Med. Chem. Univ. Szeged*, **2**, 3 (1942); **3**, 23 (1943).
111 F. B. Straub and G. Feuer, *Biochim. Biophys. Acta*, **4**, 455 (1950).

112 A. Szent-Györgyi, *Stud. Inst. Med. Chem. Univ. Szeged*, **3**, 76 (1943).
113 A. Szent-Györgyi, "Chemistry of Muscular Contraction," 1st ed., Academic Press, New York (1947).
114 M. Dainty, A. Kleinzeller, A. S. C. Lawrence, M. Miall, J. Needham, D. M. Needham and S. C. Shen, *J. Gen. Physiol.*, **27**, 355 (1944).
115 S. S. Spicer, *J. Biol. Chem.*, **199**, 289 (1952).
116 A. Szent-Györgyi, *Biol. Bull.*, **96**, 140 (1949).
117 H. Portzehl, *Z. Naturforsch.*, **7b**, 7 (1952).
118 H. H. Weber and H. Portzehl, *Progr. Biophys. Biophys. Chem.*, **4**, 60 (1954).
119 H. Portzehl, *Z. Naturforsch.*, **6b**, 355 (1951).
120 T. Hayashi, *J. Gen. Physiol.*, **36**, 139 (1952).

2
THE STRUCTURE OF THE MYOSIN MOLECULE*

'Myosin,' a structural protein of muscle, was discovered and named by Kühne[1] in 1859. However, it is fair to say that systematic research on 'myosin' with reference to the mechanism of energy transformation in living organisms did not start until the 1930's, when Edsall and Muralt[2-4] and Weber[5] made physico-chemical studies on myosin and Engelhardt and Ljubimova[6] discovered its ATPase activity. The subsequent discovery by Banga and Szent-Györgyi[7,8] that the substance previously called 'myosin' is a complex, actomyosin, composed of both myosin and actin, another important muscle structural protein, further intensified research on muscle contraction. Myosin constitutes 54% of the total protein of the myofibrils[9-11], and H. E. Huxley[12] and Kaminer and Bell[13] showed that upon isolation it aggregates under normal physiological conditions to the thick filament form observed in living muscle. Several proposed molecular mech-

* Contributor: Yutaro Hayashi

anisms have postulated that contraction is caused by either myosin or actin alone. However, in 1958 Hayashi et al.[14] showed that a muscle model containing both myosin and actin shortens anisodimensionally upon the addition of ATP, while filaments consisting only of myosin do not shorten. This demonstrated decisively that both myosin and actin play roles in muscle contraction.

Myosin has been shown to perform three important physiological functions: (1) hydrolysis of ATP to P_i and ADP (ATPase); (2) combination with actin, and (3) formation of the filaments observed in living organisms, through polymerisation under normal physiological conditions. In consequence, the study of myosin has concentrated on these three functions, since an understanding of its interactions with actin and ATP is essential for the elucidation at a molecular level of muscle contraction and of the conversion of chemical energy to mechanical energy in living organisms. This chapter describes the molecular structure of myosin, emphasising its three functions.

1. The Preparation of Myosin

The basic procedures for the preparation of myosin were discovered by Banga and Szent-Györgyi[7] and Guba and Straub[15]. The methods which are currently used by many researchers are improvements of these original procedures[16] and usually contain four steps: (1) extraction of muscle for about 10 min with a solution of neutral pH and high ionic strength (longer contact times result in the additional extraction of a large amount of actomyosin, a complex of myosin with actin); (2) removal of actomyosin by centrifuging with 0.3M KCl; (3) repeated precipitation at low ionic strength and dissolution at high ionic strength, to remove those components soluble at low ionic strength, and (4) clarification of the solution by ultracentrifugation. These four steps are all carried out at low temperature, and contamination by metal ions is avoided. Myosin specimens prepared even in this manner may still contain impurities such as actomyosin and myosin aggregates. Actomyosin can be dissociated into F-actin and myosin by the addition of ATP in the presence of Mg^{2+}, and the F-actin can then be removed easily by ultracentrifuging[17]. Minor impurities include traces of 5′-adenylic acid deaminase, myokinase and ribonucleic acid [18-20]. The 5′-adenylic acid deaminase and most of the RNA can be removed from the myosin by chromatography on phospho-cellulose[21] and DEAE-cellulose[22], respectively. Chromatography on DEAE-Sephadex A-50[20] is very useful to remove impurities after the customary preparation of myosin [22a], although the elimination of myokinase is difficult even by this procedure[20]. The irreversible aggregates of myosin are formed spontaneously and are difficult to remove completely from the specimen by usual methods. However, these aggregates can be separated from the myosin monomer by chromatography on DEAE-Sephadex A-50[20] or agarose[23].

2. Methods for Determination of Molecular Weight

The molecular weight of myosin was reported in 1948 to be 1.5×10^6 [24]. Since then many papers have been published on this subject, as shown in Table I, which covers nearly all the methods used in the determination of molecular weights of proteins. A brief digression on the measurement of weight-average molecular weight will now be made. The sedimentation coefficient, s, and the diffusion coefficient, D, are used to determine the molecular weight, M, from ultracentrifuge studies using the equation derived by Svedberg:*

$$M = \frac{RTs}{D(1-\bar{v}\rho)}, \qquad (1)$$

where R is the gas constant, T the absolute temperature, $\bar{v}$ the partial specific volume of the solute molecule, and ρ the density of the solvent. This sedimentation-diffusion method requires four parameters, s, D, $\bar{v}$ and ρ, of which the accurate measurement of D is particularly difficult. Alternatively, if the distribution of the solute molecules reaches equilibrium in the gravitational field, then the sedimentation rate of the solute molecules becomes equal to their centripetal diffusion rate and, at an initial concentration, c, and an angular velocity, ω, the following equation is obtained:

$$\bar{M}_w(c) = \frac{2RT}{(1-\bar{v}\rho)\omega^2(x^2_b - x^2_m)} \cdot \frac{c_b - c_m}{c}, \qquad (2)$$

where c_m and c_b are the concentrations at the meniscus, x_m, and the cell bottom, x_b, respectively, and are determined optically using an analytical ultracentrifuge. Thus this is a relatively direct method not requiring the measurement of D or s. Previously this sedimentation-equilibrium method had the disadvantage of needing a long time to reach equilibrium, but the recent introduction of a short column has substantially reduced the time required and may make its use more widespread[26]. A simplified version

* cf. Ref. 25 for theories and procedures of ultracentrifugation.

is Archibald's method[27], which takes advantage of the lack of effective transfer of solute in the centrifuge through the meniscus or the cell bottom, even during the approach towards equilibrium. If the concentration at either the meniscus or cell bottom, x_i, be c_i, and the concentration gradient be $(dc/dx)_i$, one obtains

$$\bar{M}_w(c) = \frac{RT(dc/dx)_i}{(1-\bar{v}\rho)\omega^2 x_i c_i}. \tag{3}$$

This method is simpler but requires some expertise to operate successfully.

At a weight concentration, c, the intensity of light, R_θ, scattered at an angle θ to the incident light is given by

$$R_\theta = Kc\bar{M}_w(c)(1+\cos^2\theta)P(\theta). \tag{4}$$

K is a constant depending on the refractive indices of the solvent and the solute and the wavelength of the light used, and the function $P(\theta)$ depends on the molecular shape and equals 1 when θ tends to 0. The above equation leads to $R_0 = Kc\bar{M}_w(c)$ when θ tends to 0, and the molecular weight can be found by measuring the intensity of light-scattering[28]. To obtain an accurate molecular weight value uninfluenced by intermolecular interaction, measurements must be made at several concentrations and extrapolated to infinite dilution: $1/\bar{M}_w(c) = 1/\bar{M}_w(0) + 2B_2 c$.

The above methods determine weight-average molecular weights ($\bar{M}_w$). Number-average molecular weights ($\bar{M}_n$) are measured only by osmometry, as is now described. When a high polymer solution is separated from its solvent by a semi-permeable membrane, the application from the solution side of a pressure corresponding to the osmotic pressure (π) of the solution stops the transfer of solvent into the solution. When the solution is of low concentration, the equation

$$\pi/RTc = 1/\bar{M}_n + B_2 c \tag{5}$$

is valid, where B_2 is the second virial coefficient and corrects for the non-ideality of the solution. π/c is found experimentally at several concentrations and then extrapolated to infinite dilution. The number-average molecular weight is then given by $\bar{M}_n = RT/(\pi/c)_{c\to 0}$. This is the most direct procedure for finding $\bar{M}_n$ and avoids the many parameters involved in other methods. However, it gives a number-average molecular weight which is markedly affected by small amounts of contamination by molecules of relatively low molecular weight.

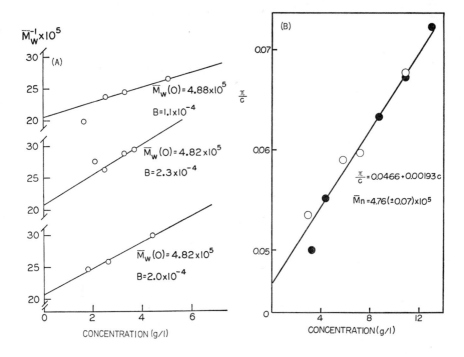

Fig. 1. Determination of molecular weight of myosin by sedimentation equilibrium and osmometry. 0.6M KCl, 0.05M Tris buffer, pH 7.0, 5°C. A: Sedimentation equilibrium method. Three kinds of preparation were used for the measurement. B: Osmometry. From $(\pi/c)_{c \to 0}$, $\bar{M}_n$ is determined as $4.76 \, (\pm 0.07) \times 10^5$.

3. The Molecular Weight of Myosin

For a monodisperse system, $\bar{M}_w = \bar{M}_n$, but for a polydisperse system, $\bar{M}_w$ is always greater than $\bar{M}_n$. As described above, it is difficult to obtain a completely monodisperse myosin specimen, and hence comparison of $\bar{M}_w$ with $\bar{M}_n$ should show a range for the molecular weight. Because of this Tonomura et al.[29] measured $\bar{M}_w$ by the sedimentation equilibrium method and $\bar{M}_n$ by osmometry (Fig. 1), and found an $\bar{M}_n$ of 4.67–4.76×10^5 and an $\bar{M}_w$ of 4.82–5.15×10^5. After correction for the effect of contaminants in the myosin specimen, which were estimated to account for 2.6% of the total protein and to have a molecular weight of about 1×10^5, the $\bar{M}_n$ value of 4.6–4.8×10^5 was obtained compared to the $\bar{M}_w$ value of 4.8–5.1×10^5.

TABLE I. Molecular Weight of Myosin Prepared from Rabbit Skeletal Muscle

Method	Year	Molecular weight ($\times 10^{-5}$)	Measured by	Reference
Sedimentation-diffusion	1950	8.58	Portzehl et al.	142
	1955	5.00	Laki and Carroll	143
	1956	5.30	Holtzer and Lowey	144
	1959	4.90	Holtzer and Lowey	30
	1960	5.40	Johnson and Rowe	145
	1964	5.40	Kaldor et al.	146
	1966	5.04	Trayer and Perry	147
Sedimentation-equilibrium	1963	5.95	Woods et al.	148
	1966	4.82–5.18	Tonomura et al.	29
	1969	4.85–5.15	Gershman et al.	85
	1970	4.65	Rossomando and Piez	23
	1970	4.58	Godfrey and Harrington	120
Archibald	1958	3.80	Mommaerts and Aldrich	149
	1958	4.20	von Hippel et al.	150
	1960	6.00–6.19	Kielley and Harrington	68
	1962	4.97	Lowey and Cohen	50
	1963	5.87–6.27	Woods et al.	148
	1964	5.24	Mueller	151
	1964	4.60–5.00	Connell and Mackie	152
Light-scattering	1959	4.30	Brahms	153
	1959	4.93	Holtzer and Lowey	30
	1962	5.25	Holtzer et al.	154
	1963	5.20	Gellert and Englander	155
		(6.20)	(as corrected and recalculated by Tomimatsu)	156
	1963	4.70	Asai	157
Osmometry	1950	8.40	Portzehl	158
	1966	4.67–4.80	Tonomura et al.	29

Hence one can conclude that the true molecular weight of myosin is about 4.8×10^5. This was the first time that $\bar{M}_w$ and $\bar{M}_n$ were both measured using the same myosin specimen. Many recent molecular weight measurements have tended towards this value (see Table I), and so, unless stated otherwise, in this monograph the molecular weight of myosin is taken to be 4.8×10^5.

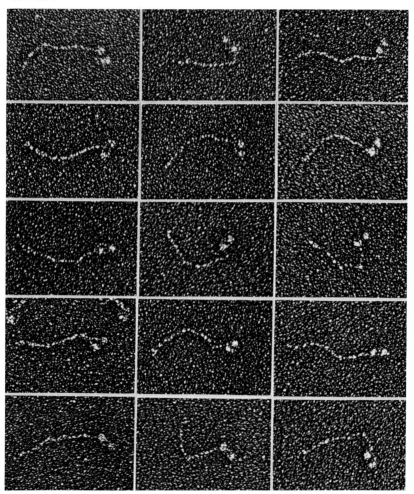

Fig. 2. Electron micrograph of the myosin molecule by the rotary shadow-casting method. 1M ammonium acetate, pH 7.5. × 175,000. (Courtesy of Dr. K. Takahashi, Faculty of Agriculture, Hokkaido University.)

4. The Molecular Shape of Myosin

The function $P(\theta)$ obtained from the light-scattering measurements described above gives a radius of gyration $<s^2>^{1/2}$ which is a measure of molecular size in solution. In the case of a rigid rod-shaped molecule, the molecular length, L, is given by $L = \sqrt{12}<s^2>^{1/2}$. Using this method, Holtzer and Lowey[30] found L for myosin to be 1,620Å, while the molecular diameter was calculated to be 23–28Å by Kirkwood and Riseman's equation[31]. However myosin does not have a simple rod-like shape. Electron microscopy shows clearly that it is rod-like but with a globular tip. Rice[32] first used a shadow-casting technique and observed a myosin image comprising a relatively thin portion, 10–30Å wide, and a thick portion 20–40Å wide. The total length was about 1,100Å. However, Rice[33], Zobel and Carlson[34] and H. E. Huxley[12] later demonstrated that the total length of the myosin molecule is about 1,600Å with a head portion 35–60Å in diameter, connected to a tail of diameter 20Å. The reported values for the length of the head have differed between researchers, varying from 150 to 440Å. Recently Slayter and Lowey[35] using a rotary shadow-casting technique observed an image in which the wide portion consisted of two spherical units with a diameter 90Å, as shown in Fig. 2. More recently, Pepe and Drucker[36] and particularly Takahashi[36a] (see the electron micrograph in the frontispiece) have been able to get reasonably good pictures of myosin molecules negatively stained with uranyl acetate and uranyl oxalate, respectively. The observed structure is consistent with that found by the shadow technique. Hence it has been accepted, from electron microscopy, that myosin is composed of two fibrous heavy chains, each with a molecular weight of about 2.1×10^5, as described later.

5. The Subunit Structure of the Myosin Molecule

A) Subfragments formed by proteolysis

Perry[37,38] and Gergely and his co-workers[39–41] showed that digestion of myosin with trypsin or chymotrypsin yields a subfragment of low viscosity, which dissolves in a low ionic strength solution and which retains its ATPase activity. Mihályi and A. G. Szent-Györgyi[42] found that digestion with trypsin gave two products, one sedimenting more slowly and the other more rapidly than the parent molecule in an ultracentrifuge. These are called light meromyosin (L-MM) and heavy meromyosin (H-

TABLE II. Molecular Weight of H-MM and L-MM

| Measured by | Year | Digestion condition | | | Methods | Molecular weight | | Reference |
		Trypsin/ protein (w/w)	Digestion time (min)	Temperature (°C)		H-MM ($\times 10^{-5}$)	L-MM ($\times 10^{-5}$)	
Gergely and Kohler	1957		4	25	Light-scattering	3.40	1.40	*159*
Lowey and Holtzer	1959	0.41/100	10	25	Archibald	3.24	1.26	*49*
Szent-Györgyi et al.	1960	1/200	10–12	23	Archibald	—	1.20	*44*
Lowey and Cohen	1962	1/300	5	room temp.	Archibald	3.20		*50*
Holtzer et al.	1962		10	25	Light-scattering	3.40	1.40	*154*
Mueller	1964	1/300	5	25	Archibald	3.50	1.51	*151*
Young et al.	1964	1/255	1.25	20	Sed. equil.	3.34	1.37	*160*
						—	—	
						3.62	1.62	
Young et al.	1965	1/255	1.25	20	Sed. equil.	3.86		*51*
						—		
						4.05		

MM) respectively[43]. L-MM does not exhibit ATPase activity, but is similar to myosin in that it does not dissolve in low ionic strength solutions. In contrast, H-MM shows ATPase activity and also dissolves in low ionic strength solutions. After ultracentrifugation and column chromatography, H-MM was found to be monodisperse, having a molecular weight of about 3.4×10^5 (Table II). However, L-MM is not homogeneous and its main component, L-MM fraction 1, has a molecular weight of 1.2×10^5 and is 90–100% helical, as measured by optical rotatory dispersion[44] (chapter 4 discusses filament formation by the polymerisation of L-MM). The helical content of H-MM is 40–50%[45]. Comparison of the amino acid compositions of H-MM and L-MM fr. 1 shows that the former contains 98 to 109 proline residues per molecular weight of $3.4 \times 10^{5\,[46,47]}$, while the latter has none, indicating that the proline residue prevents helix formation.

Electron microscopy studies show that H-MM contains a portion identical in size to the globular region observed in myosin, and also has a short tail, the overall length of the H-MM molecule being 600–900Å. In contrast, L-MM is 900Å in total length and consists of a simple rod of diameter 15–20Å[33,34,48]. H-MM and L-MM are obtained in yields of 70% and 25–30% respectively from myosin[49,50]. The evidence from morphology, size, and yield presented above shows beyond doubt that the myosin molecule is composed of 1 molecule of H-MM and 1 molecule of L-MM, joined end to end.

Mueller and Perry[47] succeeded in obtaining a smaller active fragment by further digestion of H-MM with trypsin. This was called subfragment-1 (S-1). As shown in Table III, its molecular weight is $1.0–1.8 \times 10^5$, but this may depend on the conditions of digestion. The molecular weight of the S-1 obtained by trypsin digestion under the conditions used by many researchers is $1.1–1.2 \times 10^5$. S-1 is obtained in fairly high yield from H-MM[51,52]. It is spherical in shape, with no tail, its diameter being 90 Å[33]. The most reasonable conclusion from these observations is that one molecule of myosin or of H-MM contains two S-1 components, since we already know that the globular portion of myosin consists of two fragments with a diameter of 90Å. Digestion of H-MM with trypsin gives, besides S-1, a subfragment-2 which is assumed to be the tail portion of H-MM observed by electron microscopy. Subfragment-2 was also prepared by Lowey and her colleagues by digestion of myosin with a water-insoluble polyanionic derivative of trypsin[53] or papain[54]. It has a molecular weight of 6.1×10^4, a weight-average length of 475Å and a helical content of 80–90%. Furthermore, under suitable conditions, papain could split

TABLE III. Molecular Weight of Subfragment-1

| Measured by | Year | Protein digested | Digestion condition | | | Method | Molecular weight ($\times 10^{-5}$) | Reference |
			Trypsin/protein (w/w)	Digestion time (min)	Temperature (°C)			
Young et al.	1964	H-MM	1/15	about 10	25	Sed. equil.	1.17–1.21	160
Mueller	1965	H-MM	1/5–1/20	4	23	Archibald	1.70	161
				5			1.48	
				8			1.29	
				20			1.12	
Jones and Perry	1966	H-MM	1/20	90	25	Sed. diffusion	1.29	98
Tokuyama et al.	1966	H-MM	1/20	10	18	Archibald	1.80	162
Trotta et al.	1968	H-MM	1/20	25	25	Sed. equil.	1.08	60
		Myosin	1/200	15	25		1.043	
			(insoluble papain)					
Lowey et al.	1969	Myosin	1/70–1/100	10	room temp.	Sed. equil.	1.15	54
			(insoluble papain)					
Hayashi	1972	H-MM	1/20	17	25	Gel filtration	1.04–1.09	61
						Osmometry	1.09–1.25	

TABLE IV. ATPase Activity of Myosin and Its Subfragments[a]

	Myosin (4.8×10^5)	H-MM (3.4×10^5)	S-1 (1.2×10^5)[b]
	ATPase activity (moles P_i/min/mole of protein)		
Ca^{2+} (7mM)	90.7	80.2	43.6
EDTA (3mM)	725	749	465
Mg^{2+} (20mM)	1.20	1.10	0.582

[a] 1.0mM ATP, 1.0M KCl, 0.05M Tris-maleate, pH 7.0, 25°C.
[b] Values in parentheses indicate molecular weights.

the enzymic globular regions of myosin, leaving the long rod intact along its length[45]. The rod has a length of 1,360–1,386Å, a molecular weight of 22×10^4 and is 94% helical. It contains no proline residue.

We[55] have compared the Ca^{2+}, Mg^{2+} and EDTA-ATPase activities of a myosin specimen (see chapter 3) with the activities of the H-MM and S-1 prepared from it (Table IV). If the respective molecular weights were assumed to be 4.8, 3.4 and 1.2×10^5, then the relative specific activities per mole under all conditions investigated were approximately myosin: H-MM: S-1 = 1 : 1 : 0.5. Furthermore, Nauss et al.[56] observed that there was no significant alteration in either the Ca^{2+}- or EDTA-activated total ATPase activity during the digestion of H-MM by trypsin. The constant ATPase activity throughout the entire digestion period eliminated the possibility of an initial activation followed by destruction of the enzymatic site. The yield of S-1 from the tryptic digestion of H-MM was 65%, two molecules of S-1 being obtained from one molecule of H-MM. The characteristic shape of the pH-activity curve of myosin (see chapter 3) is also shown by H-MM and S-1[57]. From the above retention of total ATPase activity and the identical influence of pH upon this activity one can reasonably conclude that the active site of ATPase in myosin lies entirely in the S-1.

We may now consider the locations in the myosin molecule of its other two functions—formation of filaments and ability to combine with actin. As will be described in section 7, the former is in the L-MM. When the ionic strength of a solution of L-MM is low, it aggregates in a similar manner to myosin, forming the backbone of the thick filament. In contrast, H-MM and S-1 remain in solution at low ionic strength. The ability to combine with actin is present in both H-MM and S-1, but not in L-MM. However, the association constant of S-1 with actin seems to be smaller than that of H-MM[58,59]. Chapter 6 will describe detailed measurements of the combination of F-actin with both H-MM and S-1.

Recently, Trotta et al.[60] reported that the S-1 obtained by tryptic digestion (molecular weight, 1.1×10^5) consists of a peptide chain with a molecular weight of 6.7×10^4, a light chain of molecular weight 1.8×10^4 (to be described later) and 16% of a peptide of molecular weight 2.1×10^3. Our gel filtration of S-1 in 5M guanidine-HCl also indicates that 10–30% of the components of S-1 have a molecular weight substantially smaller than $1.2 \times 10^{4[55]}$. The involvement of small peptides in S-1 makes its structural analysis difficult. Consequently we took a solution of S-1 obtained by the gel filtration of the tryptic digest of H-MM, stood it at pH 11.0 at 0°C for 60 min and, after neutralisation, chromatographed it on Sephadex G-200 and DEAE-Sephadex A-50, in order to obtain a sample of S-1 with an ATPase activity approximating that of the original specimen[61]. This alkali-treated S-1 showed, on gel filtration through Sephadex G-200 in the presence of guanidine-HCl and EDTA, the presence of two components, a and b, with relative area ratios of 0.92 : 1.0 (Fig. 3). Figure 4 shows the relationship of molecular weight to elution volume, in the presence of guanidine-HCl, of polypeptide chains from various proteins containing no S-S bridges. This enables the molecular weights of fractions a and b to be estimated as 5.5×10^4 and 2.6×10^4 respectively. In contrast, the

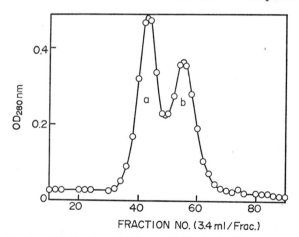

Fig. 3. Gel filtration of alkali-treated S-1. Gel filtration of 6 ml of the alkali-treated S-1 (5.9 mg/ml), obtained by the procedures given in the text on a Sephadex G-200 (3.7 ×31 cm) column equilibrated with 5M guanidine-HCl and 2mM EDTA. Flow-rate 11 ml/hr.

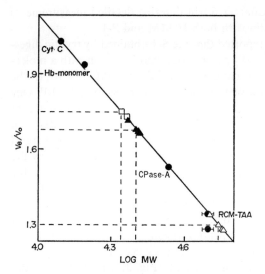

Fig. 4. The relationship of the relative elution volume (V_e/V_0) to molecular weight (MW) in gel filtration on Sephadex G-200 of various proteins in 5M guanidine-HCl. RCM-TAA: reduced carboxymethyl Taka-amylase; CPase A: carboxypeptidase A; Hb: hemoglobin; Cyt c: cytochrome c. △, Fraction a; ▲, fraction b of alkali-treated S-1; □, light chains of myosin.

molecular weight of alkali-treated S-1 as measured by osmometry and gel filtration (Fig. 5) was 11.7 and 10.7×10^4, respectively.

Recently, electrophoresis in polyacrylamide gels in the presence of the anionic detergent sodium dodecyl sulphate (SDS) has proved to be a useful tool for the rapid and simple estimation of the molecular weights of proteins and their subunits[62-64]. Hydrodynamic studies made by Reynolds and Tanford[65] suggested that the complex of protein with SDS is a rod-like particle, the length of which varies uniquely with the molecular weight of the protein moiety. These results explain the empirical observation that, when the electrophoretic mobilities are plotted against the logarithm of the known polypeptide chain molecular weights, a smooth curve is obtained. We[61] have recently separated polypeptide chains contained in S-1 using this method. Figure 6 shows densitometer tracing patterns of S-1. Fraction a was shown to contain only one peptide chain, of molecular weight 5.2×10^4. As will be described in the next section, the myosin molecule consists of two heavy chains of molecular weight 2.1×10^5 and four light chains of molecular weights from 1.4 to 2.5×10^4. Therefore, fraction a must be

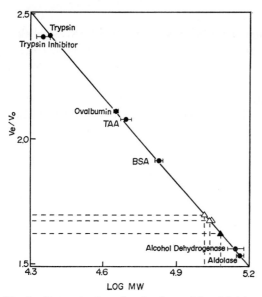

Fig. 5. Determination of molecular weight of S-1 by gel filtration. Trypsin inhibitor: trypsin inhibitor (soy bean); TAA: Taka-amylase; BSA: bovine serum albumin; alcohol dehydrogenase: alcohol dehydrogenase (yeast); aldolase: aldolase (muscle). △, alkali-treated S-1; ▲, control S-1.

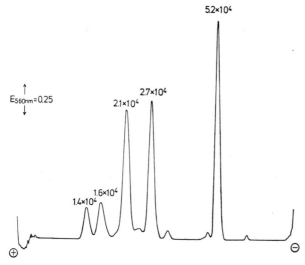

Fig. 6. Densitometer trace of the SDS-gel electrophoresis of subfragment-1. 7.5 μg of alkali-treated S-1 which binds to F-actin. 10% acrylamide, 0.27% methylene-bisacrylamide. 0.1% SDS, 0.05M phosphate buffer, pH 7.3. 8 mA/tube. Stained by 0.2% coomassie brilliant blue for 1hr at 45°C.

derived from the heavy chains. Fraction b was separated by SDS-gel electrophoresis into four components, whose molecular weights were 2.7, 2.1, 1.6 and 1.4×10^4, respectively. A detailed discussion of the submolecular structure of S-1 will be given in the next section after describing the properties of the heavy and light chains of myosin, but our conclusion on the submolecular structure of S-1 is summarised as follows: Subfragment-1 consists of two different molecular species, S-1A and S-1B. S-1A consists of four polypeptide chains of molecular weights 5.2–5.5, 2.7, 1.6 and 1.4×10^4 and has no ATPase activity, while S-1B consists of three polypeptide chains of molecular weights 5.2–5.5, 2.7 and 2.1×10^4 and is responsible for the ATPase activity of myosin.

Yagi et al.[66] have treated S-1 with the proteolytic enzyme Nagarse to obtain an active fragment S-1N with a molecular weight of 1×10^5, as measured by light-scattering. It is anticipated that future progress will allow us to obtain a much smaller active fragment from myosin.

B) Heavy and light chains

Myosin is degraded to smaller subunits by urea or guanidine-HCl. Tsao[67] first reported that the myosin molecule in urea dissociates into a polypeptide chain of molecular weight 1.65×10^5 and a light chain of molecular weight 1.6×10^4, which latter accounts for 8% of the total weight. Kielley, Harrington and their co-workers[68-70], after a detailed study of the subunits produced by treatment with 5M guanidine-HCl, concluded that myosin contains subunits of molecular weight 2.06×10^5 and these were shown by electrophoresis to be monodisperse. On the assumption that the molecular weight of myosin is 6×10^5, they deduced that myosin consists of three peptide chains with identical molecular weights of 2.06×10^5. The SH peptide analyses of myosin made by Kimura and Kielley[71] and Weeds and Hartley[72] also supported this conclusion of the identity of the main chains. More recently, Huszar and Elzinga[73] also supported the conclusion from a study of the amino acid sequence around 3-methylhistidine (*cf.* p. 117). However, it is undoubtedly an enormous problem to determine whether the two main polypeptide chains composing the myosin molecule have identical primary structures. Recent studies of the N-terminal amino acid sequence of the polypeptide chains by Offer and Starr[73a] suggest that their primary chemical structures are very similar, but may not be identical. Bechtel et al.[73b] and Hale and Beecher[73c] have also presented evidence for two non-identical heavy chains in myosin by isoelectric focusing and disc gel electrophoresis in 11M urea, but no conclusive evidence exists for non-identity of the two heavy chains in the myosin molecule.

Seven years after Tsao's work[67], Wetlaufer and Edsall[74] reported the dissociation of light components from myosin in urea or guanidine-HCl. Dreizen et al.[75] found that there is about 17% of lower molecular weight components (molecular weight 4.6×10^4) in myosin in addition to the 83% composed of peptide chains of molecular weight 2.0–2.6×10^5, as argued by Kielley and Harrington. Treatment of myosin with an alkaline solution (pH 10 or pH 11–12.5) gave 12–18% of a low molecular weight fraction in the sedimentation pattern[76,77]. We shall call a main peptide chain of molecular weight 2.0–2.6×10^5 a heavy chain or a fibrous subunit (f subunit), and a fraction of molecular weight less than 4.6×10^4 a light chain or a globular subunit (g subunit), after the nomenclature of Stracher. Besides the alkali treatment, other methods such as acetylation, carboxymethylation and succinylation[78,79] have permitted the isolation of light chains from myosin. These treatments all increase the negative charge on the protein and the electrostatic repulsion is thought to separate the light chains which are bound by weak secondary forces. Hence it appears that myosin molecules prepared by the procedures given in section 1 consist of two heavy chains of molecular weight 2.0–2.6×10^5 and several light chains of molecular weight less than 4.6×10^4.

There is a divergence of opinion on the properties and functions of the light chains in the myosin molecule.* Frederiksen and Holtzer[84] reported recently that myosin consists of two heavy chains of molecular weight 2.2×10^5 and two light chains of molecular weight 3.0×10^4. They[84] made the interesting observation that subunits isolated by fractionation with ammonium sulphate from an alkaline solution of myosin have neither ATPase activity nor the ability to combine with actin. Dissociation of light chains from myosin with alkali followed by subsequent neutralization resulted in the recombination of the light chains and the recovery of the major portion of both activities. Gershman and his colleagues reported that when myosin is kept at pH 11 and in 2M guanine, or in concentrated salt solution after thiol protection, three light chains of average molecular weight 2×10^4 dissociate from the myosin molecule and ATPase activity is lost[85,86]. Stracher[87] separated the heavy and light chains by gel filtra-

---

* The light chains vary among species, between adult and foetus, between white and red muscles, and between skeletal and cardiac muscles. In the following, we will describe only the properties of the light chains of myosin from rabbit white muscles. For the properties of the light chains of myosins from other muscles see Refs. *80–83c.*

tion of myosin in 4M LiCl and 1mM Mg-ATP on Sephadex G-200, and reported that the heavy and light chains themselves are devoid of ATPase activity, but that the ATPase can be restored upon remixing, although to a rather low extent. Later, Dreizen and Gershman[88] separated the heavy and light chains by rapid LiCl-citrate fractionation with thiol protection, and succeeded in reconstituting from two subunits myosin with 70% of the ATPase activity of the original myosin, although the reconstitution of myosin from isolated subunits is very difficult and has not yet been confirmed by others. They concluded that myosin comprises two enzymatically active protomers, each containing one heavy chain and one essential light chain, since the ATPase activity was unaffected by removal of one light chain out of the three chains by treatment with 4M NH_4Cl. However, this conclusion is inconsistent with their own observations that the ATPase activity disappeared when one mole of the light chain was removed from myosin which had been treated by NH_4Cl and contained two moles of the light chain, and that the activity restored when one mole of the light chain combined again with myosin which had lost its ATPase activity.

In contrast, Gaetzens et al.[89] reported that 5.0×10^5g of purified myosin contained only one mole of light chains with a molecular weight of 2.4×10^4, and treatment to remove this component led to a concomitant loss in ATPase activity. Gazith et al.[90] reported that the low molecular weight protein associated with myosin comprises 9% of the myosin. Removal of 40–50% of this caused no decrease in ATPase activity, but any further removal resulted in an accompanying loss of activity. Paterson and Strohman[91] succeeded in the simultaneous resolution of the sulphonyl derivatives of the light and heavy chains of myosin on SDS acrylamide gels, and concluded that the heavy chain is homogeneous on gel electrophoresis, but that the light component is composed of two chains of molecular weights 1.85–1.95 and $3.21–3.30 \times 10^4$. More recently, Sarker and Cooke[92] and Low et al.[93] made the very interesting observations that the heavy chain is synthesised on polysomes which contain 50–60 ribosomes, while the three light chains of molecular weights 2.6, 1.7 and 1.5×10^4 are synthesised by polysomes of 5–9 ribosomes.*

These findings suggest that the light chains are essential components for the ATPase activity and the ability to bind with actin. However, there are still problems remaining as to the homogeneity of the light chains, and it is not certain how many light chains related to physiological functions are

* For the synthesis of myosin in cell-free systems see the very interesting and important papers by Heywood and his co-workers[94–97].

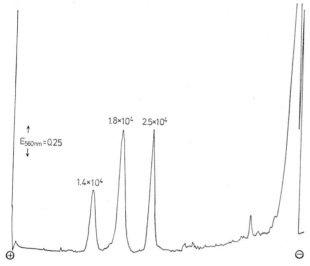

Fig. 7. Densitometer trace of the SDS-gel electrophoresis of myosin. 23 μg myosin. 10% acrylamide, 0.27% methylene-bisacrylamide. 0.1% SDS, 0.05M phosphate buffer, pH 7.3. 8 mA/tube. Stained by 0.2% coomassie brilliant blue for 1 hr at 45°C.

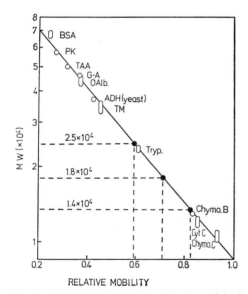

Fig. 8. Determination of the molecular weight (MW) of the light chains by SDS-gel electrophoresis. BSA: bovine serum albumin; PK: pyruvate kinase (muscle); TAA: Taka-amylase; G-A: g-actin; OAlb.: ovalbumin; ADH (yeast): alcohol dehydrogenase from yeast; TM: tropomyosin; Tryp.: trypsin; Chymo. B: B chain of chymotrypsin; Cyt c: cytochrome c; Chymo. C: C chain of chymotrypsin.

TABLE V. Properties of Fragments of Myosin

Subfragment or subunit	Function	Molecular weight	Number existing in myosin	Dimension (Å)	Helix content (%)
H-Meromyosin (H-MM)	ATPase; actin-binding	3.4×10^5	1	600–900	46
L-Meromyosin (L-MM)	Filament formation	1.2×10^5	1	900	90
Subfragment-1 (S-1)	ATPase; actin-binding	$1.1–1.2 \times 10^5$	2	90	33
Subfragment-2 (S-2)	—	6.1×10^4	1	475	87
Heavy chain	—	$2.0–2.2 \times 10^5$	2	—	44
Light chain	—	$1.4, 1.8, 2.5 \times 10^4$	3–4	—	33

contained in a myosin molecule. Therefore, we[61] separated the light chains by SDS-gel electrophoresis. The content of light chains in our myosin preparation, estimated by Sephadex G-200 column chromatography in 5 M guanidine-HCl, was 12%. As shown in Fig. 7, the densitometer trace of the SDS-gel electrophoresis of the light chains revealed the presence of three components, of molecular weights 2.5, 1.8 and 1.4×10^4 (Fig. 8). This result agreed well with that obtained by Sarker and Cooke[92], mentioned above. We will call these three components g_1, g_2 and g_3 respectively, in decreasing order of their molecular weights. The molar ratio of g_1, g_2 and g_3 was estimated from areas under peaks in densitometer tracing to be 1.0:2.1:1.1. These ratios agree with those reported recently by Weeds and Lowey[97a], using a more quantitative method.

The relationship between the physiological functions of myosin and the light chains must be further explored and should give key information for elucidating the role of myosin in muscle contraction at a molecular level. We will discuss this point further in chapter 4 in relation to the chemical modification of ATPase. Properties of the subfragments of myosin discussed in this section are summarised in Table V.

6. The Submolecular Structure of the Myosin Molecule

The investigations of the structure of myosin described above have shown that the molecule comprises two heavy chains of molecular weight 2.0–2.2×10^5, and 1 to 4 light chains of molecular weight 1.4–4.6×10^4. As described in section 5, there is no dispute that myosin contains two heavy chains, since its molecular weight is 4.8×10^5, but the number of light chains is not yet certain, although four have recently been reported. However, recent work suggests that the light chains are responsible for the ATPase and actin-combining activities of myosin. Hence it is increasingly important to know accurately the number, properties and binding sites of the light chains. The proteolysis of myosin to determine its substructure shows that the ATPase and actin-combining activities are completely localised in the head portion, S-1. Therefore if the light chains are the structural units necessary for these physiological functions, they must be present in the S-1. A schema of the myosin molecule and the structure of its various subunits based on this concept is shown in Fig. 9.

Trotta et al.[60] treated S-1 with alkali and isolated a fragment of molecular weight about 2×10^4, suggesting that fractions corresponding to the light chains are present in S-1. However, as described already, there is sig-

48

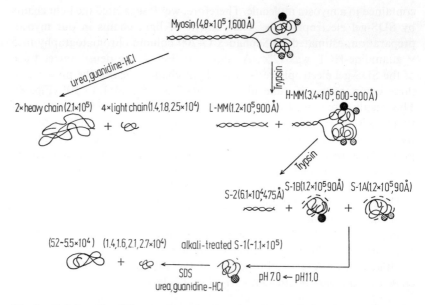

Fig. 9. Relationship of the myosin molecule to its subunits. Values in parentheses indicate molecular weight and length.

nificant cleavage of peptide links within the S-1 during isolation by the usual method, and it is consequently unsound to conclude that light chains are present in S-1 simply because there is a similarity in the molecular weights of the isolated fractions. Other methods of identification of the light chains are required to resolve this question, and Lowey and Steiner[97b] have recently shown, using fluorescein-labelled antibodies against the light chains, that the light chains are a structural component of S-1 of myosin and also of the thick filament in the myofibril.

As described in detail above, it is thought, from electron microscopy and from the yield of S-1 from H-MM, that the head portion of myosin is divided into two sections. Digestion of H-MM with trypsin separates the two S-1 fractions (S-1A and S-1B). It is important to know whether these fractions are identical, especially whether both contain active sites for the ATPase and, if so, whether their properties are the same. Similar questions are relevant to the sites for combining with actin. As described in detail in chapters 3 and 4, all our studies of the reaction mechanism of myosin-ATPase, the mechanism of binding of myosin with PP_i and ADP, and the chemical modification of active sites, indicate that one molecule of myosin contains one site for the simple hydrolysis of ATP and another for phosphorylation. Therefore, it is likely that these sites are located in either one of S-1A or S-1B, but not in both. This means at least that the ATPase activities of S-1A and S-1B are not identical. On the other hand, as described in chapter 6, all of the S-1 (S-1A and S-1B) combines with F-actin in the presence of Mg^{2+} but without ATP, and hence the actin-combining sites must be present in both fractions[98].

These conclusions have been supported by our studies[61] on the submolecular structure of subfragment-1. We first investigated the subunit structure of fractions a and b from S-1 using disc electrophoresis. Fraction a contained one main component in 7M urea at pH 8.0. Furthermore, fraction b in 7M urea at pH 4.0 contained at least three components, and the mobilities of two out of the three components coincided with those of the light chains isolated from myosin.

As shown in Fig. 6, the densitometer trace of alkali-treated S-1 in SDS-gel electrophoresis showed five polypeptide chains of molecular weights 5.2–5.5,* 2.7, 2.1, 1.6 and 1.4×10^4. Fraction a showed only one component of molecular weight 5.2×10^4 in SDS-gel electrophoresis, and other components were contained in fraction b. The chains of molecular weights 5.2–5.5 and 2.7×10^4 must be derived from the heavy chains of myosin and are called f' and f", respectively, since they are larger than intact light chains. H-Meromyosin contained components of molecular weights 5.2, 2.7, 2.3, 1.6 and 1.4×10^4, besides components whose molecular weights are larger than 7×10^4 (Fig. 10A). Furthermore, the changes in components of low molecular weights contained in myosin and H-meromyosin were followed during trypsin digestion by SDS-gel electrophoresis (Fig. 10). These results suggested strongly that the component of molecular weight 2.3×10^4 (g_1') in H-MM and that of a molecular weight 2.1×10^4 (g_1'') in S-1 are derived from g_1, while the component of molecular weight 1.6×10^4 (g_2') is derived from g_2 and the component of molecular weight 1.4×10^4 is g_3 itself. Myosin and H-MM molecules contain one g_1, two g_2 and one g_3 components, as described above, while the S-1 molecule contains only one g_2' chain. Thus, during digestion of H-MM with trypsin to form S-1, one of the two g_2' chains is digested into small peptides, and the other remained intact. However, the content of g_1'' in S-1 at a molar basis was usually much larger than that of g_1' in H-MM. For example, the molar ratio of f" : g_1'' : g_2' : g_3 in S-1, estimated from areas

* From gel filtration (see p. 39).

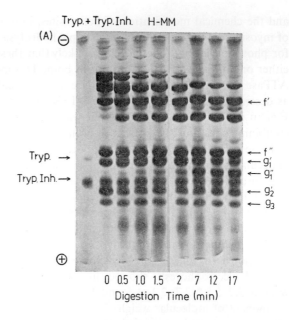
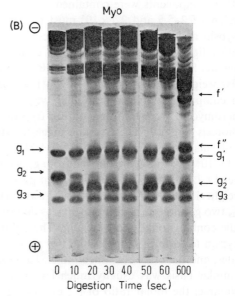

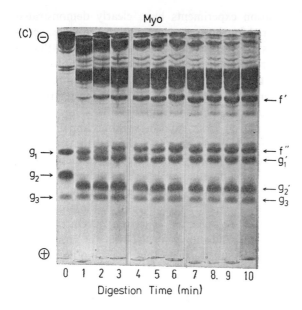

Fig. 10. Changes in submolecular structures of H-meromyosin and myosin during trypsin digestion, as revealed by SDS-gel electrophoresis. A: H-MM (17.7mg/ml) in 0.05M KCl and 0.05M Tris-HCl was digested with trypsin (20: 1 w/w) at pH 7.6 and 25°C. Note that the components of molecular weights 5.2 (f'), 2.7 (f''), 1.6 (g_2') and 1.4×10^4 (g_3) remained unchanged, while the component of molecular weight 2.3×10^4 (g_1') was converted by trypsin digestion to that of molecular weight 2.1×10^4 (g_1''). The first column shows SDS-gel electrophoresis of trypsin (Tryp.) and trypsin inhibitor (Tryp. Inh.). B and C: Myosin (16mg/ml) in 0.5M KCl and 0.05M phosphate buffer was digested with trypsin (100: 1 w/w) at pH 7.5 and 20°C. Note that the component of molecular weight 1.8×10^4 (g_2) was very rapidly converted to the component of molecular weight 1.6×10^4 (g_2'), and then the component of molecular weight 2.5×10^4 (g_1) was converted to g_1' (molecular weight 2.3×10^4). Note also that the components of molecular weight, 5.2×10^4 (f') and 2.7×10^4 (f'') appeared after a short-time digestion of myosin, while g_3 remained unchanged.

under peaks in densitometer tracing shown in Fig. 6, was 2.0: 2.5: 1.0: 0.8, while the molar ratio of f'': g_1': g_2': g_3 of H-MM was 2.0: 1.1: 2.3: 1.0. These results showed that, during digestion of H-MM to S-1 with trypsin, a considerable amount of the derivative from f-subunit was digested into a polypeptide as large as g_1''.

Recently, Dow and Stracher[98a] showed that adult chick myosin contains g_1, g_2 and g_3, while the embryonic form contains only g_1 and g_2. Recombination and hybridisation experiments have clearly demonstrated that only g_1 and g_2 are required for the expression of the full ATPase activity of myosin. These authors[98a] also gave evidence which suggests that g_3 may be derived from g_1. More recently, Weeds and Lowey[97a] showed that g_1 and g_3 are chemically related in that they both contain an identical thiol sequence, although there remains a possibility that the chemical structure of g_3 other than the thiol sequence is different from that of g_1. They[97a] also reported that g_2 can be removed from myosin without impairing the ATPase activity. Furthermore, Hayashi et al.[98b] showed that only one g_2' in the two was removed from H-meromyosin by treating H-meromyosin with PCMB and then with β-mercaptoethanol[61,90]. This did not affect the ATPase activity of H-meromyosin but it completely inhibited the substrate inhibition and the EGTA-sensitivity of H-meromyosin in the presence of F-actin and relaxing proteins[99] (cf. chapter 7, section 7). On the other hand, S-1 contained only one g_2' and this g_2' in S-1 was not removed by treatment with PCMB. These results suggest that g_1 is the component essential to the enzymatic activity, and that the two g_2 chains in the myosin molecule are not identical to each other, the one being removed from myosin by the PCMB-treatment and easily digested by trypsin, while the other, which is not removed by PCMB-treatment, being resistant to tryptic digestion and involved in S-1.

The molecular weight of S-1 was about 11×10^4, and the ratio of fraction a to b was 0.92 to 1.0. Furthermore, the two heavy chains of myosin can be assumed to be identical with each other, and the light chains account for 12% of the total myosin. It is rather difficult to construct a definite model of S-1, since g_1'' was contaminated by a polypeptide chain derived from the heavy chains, and since the possibility could not be excluded that even one myosin preparation purified from a single type of muscle contains several kinds of myosin molecules. However, we will tentatively adopt 1:1:1 as molar ratios of g_1'', g_2' and g_3 in a typical S-1 molecule. Basing on this assumption, all the results can be interpreted by assuming that S-1A, which has no ATPase activity, consists of four peptide chains of molecular weights 5.2–5.5 (f'), 2.7 (f''), 1.6(g_2') and 1.4×10^4 (g_3). The former two chains are derived from the heavy chains and the latter two chains are derived from g_2 and g_3 itself, respectively. S-1B, which has ATPase activity, comprises f', f'' and a light chain of molecular weight 2.1×10^4 (g_1'') derived from g_1. These tentative conclusions are summarised in Table VI and must be examined by determining the chemical structures not only of the components of S-1 but also of the light chains.

TABLE VI. Submolecular Structures of Myosin and Its Subfragment-1
Myosin$=2\times f(20.5\times 10^4)+g_1(2.5\times 10^4)+2\times g_2(1.8\times 10^4)+g_3(1.4\times 10^4)$

	obs.	calc.
Molecular weight	48×10^4	48.5×10^4
Heavy chain: Light chains	0.88 : 0.12	0.85 : 0.15

Subfragment-1 A$=f'(5.2-5.5\times 10^4)+f''(2.7\times 10^4)+g_2'(1.6\times 10^4)+g_3(1.4\times 10^4)$
              B$=f'(5.2-5.5\times 10^4)+f''(2.7\times 10^4)+g_1''(2.1\times 10^4)$

	obs.	calc.	
Molecular weight	10.7×10^4	S-1A	$10.9-11.2\times 10^4$
		S-1B	$10.0-10.3\times 10^4$
Fraction a : b	0.92 : 1		0.99−1.05 : 1
			($2f'$: $2f''+g_1''+g_2'+g_3$)

A detailed study of the structural and functional difference between S-1A and S-1B will be an important future problem in the elucidation of the role of myosin in muscle contraction, and we believe that subfragment-1A and B will be able to be reconstituted by self-assembly from their constituent peptide chains in the near future, since, as mentioned above, their submolecular structures are rather simple.*

7. The Formation of Myosin Filaments

After the electron micrographs of myosin filaments by Jakus and Hall[107], Noda and Ebashi[108] suggested from flow-birefringence studies of myosin solutions that filaments of about 1μ long are formed at low ionic strength. H. E. Huxley[12] showed by electron microscopy that a thick filament from rabbit psoas had a diameter of 100–120Å and a length of 1.5–1.6μ, its tip being tapered. On its surface were irregular looking projections, except for the central part of the filament where there is a bare region about 0.15–0.2μ long. At high ionic strength, negative staining methods showed no molecular image for myosin, but at ionic strengths as low as 0.1–0.2 a rod-shaped particle was observed[12], with a length of 0.2–0.5μ and a diameter

* Monod et al.[99a] suggested the presence of symmetry in the distribution of protomers in an enzyme molecule as a structural basis for allosteric regulation of enzymic activity. Their hypothesis has been supported by many experimental results[100]. However, the existence of asymmetric enzymes has recently been described by some investigators. Yeast epimerase is asymmetric; one dimer carries only one bound pyridine nucleotide, and not two[101, 102]. Tryptophan synthetase[103] and tryptophanase[104] which have covalently bound pyridoxal-phosphate show a related type of asymmetry, while lactose synthetase shows yet another type of asymmetry[105, 106].

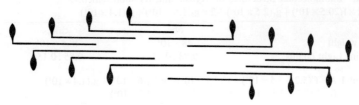

Fig. 11. The composition of the thick filament of myosin formed by aggregation. (Adapted from H. E. Huxley's data, Ref. *12*).

ranging from 60–200Å with no constant value. The irregular projections, the bare region of 0.15–0.2μ in the centre and the tapering at each end found with the thick filaments of muscle by H. E. Huxley were also observed with the myosin aggregates. Furthermore, Kaminer and Bell[13] demonstrated that myosin aggregates into its longest filaments (1.8μ) under normal physiological conditions (pH 6.5, I=0.1). Formation of synthetic myosin filaments has also been studied by Josephs and Harrington[109,110], Zobel and Carlson[34], Sanger[111], Takahashi[112] and Katsura and Noda[113]. At low ionic strength, L-MM formed filaments having no surface projections[12], showing that the backbone of the thick filament is composed of L-MM and the projections are probably part of the H-MM. The projections occur at intervals of about 400Å along a given longitudinal axis of the filament, and are staggered so that 6 projections correspond to one turn. Thus the reconstituted thick filament has projections containing H-MM on a 6-fold helical screw axis of period 60–70Å and has the same structure as a thick filament from living muscle, as described in chapter 1. The shortest filaments are 0.25–0.3μ long and contain approximately 2 molecules of myosin. Regardless of length, all filaments have a bare region 0.15–0.2μ long at the centre. The above data led H. E. Huxley[12] to propose the scheme of formation of myosin filaments shown in Fig. 11. More detailed periodic structure has been observed by Hanson, O'Brien and Bennett[113a] in assemblies of the thick filaments isolated from skeletal muscle. The structure is bipolar. The central M band (width $\sim$495Å) is flanked on either side by a heavily stained 'bare' zone (width $\sim$395Å), following which is a series of ten bands of equal width (420Å$\pm$15Å) showing a polarised structure. They concluded that the band pattern represents mainly the bipolar structure of the L-meromyosin backbone. The use of antibody staining techniques in fluorescent and electron microscopy has added considerably to our knowledge of the molecular anatomy of the myosin filament[114-119]. In particular, it was found that in different regions of the filament the length of the rod part of the myosin molecules involved in aggregation in the core of the filament is different. Pepe[115-119] derived a model for the detailed molecular packing of myosin molecules in the myosin filament. This is the most detailed model for the myosin filament yet obtained.

Another type of assembly of myosin has recently been observed by Harrison et al.[119a] In the presence of 0.075–0.1M KSCN, 0.05M $CaCl_2$ and 0.05M Tris-HCl at pH 8.2, myosin precipitates as a bipolar segment, showing a lightly stained central region about 900Å wide, on either side of which is heavily stained zone. The globular heads of the myosin molecules appear about 625Å from the light region. Harrison and his colleagues inferred that this bipolar segment consists of two oppositely oriented arrays of myosin molecules which overlap by 900Å.

Recently Godfrey and Harrington[120] measured a monomer-dimer equilibrium, and found it to have an equilibrium constant of 2.29×10^5 l/mole (at 6°C) even at high salt concentrations (0.5M KCl – 0.2M PO_4^{2-}). This explains not only the high molecular weights of myosin reported previously by Harrington and co-workers, but also why there is aggregation of myosin molecules at high salt concentrations. The investigation on the geometry of the myosin dimer is now in progress in Harrington's laboratory[120a], and it should help us to gain deeper understanding of the forces involved in the formation of myosin filaments. Furthermore, a new protein (an M line protein) with a molecular weight of 88,000 was isolated from skeletal muscle by Morimoto and Harrington[120b]. They presented evidence for its location on the M line structure and its association with myosin filaments to form large diffuse aggregates of the filaments of indefinite length and thickness.

8. Changes in Secondary Structure at the Active Sites of Myosin

As will be described in chapter 8, the development of tension in muscle fibre and its ATPase activity are associated with the interaction of the projections from the thick filaments with the thin filaments. As discussed in detail in chapter 13, it is highly probable that the driving force for muscle contraction may result from conformational changes in myosin at the position of its interaction with actin. In our pursuit of the molecular mechanism of contraction, we have come to believe that it is essential to study the structural changes in the myosin molecule.

The structure of proteins depends on four factors—the primary structure of the peptide chain, the secondary structure governing conformation of the peptide chain, the tertiary structure governing the shape of the peptide chain and the quaternary structure governing the distribution of peptide chains in the molecule. In proteins with secondary structures involving α-helices and random coils, the α-helix portions have a regular helical arrangement of asymmetric carbon atoms, resulting in an effective residue rotation, $[m']$[121,122]

$$[m'] = \frac{3}{n^2+2} \cdot \frac{M_0}{100} \, [\alpha] = \frac{a_0 \lambda^2_0}{\lambda^2 - \lambda^2_0} + \frac{b_0 \lambda^4_0}{(\lambda^2 - \lambda^2_0)^2},$$

where n is the refractive index of the solvent, λ_0 is the wavelength of the absorption maximum, $[\alpha]$ is the specific rotation, and M_0 is the average molecular weight of the amino acid residues. A plot of $[m'] \, (\lambda^2 - \lambda^2_0)$ *versus* $(\lambda^2 - \lambda^2_0)^{-1}$ (the Moffit-Yang plot) gives b_0 from its slope. According to

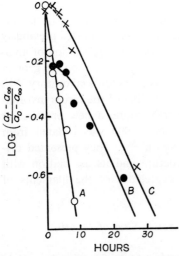

Fig. 12. The effect of PP$_i$ on the denaturation of myosin-ATPase by heat. 0.6M KCl, 0.1mM CaCl$_2$, pH 7.0, incubation at 30°C. A: Control; B: addition of 10mM PP$_i$ after incubation for 2 hr at 30°C; C: in the presence of 10mM PP$_i$. The ATPase activity was measured under the conditions of 0.6M KCl, 1mM ATP, 3mM Ca^{2+}, pH 7.0 and 21°C. a_0, a_t and a_∞ represent activities initially, at t hr, and after a sufficiently long time that no further reaction occurs at 30°C. When purified myosin was used, a_∞ was 0. On the other hand a_∞ was a definite value when myosin B was used, since myosin B was usually a mixture of myosin and actomyosin and actomyosin-ATPase was resistant to incubation at 30°C.

Yang and Doty, the absolute value of b_0 is proportional to the helical content of the polymer, and $-b_0$ is 600–700 for a perfect α-helix. The helical content of myosin indicated by this method is 55–60%[84,123]. In this way we have studied conformational changes in myosin by measuring the effects of various factors on its helical content.

Denaturation changes the helical content of myosin. The inactivation of myosin-ATPase follows first order kinetics at various temperatures, as shown in Fig. 12[124,125], and so it is a typical intramolecular change. Inactivation is affected by the addition of PP_i—a substrate analogue—which causes the appearance of an initial lag (Fig. 12). From measurements of the kinetics of inactivation under various conditions, the following mechanism has been proposed:

$$N \xrightarrow{k_1} N^* \xrightarrow{k_2} \text{denatured},$$

where the ATPase activity of the intermediate N^* is equal to that of N, the native state, in the presence of Ca^{2+} or EDTA, but is significantly lower

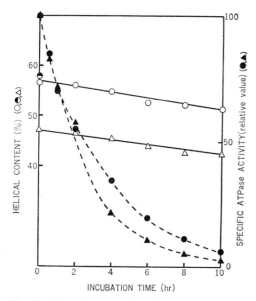

Fig. 13. The change in helical content of myosin on denaturation by acid or alkali. Incubated with 0.5M KCl and 50mM buffer at 20°C. The helical content was measured at pH 5.7 (○), 10.3 (△) and 7.0 (◓). The ATPase activity was measured at 0.5M KCl, 10mM $CaCl_2$, pH 7.0, and 21°C after incubation at pH 5.7 (●) or pH 10.3 (▲).

than N in the presence of Mg^{2+}. The results can be adequately explained by assuming that in the absence of PP_i, $k_2 \gg k_1$, while in the presence of PP_i, k_2 is reduced so that it approximates k_1. During denaturation by acid, alkali or heat the decrease in helical content is of only a few per cent, even when the activity has decreased significantly (Fig. 13)[125]. Two interpretations of the relationship between changes in helical content and activity are possible. One is that the active centre is non-helical, and hence the change in its secondary structure is not observable as a change in rotatory power. The other interpretation is that the active centre has an α-helix structure, but only this centre is changed by a local fusion accompanying denaturation. Much of the experimental work described later seems to support the latter interpretation.

Inorganic salts are of two kinds, depending on their effects on the structure of myosin[126]. One group contains KCl and NaCl, the other LiBr, LiCl, KI and KSCN. KCl is representative of the first group. An increase in KCl concentration decreases the ATPase activity, but the effect is completely reversible and subsequent reduction of KCl concentration restores the ATPase activity. Even in high concentrations of KCl, there is very little change in the viscosity or helical content of the myosin. In contrast, LiBr, which belongs to the second group, causes significant structural changes over a very narrow concentration range around 2.5M, when the helical content decreases drastically from 60% to about 10%, and η_{sp}/C decreases from 2.2 (100 ml/g) to 0.5 (Fig. 14). These changes are also apparently reversible. With LiCl, a similarly abrupt large structural change is observed around a concentration of 3.3M, and is also reversible. LiBr at a concentration of about 1.5M irreversibly decreases the ATPase activity before this large structural change takes place, but with LiCl, the decrease in ATPase activity is mostly reversible at concentrations up to 2.5M, although it is irreversible once the structural changes have completely occurred. Thus in 2.5M LiCl, the ATPase activity disappears with retention of the structure, and is recovered when the LiCl concentration is decreased. The reversibility of ATPase activity with LiCl treatment is improved by thiol protection[88].

Treatment of myosin with 1.5M LiBr followed by transfer to 0.6M KCl causes little change in helical content, but completely destroys ATPase activity. The bonding of trinitrobenzenesulphonate (TBS) to this myosin is intrinsically different from that to an intact myosin[127]. The binding of TBS to one specific lysine residue located near the active site of ATPase is inhibited by treatment of myosin with LiBr. This indicates that LiBr

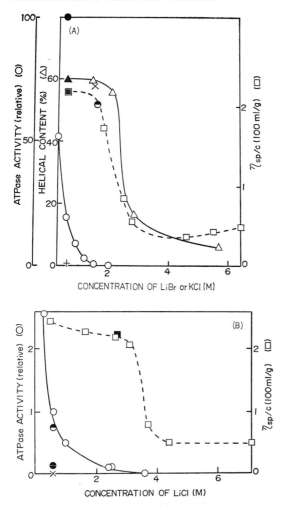

Fig. 14. Dependence of the helical content, reduced viscosity and ATPase activity of myosin on LiBr and LiCl concentration. A: LiBr, pH 7.0, 20°C. △, helical content; □, reduced viscosity; ○, ATPase activity relative to that in 0.6M KCl and 7mM Ca^{2+}. ×, helical content in 1.5M LiBr after incubation in 3M LiBr. ◓, reduced viscosity in 1.6M LiBr after incubation in 3.5M LiBr for 15–30 min. ▲, ■, ● show helical content, reduced viscosity and ATPase activity, respectively, in 0.6M KCl. +is the ATPase activity in 0.6M KCl after 5 min incubation in 2M LiBr. B: LiCl, pH 7.0, 20°C, □, reduced viscosity; ○, ATPase activity relative to 0.6M LiCl, 7mM Ca^{2+}. ■, reduced viscosity in 2.68M LiCl after 5 min incubation in 4.82M LiCl. ◓, ●, and ×, all ATPase activities in 0.6M LiCl after 5 min incubation in 2.45, 3.6 and 4.8M LiCl, respectively.

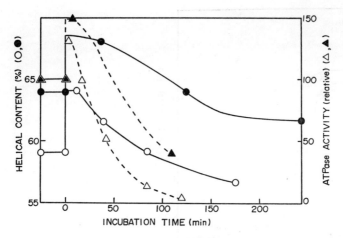

Fig. 15. Change in the ATPase activity and helical content of myosin on addition of dioxane. 0.6M KCl, 7mM $CaCl_2$, pH 7.0 and 20°C. Addition of 8% dioxane at time 0: ●, change of helical content; ▲, change in ATPase activity. Addition of 10% dioxane at time 0: ○, change of helical content; △, change in ATPase activity. The ATPase activity was measured in the presence of 7mM Ca^{2+} and 0.5mM ATP.

causes a characteristic change in the TBS binding site and hence in the secondary and tertiary structure near the active sites.

Dissolution in an organic solvent can change the structure and reactivity of myosin[128,129]. For example when myosin is dissolved in 8% dioxane, the ATPase activity immediately increases by about 50% and then gradually decreases after about 10 min. The helical content concurrently increases by about 10% and then gradually decreases, while the reduced viscosity decreases by about 17% and then later gradually increases (Fig. 15). Both the Michaelis constant (K_m) and the maximum velocity (V_{max}) of the ATPase increase in 8% dioxane, while at the same time the maximum and minimum, found to be near neutrality in the pH-activity curve of intact myosin, both disappear. The effect of dioxane is very similar to the effect of chemical modification of an SH group with p-chloromercuribenzoate (PCMB), as described in chapter 4. However, this is not due to attack on the SH group by a peroxide present in dioxane, but is a solvent effect[128,129]. The change in helical content induced by the addition of dioxane is prevented by PP_i, which is a competitive inhibitor for the ATPase. Addition of PP_i after the addition of dioxane causes no further change. When the helical content is reduced by several per cent by preliminary denaturation with heat, alkali or acid, there is no change on the subsequent addition of dioxane, showing that dioxane acts locally near the active site of the ATPase. The effect on H-MM is qualitatively similar but even more marked than that on myosin, but in support of the above conclusion there is no effect on L-MM.

It is particularly important to know the structural changes of enzymes upon reaction with their substrates. The structural change on bonding of myosin with actin must be determined to help elucidate the mechanism of contraction. The overall structure of the myosin molecule hardly changes upon addition of ATP, as is shown by the lack of alteration in intrinsic viscosity, sedimentation coefficient and intensity of light-scattering. In particular, no change in the sedimentation coefficient of myosin could be detected on adding ATP, even by the differential velocity sedimentation method[130]. We have observed a change induced by ATP of -5 to $+5\%$ in the helical content of myosin, although the direction of change depends on the season in which the preparation was made [123]. Thus in winter and early spring, $-b_0$ decreases with addition of ATP, while in other seasons it increases. The absolute amount of change is maximum at neutral pH, and is greater at lower temperatures, and the changes are inhibited by treatment of the myosin with PCMB and then with β-mercaptoethanol. Alteration in the helical content is observed even in the presence of EDTA, and hence it is difficult to attribute it to the formation of myosin-phosphate-ADP complex, since as shown in chapter 3, the complex is not formed when EDTA is present. There is no change with ADP, which is a substrate analogue, but PP_i always causes several per cent decrease in the helical content. As will be described in chapter 6, a decrease of several per cent in the helical content was observed when myosin combined with actin to form actomyosin. This is more clearly seen with acto-H-meromyosin, when complex formation decreases the helical content by 19%[131]. However, the amount of change in helical content of myosin on its binding with ATP or F-actin is very small, and very recently Gratzer and Lowey[132] reported that ATP does not change the helical content. Further investigation by the method of optical rotatory dispersion (ORD) is necessary to establish the change induced by ATP in the secondary structure of myosin.

The local conformational changes around the active site induced by the substrate or its analogues have recently been detected by methods other than ORD, such as the change in UV absorption spectra, the change in reactivity of myosin with some reagents, and spin-labelling. As will be described in the next chapter, changes in the absorption of H-meromyosin of around 280nm are induced by the addition of ATP or ADP[133-135]. A conformational change around the active site induced by ATP was recently indicated by the changes in the intrinsic fluorescence of the tryptophane residues of H-meromyosin induced by ATP and other analogues[135a,b]. A conformational change around the active site was also suggested by chemical modifications of the ATPase. For example, *p*-nitroaniline diazonium fluoroborate strongly inhibited the Ca^{2+}-ATPase activity, but slightly depressed the EDTA-ATPase activity. Conversely, when ATP or ADP was present in the coupling reaction mixture, the diazonium compound suppressed the EDTA-ATPase, but only slightly inhibited the Ca^{2+}-ATPase[136]. More recently, Seidel *et al.*[137] reported that the spin label, N-2,2,6,6-tetramethyl piperidine nitroxide iodoacetamide, reacted to two rapidly reacting thiol groups per mole of myosin (see chapter 4, section 3).* The addition of ATP, ITP, ADP or PP_i increased the mobility of spin labels bound to the thiol groups. Lately, Seidel and Gergely[137a] have reported that the EPR spectrum of the spin labels during the steady-state of ATP-hydrolysis shows a marked increase in their mobility. A similar change induced by ADP in the EPR spectrum of the spin-labelled myosin was also reported by Stone[139]. Furthermore, the fluorescence of a macromolecular probe, 8-anilino-1-naphthalenesulfonate, bound to myosin suffered a small reversible decrease in the presence of ATP, which was correlated with the kinetics of hydrolysis[140]. Our kinetic analysis of the ATPase[141] also indicates a conformational change of myosin induced by Mg-ATP (chapter 3, section 5). On adding Mg-ATP, myosin changed to a new conformational state, and this change accompanied a dramatic acceleration of the conversion of the myosin-ATP complex from $E_2{}^{1S}$ to E_2S^1 (see p. 90).

The above experimental results suggest that the conformational change in the myosin molecule occurs rather locally near the active site upon the addition of ATP. It has been suggested recently that when the light chains are dissociated from myosin, there is about a 10% decrease in helical content[84]. It seems reasonable to assume that the ATPase activity is controlled by induction of a conformational change at the active site through

---

* An excellent introduction to nitroxide spectra and spin-labelling is given by McConnell and his co-workers[138, 138a]. Briefly, the protein is labelled with a molecule containing the nitroxide group ($>$N-O), which has a simple three line EPR spectrum which is sensitive to the rotational rate of the spin label. Changes in protein structure induced by chemical reagents are detected from changes in the EPR spectrum of the spin label.

the interaction of light and heavy chains (see chapter 13). In the light of these ideas we believe that re-investigation of the change in helical content using the smallest active fragment (S-1) currently known, instead of myosin itself, will clarify the conformational changes in myosin and help elucidate the molecular mechanism of muscle contraction.

REFERENCES

1 W. Kühne, *Arch. Anat. u. Physiol.*, 748 (1859).
2 J. T. Edsall, *J. Biol. Chem.*, **89**, 289 (1930).
3 A. von Muralt and J. T. Edsall, *J. Biol. Chem.*, **89**, 315 (1930).
4 A. von Muralt and J. T. Edsall, *J. Biol. Chem.*, **89**, 351 (1930).
5 H. H. Weber, *Ergeb. Physiol.*, **36**, 109 (1933).
6 W. A. Engelhardt and M. N. Ljubimova, *Nature*, **144**, 668 (1939).
7 I. Banga and A. Szent-Györgyi, *Stud. Inst. Med. Chem. Univ. Szeged*, **1**, 5 (1941).
8 A. Szent-Györgyi, "Chemistry of Muscular Contraction," 1st ed. (1947) & 2nd ed. (1951), Academic Press, New York.
9 W. Hasselbach and G. Schneider, *Biochem. Z.*, **321**, 462 (1951).
10 H. E. Huxley and J. Hanson, *Biochim. Biophys. Acta*, **23**, 229 (1957).
11 J. Hanson and H. E. Huxley, *Biochim. Biophys. Acta*, **23**, 250 (1957).
12 H. E. Huxley, *J. Mol. Biol.*, **7**, 281 (1963).
13 B. Kaminer and A. L. Bell, *J. Mol. Biol.*, **20**, 391 (1966).
14 T. Hayashi, R. Rosenbluth, P. Satir and M. Vozick, *Biochim. Biophys. Acta*, **28**, 1 (1958).
15 F. Guba and F. B. Straub, *Stud. Inst. Med. Chem. Univ. Szeged*, 3, 46 (1943).
16 S. V. Perry, *in* "Methods in Enzymology," ed. by S. P. Colowick and N. O. Kaplan, Academic Press, New York, Vol. 2, p. 582 (1955).
17 A. Weber, *Biochim. Biophys. Acta*, **19**, 345 (1956).
18 E. Mihályi, K. Laki and M. I. Knoller, *Arch. Biochem. Biophys.*, **68**, 130 (1957).
19 S. V. Perry and M. Zydowo, *Biochem. J.*, **72**, 682 (1959).
20 E. G. Richards, C.-S. Chung, D. B. Menzel and H. S. Olcott, *Biochemistry*, **6**, 528 (1967).
21 M. Harris and C. H. Suelter, *Biochim. Biophys. Acta*, **133**, 393 (1967).
22 S. V. Perry, *Biochem. J.*, **74**, 94 (1960).
22a R. Starr and G. Offer, *FEBS Letters*, **15**, 40 (1971).
23 E. F. Rossomando and K. A. Piez, *Biochem. Biophys. Res. Commun.*, **40**, 800 (1970).
24 O. Snellman and T. Erdös, *Biochim. Biophys. Acta*, **2**, 650 (1948).

25 H. K. Schachman, "Ultracentrifugation in Biochemistry," Academic Press, New York (1959).
26 K. F. Van Holde and R. L. Baldwin, *J. Phys. Chem.*, **62**, 734 (1958).
27 W. J. Archibald, *J. Phys. Colloid Chem.*, **51**, 1204 (1947).
28 B. H. Zimm, *J. Polymer Sci.*, **14**, 29 (1948).
29 Y. Tonomura, P. Appel and M. Morales, *Biochemistry*, **5**, 515 (1966).
30 A. Holtzer and S. Lowey, *J. Amer. Chem. Soc.*, **81**, 1370 (1959).
31 J. G. Kirkwood and J. Riseman, *J. Chem. Phys.*, **18**, 512 (1950).
32 R. V. Rice, *Biochim. Biophys. Acta*, **52**, 602 (1961).
33 R. V. Rice, in "Biochemistry of Muscle Contraction," ed. by J. Gergely, Little Brown and Co., Boston, p. 41 (1964).
34 C. R. Zobel and F. D. Carlson, *J. Mol. Biol.*, **7**, 78 (1963).
35 H. S. Slayter and S. Lowey, *Proc. Natl. Acad. Sci. U. S.*, **58**, 1611 (1967).
36 F. A. Pepe and B. Drucker, cited in F. A. Pepe, *Progr. Biophys. Mol. Biol.*, **22**, 77 (1971).
36a K. Takahashi, unpublished.
37 S. V. Perry, *Biochem. J.*, **47**, XXXVIII (1950).
38 S. V. Perry, *Biochem. J.*, **48**, 257 (1951).
39 J. Gergely, *Federation Proc.*, **9**, 176 (1950).
40 J. Gergely, *J. Biol. Chem.*, **200**, 543 (1953).
41 J. Gergely, M. A. Gouvea and D. Karibian, *J. Biol. Chem.*, **212**, 165 (1955).
42 E. Mihályi and A. G. Szent-Györgyi, *J. Biol. Chem.*, **201**, 189 (1953).
43 A. G. Szent-Györgyi, *Arch. Biochem. Biophys.*, **42**, 305 (1953).
44 A. G. Szent-Györgyi, C. Cohen and D. E. Philpott, *J. Mol. Biol.*, **2**, 133 (1960).
45 S. Lowey, *Science*, **145**, 597 (1964).
46 D. R. Kominz, A. Hough, P. Symonds and K. Laki, *Arch. Biochem. Biophys.*, **50**, 148 (1954).
47 H. Mueller and S. V. Perry, *Biochem. J.*, **85**, 431 (1962).
48 R. V. Rice, A. C. Brady, R. H. Depue and R. E. Kelly, *Biochem. Z.*, **345**, 370 (1966).
49 S. Lowey and A. Holtzer, *Biochim. Biophys. Acta*, **34**, 470 (1959).
50 S. Lowey and C. Cohen, *J. Mol. Biol.*, **4**, 293 (1962).
51 D. M. Young, S. Himmelfarb and W. F. Harrington, *J. Biol. Chem.*, **240**, 2428 (1965).
52 K. M. Nauss and J. Gergely, *Federation Proc.*, **26**, 727 (1967).
53 S. Lowey, L. Goldstein, C. Cohen and S. M. Luck, *J. Mol. Biol.*, **23**, 287 (1967).
54 S. Lowey, H. S. Slayter, A. G. Weeds and H. Barker, *J. Mol. Biol.*, **42**, 1 (1969).
55 Y. Hayashi and Y. Tonomura, *J. Biochem.*, **68**, 665 (1970).
56 K. M. Nauss, S. Kitagawa and J. Gergely, *J. Biol. Chem.*, **244**, 755 (1969).
57 K. Yagi and Y. Yazawa, *J. Biochem.*, **60**, 450 (1966).

58 E. Eisenberg and C. Moos, *Biochemistry*, **7**, 1486 (1968).
59 E. Eisenberg, C. R. Zobel and C. Moos, *Biochemistry*, **7**, 3186 (1968).
60 P. P. Trotta, P. Dreizen and A. Stracher, *Proc. Natl. Acad. Sci. U. S.*, **61**, 659 (1968).
61 Y. Hayashi, *J. Biochem.*, **72**, 83 (1972).
62 A. L. Shapiro, E. Viñuela and J. V. Maizel, *Biochem. Biophys. Res. Commun.*, **28**, 815 (1967).
63 K. Weber and M. Osborn, *J. Biol. Chem.*, **244**, 4406 (1969).
64 A. K. Dunker and R. R. Rueckert, *J. Biol. Chem.*, **244**, 5074 (1969).
65 J. A. Reynolds and C. Tanford, *J. Biol. Chem.*, **245**, 5161 (1970).
66 K. Yagi, Y. Yazawa and T. Yasui, *Biochem. Biophys. Res. Commun.*, **29**, 331 (1967).
67 T. -C. Tsao, *Biochim. Biophys. Acta*, **11**, 368 (1953).
68 W. W. Kielley and W. F. Harrington, *Biochim. Biophys. Acta*, **41**, 401 (1960).
69 P. A. Small, W. F. Harrington and W. W. Kielley, *Biochim. Biophys. Acta*, **49**, 462 (1961).
70 W. W. Kielley and L. M. Barnett, *Biochim. Biophys. Acta*, **51**, 589 (1961).
71 M. Kimura and W. W. Kielley, *Biochem. Z.*, **345**, 188 (1966).
72 A. G. Weeds and B. S. Hartley, *Biochem. J.*, **107**, 531 (1968).
73 G. Huszar and M. Elzinga, *Biochemistry*, **10**, 229 (1971).
73a G. W. Offer and R. Starr, presented at the Symposium on Molecular Biophysics of the International Union of Pure and Applied Chemistry, Cambridge (1968).
73b P. J. Bechtel, A. P. Pearson and C. E. Bodwell, *Federation Proc.*, **30**, 1310 (1971).
73c R. G. Hale and G. R. Beecher, *FEBS Letters*, **18**, 245 (1971).
74 D. B. Wetlaufer and J. T. Edsall, *Biochim. Biophys. Acta*, **43**, 132 (1960).
75 P. Dreizen, D. J. Hartshorne and A. Stracher, *J. Biol. Chem.*, **241**, 443 (1966).
76 D. R. Kominz, W. R. Carroll, E. N. Smith and E. R. Mitchell, *Arch. Biochem. Biophys.*, **79**, 191 (1959).
77 L. C. Gershman, P. Dreizen and A. Stracher, *Proc. Natl. Acad. Sci. U. S.*, **56**, 966 (1966).
78 H. Oppenheimer, K. Bárány, G. Hamoir and J. Fenton, *Arch. Biochem. Biophys.*, **120**, 108 (1967).
79 R. H. Locker and C. J. Hagyard, *Arch. Biochem. Biophys.*, **120**, 454 (1967).
80 R. H. Locker and C. J. Hagyard, *Arch. Biochem. Biophys.*, **122**, 521 (1967).
81 R. H. Locker and C. J. Hagyard, *Arch. Biochem. Biophys.*, **127**, 370 (1968).
82 W. T. Perrie and S. V. Perry, *Biochem. J.*, **119**, 31 (1970).
83 F. J. Samaha, L. Guth and R. W. Albers, *J. Biol. Chem.*, **245**, 219 (1970).
83a S. Lowey and D. Risby, *Nature*, **234**, 81 (1971).
83b A. G. Weeds and B. Pone, *Nature*, **234**, 85 (1971).
83c S. Sarkar, F. A. Sreter and J. Gergely, *Proc. Natl. Acad. Sci. U. S.*, **68**, 946 (1971).

84 D. W. Frederiksen and A. Holtzer, *Biochemistry*, **7**, 3935 (1968).
85 L. C. Gershman, A. Stracher and P. Dreizen, *J. Biol. Chem.*, **244**, 2726 (1969).
86 L. C. Gershman and P. Dreizen, *Biochemistry*, **9**, 1677 (1970).
87 A. Stracher, *Biochem. Biophys. Res. Commun.*, **35**, 519 (1969).
88 P. Dreizen and L. C. Gershman, *Biochemistry*, **9**, 1688 (1970).
89 E. Gaetzens, K. Bárány, G. Bailin, H. Oppenheimer and M. Bárány, *Arch. Biochem. Biophys.*, **123**, 82 (1968).
90 J. Gazith, S. Himmelfarb and W. F. Harrington, *J. Biol. Chem.*, **245**, 15 (1970).
91 B. Paterson and R. C. Strohman, *Biochemistry*, **9**, 4094 (1970).
92 S. Sarker and P. H. Cooke, *Biochem. Biophys. Res. Commun.*, **41**, 918 (1970).
93 B. B. Low, J. N. Vournakis and A. Rich, *Biochemistry*, **10**, 1813 (1971).
94 S. M. Heywood, R. Dowben and A. Rich, *Proc. Natl. Acad. Sci. U. S.*, **57**, 1002 (1967).
95 S. M. Heywood and A. Rich, *Proc. Natl. Acad. Sci. U. S.*, **59**, 590 (1968).
96 S. M. Heywood, R. Dowben and A. Rich, *Biochemistry*, **7**, 3289 (1968).
97 S. M. Heywood and M. Nwagwa, *Proc. Natl. Acad. Sci. U. S.*, **60**, 229 (1968).
97a A. G. Weeds and S. Lowey, *J. Mol. Biol.*, **61**, 701 (1971).
97b S. Lowey and L. A. Steiner, *J. Mol. Biol.*, **65**, 111 (1972).
98 J. M. Jones and S. V. Perry, *Biochem. J.*, **100**, 120 (1966).
98a J. Dow and A. Stracher, *Proc. Natl. Acad. Sci. U. S.*, **68**, 1107 (1971).
98b Y. Hayashi, H. Takenaka and Y. Tonomura, unpublished.
99 Y. Tonomura, J. Yoshimura and S. Kitagawa, *J. Biol. Chem.*, **236**, 1968 (1961).
99a J. Monod, J. Wyman and P. Changeux, *J. Mol. Biol.*, **12**, 88 (1965).
100 A. Engström and B. Strandberg, "Symmetry and Function of Biological Systems at the Macromolecular Level (Nobel Symposium 11)," Almqvist & Wiksell, Stockholm (1969).
101 R. A. Darrow and R. Rodstrom, *Proc. Natl. Acad. Sci. U. S.*, **55**, 205 (1966).
102 A. U. Bertland and H. M. Kalckar, *Proc. Natl. Acad. Sci. U. S.*, **61**, 629 (1968).
103 T. Creighton and C. Yanofsky, *J. Biol. Chem.*, **241**, 980 (1966).
104 Y. Morino and E. E. Snell, *J. Biol. Chem.*, **242**, 5602 (1967).
105 U. Brodbeck and K. E. Ebner, *J. Biol. Chem.*, **241**, 762 (1966).
106 K. Brew, T. C. Vanaman and R. L. Hill, *Proc. Natl. Acad. Sci. U. S.*, **59**, 491 (1968).
107 M. A. Jakus and C. E. Hall, *J. Biol. Chem.*, **167**, 705 (1947).
108 H. Noda and S. Ebashi, *Biochim. Biophys. Acta*, **41**, 386 (1960).
109 R. Josephs and W. F. Harrington, *Biochemistry*, **5**, 3474 (1966).
110 R. Josephs and W. F. Harrington, *Biochemistry*, **7**, 2834 (1968).
111 J. W. Sanger, *J. Gen. Physiol.*, **50**, 2493 (1967).
112 K. Takahashi, *Protein, Nucleic Acid, Enzyme*, **12**, 61 (1967) (in Japanese).

113   I. Katsura and H. Noda, *J. Biochem.*, **69**, 219 (1971).
113a J. Hanson, E. J. O'Brien and P. M. Bennett, *J. Mol. Biol.*, **58**, 865 (1971).
114   F. A. Pepe and H. E. Huxley, in "Biochemistry of Muscle Contraction," ed. by J. Gergely, Little Brown and Co., Boston, p. 320 (1963).
115   F. A. Pepe, *J. Cell Biol.*, **28**, 505 (1966).
116   F. A. Pepe, *J. Mol. Biol.*, **27**, 203 (1967).
117   F. A. Pepe, *J. Mol. Biol.*, **27**, 227 (1967).
118   F. A. Pepe, *Int. Rev. Cytol.*, **24**, 193 (1968).
119   F. A. Pepe, *Progr. Biophys. Mol. Biol.*, **22**, 77 (1971).
119a R. G. Harrison, S. Lowey and C. Cohen, *J. Mol. Biol.*, **59**, 531 (1971).
120   J. E. Godfrey and W. F. Harrington, *Biochemistry*, **9**, 894 (1970).
120a W. F. Harrington and M. Burke, *Biochemistry*, **11**, 1448 (1972).
120b K. Morimoto and W. F. Harrington, *J. Biol. Chem.*, **247**, 3052 (1972).
121   W. Moffit, *J. Chem. Phys.*, **25**, 467 (1956).
122   P. Urnes and P. Doty, *Advan. Protein Chem.*, **16**, 402 (1961).
123   K. Sekiya, S. Mii, K. Takeuchi and Y. Tonomura, *J. Biochem.*, **59**, 584 (1966).
124   T. Yasui, Y. Hashimoto and Y. Tonomura, *Arch. Biochem. Biophys.*, **87**, 55 (1960).
125   K. Takahashi, T. Yasui, Y. Hashimoto and Y. Tonomura, *Arch. Biochem. Biophys.*, **99**, 45 (1962).
126   Y. Tonomura, K. Sekiya and K. Imamura, *J. Biol. Chem.*, **237**, 3110 (1962).
127   Y. Tonomura, J. Yoshimura and T. Ohnishi, *Biochim. Biophys. Acta*, **78**, 698 (1963).
128   Y. Tonomura, S. Tokura, K. Sekiya and K. Imamura, *Arch. Biochem. Biophys.*, **95**, 229 (1961).
129   Y. Tonomura, K. Sekiya and K. Imamura, *Biochim. Biophys. Acta*, **69**, 296 (1963).
130   J. E. Godfrey and W. F. Harrington, *Biochemistry*, **9**, 886 (1970).
131   Y. Tonomura, K. Sekiya and K. Imamura, *Biochim. Biophys. Acta*, **78**, 690 (1963).
132   W. B. Gratzer and S. Lowey, *J. Biol. Chem.*, **244**, 22 (1969).
133   F. Morita and K. Yagi, *Biochem. Biophys. Res. Commun.*, **22**, 297 (1966).
134   F. Morita, *J. Biol. Chem.*, **242**, 4501 (1967).
135   K. Sekiya and Y. Tonomura, *J. Biochem.*, **61**, 787 (1967).
135a F. Morita, in "Molecular Mechanisms of Enzyme Action," ed. by Y. Ogura, Y. Tonomura and T. Nakamura, University of Tokyo Press, Tokyo, p. 281 (1972).
135b M. M. Werber, A. G. Szent-Györgyi and G. D. Fasman, *Biochemistry*, **11**, 2872 (1972).
136   T. Yamashita, S. Kobayashi and T. Sekine, *J. Biochem.*, **65**, 869 (1969).
137   J. C. Seidel, M. Chopek and J. Gergely, *Biochemistry*, **9**, 3265 (1970).

137a J. C. Seidel and J. Gergely, *Biochem. Biophys. Res. Commun.*, **44**, 826 (1971).
138 C. L. Hamilton and H. M. McConnell, *in* "Structural Chemistry and Molecular Biology," ed. by A. Rich and N. Davidson, W. H. Freeman and Co., San Francisco, p. 115 (1968).
138a H. M. McConnell and B. G. McFarland, *Quart. Rev. Biophysics*, **3**, 91 (1970).
139 D. B. Stone, *Arch. Biochem. Biophys.*, **141**, 378 (1970).
140 H. C. Cheung, *Biochim. Biophys. Acta*, **194**, 478 (1969).
141 A. Inoue, K. Sekiya-Shibata and Y. Tonomura, *J. Biochem.*, **71**,115(1972).
142 H. Portzehl, G. Schramm and H. H. Weber, *Z. Naturforsch.*, **5b**, 61 (1950).
143 K. Laki and W. R. Carroll, *Nature*, **175**, 389 (1955).
144 A. Holtzer and S. Lowey, *J. Amer. Chem. Soc.*, **78**, 5954 (1956).
145 P. Johnson and A. J. Rowe, *Biochem. J.*, **74**, 432 (1960).
146 G. Kaldor, J. Gitlin, F. Westley and B. W. Volk, *Biochemistry*, **3**, 1137 (1964).
147 I. P. Trayer and S. V. Perry, *Biochem. Z.*, **345**, 87 (1966).
148 E. F. Woods, S. Himmelfarb and W. F. Harrington, *J. Biol. Chem.*, **238**, 2374 (1963).
149 W. F. H. M. Mommaerts and B. B. Aldrich, *Biochim. Biophys. Acta*, **28**, 627 (1958).
150 P. H. von Hippel, H. K. Schachman, P. Appel and M. F. Morales, *Biochim. Biophys. Acta*, **28**, 504 (1958).
151 H. Mueller, *J. Biol. Chem.*, **239**, 797 (1964).
152 J. J. Connell and I. M. Mackie, *Nature*, **201**, 78 (1964).
153 J. Brahms, *J. Amer. Chem. Soc.*, **81**, 4997 (1959).
154 A. Holtzer, S. Lowey and T. M. Schuster, *in* "Molecular Basis of Neoplasia," Univ. of Texas Press, Austin, Texas, p. 259 (1962).
155 M. F. Gellert and S. W. Englander, *Biochemistry*, **2**, 39 (1963).
156 Y. Tomimatsu, *Biopolymers*, **2**, 275 (1964).
157 H. Asai, *Biochemistry*, **2**, 458 (1963).
158 H. Portzehl, *Z. Naturforsch.*, **5b**, 75 (1950).
159 J. Gergely and H. Kohler, *in* "Conf. Chem. Muscular Contraction," Igaku Shoin, Tokyo, p. 14 (1957).
160 D. M. Young, S. Himmelfarb and W. F. Harrington, *J. Biol. Chem.*, **239**, 2822 (1964).
161 H. Mueller, *J. Biol. Chem.*, **240**, 3816 (1965).
162 H. Tokuyama, S. Kubo and Y. Tonomura, *Biochem. Z.*, **345**, 57 (1966).

3
REACTION MECHANISM OF MYOSIN-ATPase*

The interaction between myosin and ATP is the key reaction in muscle contraction. Therefore, to clarify the molecular mechanism of muscle contraction it is essential to clarify the kinetic properties of the reaction intermediates which are formed in the ATPase reaction. In this chapter we will describe the reaction mechanism of ATPase, and the effects of F-actin on the myosin-ATP system will be described in chapters 6 and 7.

1. Energy Changes Accompanying the Hydrolysis of ATP

Adenosine triphosphate (ATP) was isolated from muscle in 1929 by Lohmann[1,2] and Fiske and SubbaRow[3], who suggested its structure to be that shown in Fig. 1. Baddiley et al.[4] confirmed this structure by synthesis in 1949. Recently Kennard and his colleagues[5] reported the three

* Contributors: Masato Ohe and Hiroshi Nakamura

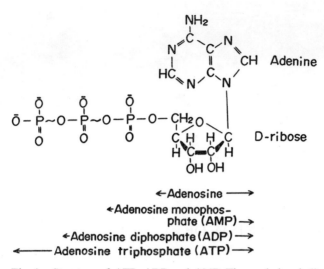

Fig. 1. Structure of ATP, ADP and AMP. The symbol ~ indicates the high energy phosphate bond.

dimensional structure of the hydrated disodium salt of ATP at a resolution of 0.9Å. It is well known that the synthesis of ATP in living organisms is coupled with a redox reaction, and occurs mainly in the three processes of glycolysis, oxidative phosphorylation and photosynthetic phosphorylation (cf. chapter 14).

Lipmann[6] pointed out that the particular importance of ATP derives from the large amount of energy liberated during its hydrolysis, even though it is relatively stable under normal physiological conditions. Podolsky and Morales[7] and Kitzinger and Benzinger[8] measured the heat generated by the reaction of ATP with myosin and determined the standard enthalpy change of hydrolysis of ATP to be -4.7 and -4.8 kcal/mole at pH 8.0 and 20°C in 0.6M KCl, respectively. The Gibbs free energy of hydrolysis, ΔG_{obs}, was calculated by Benzinger et al.[9] to be -8.16 kcal/mole at pH 7.0, 37°C, I=0.2 and 35mM Mg^{2+}, from the equilibrium constant for the glutamine synthetase

glutamate + ATP + ammonia ⇌ glutamine + ADP + P_i

and the equilibrium constant for the hydrolysis of glutamine

glutamine + H_2O ⇌ glutamate + ammonia.

When the correction for the binding of glutamate by Mg^{2+} is included,

ΔG_{obs} becomes to be -8.98 kcal/mole[10]. More recently, the values for the equilibrium of the glutamine synthetase reaction and for the stability constant of the Mg^{2+}-glutamate were re-examined by Rosing and Slater[10a]. The value for ΔG_{obs} of the ATP-hydrolysis calculated from these new data is -7.36 kcal/mole under the same conditions. The dissociation constants of H^+ from ATP have been measured by Alberty and his co-workers[11,12] and Phillips et al.[13,14] The dissociation constants of the complexes of ATP with divalent cations have been determined by many researchers[11,15-23].* Recently, Phillips et al.[10], Alberty[24] and Rosing and Slater[10a] calculated the Gibbs free energy change as a function of pH and Mg^{2+} concentration, using these dissociation constants. The rate constants of formation and dissociation of ATP and ADP complexes with Ca^{2+} and Mg^{2+} were determined by Eigen and Hammes[25-28]. Since they are very large, these reaction steps cannot be the rate-determining ones in the ATPase reaction.

The origin of the large decrease in ΔG resulting from hydrolysis of the pyrophosphate bond has been investigated by Pullman and Pullman[29] who performed quantum mechanical calculations of the energy of hydrolysis of ATP and similar compounds. These calculations suggested that both the 'opposing resonance' proposed by Kalckar[30] and the intramolecular electrostatic repulsion proposed by Hill and Morales[31,32] are important factors. More recently, George et al.[33] have emphasised differences between solvation energies for reactant and product species as a major factor contributing to the greater thermodynamic instability of 'high energy' phosphate compounds. Two routes for the hydrolysis of the terminal phosphate group of ATP are possible, since cleavage could occur at either the —O$\underset{\uparrow}{-}$P or $\underset{\uparrow}{-}$O–P positions. This question has been much investigated in the analogous case of ester hydrolysis, for which diagnostic techniques have been developed. Koshland et al.[34] used the identical procedures to study the hydrolysis of ATP by myosin in solutions containing Ca^{2+} and $H_2^{18}O$, in order to determine which of the reaction products, ADP and P_i, incorporated ^{18}O. They found that only the P_i incorporated one mole of ^{18}O per mole, showing that cleavage takes place at the O–P bond.**

---

* See chapter 4, section 6 for the structure of the complexes of ATP with divalent cations.
** As described in section 8, in the presence of Mg^{2+} one mole of P_i incorporates 2 moles or more of ^{18}O.

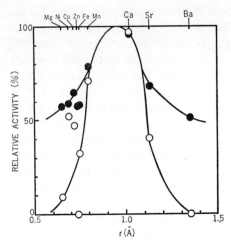

Fig. 2. Relationship between the ionic radius of added divalent metal ions and ATPase activity. The abscissa shows the ionic radius (r) of Me^{2+}, the divalent metal ion. The ordinate shows the ATPase activity relative to 100 when $r=0.95$Å. 1mM ATP, pH 8.2, 22°C. Myosin B was used as the enzyme, so that myosin-type ATPase activity was observed in 0.6M KCl, and actomyosin-type ATPase activity in 0.14M KCl. ○, 0.6M KCl, 5mM Me^{2+}; ●, 0.14M KCl, 1mM Me^{2+}.

2. The Steady-state Reaction of Myosin with ATP

Myosin-ATPase reacts in a complex and characteristic manner with ATP. Initiation of the reaction followed by quenching with trichloroacetic acid (TCA) results in the rapid initial discharge of P_i followed by a further slower constant discharge. Thus there is a pre-steady-state phase and then a steady-state. The pH-activity curve for myosin-ATPase is also complicated, indicating that the reaction mechanism is involved. Therefore, this chapter first describes the steady-state reaction of myosin-ATPase, then later discusses a more detailed mechanism based on studies of the pre-steady-state. The actomyosin-ATPase, which differs considerably from the myosin-ATPase, will not be discussed until a later chapter.

We have followed the reaction of myosin with ATP by three methods. In the first, the hydrolysis was quenched by the addition of TCA after a certain lapse of time, and the P_i liberated was determined, usually by the Martin-Doty technique[35]. In the second method phosphoenol pyruvate (PEP) and pyruvate kinase were added, and made to react quantitatively with the ADP formed to produce pyruvate. The pyruvate was converted to the 2,4-dinitrophenyl hydrazone on reaction with 2,4-dinitrophenyl hydrazine, and on subsequent addition of NaOH, formed a stable brown colour which can be determined colourimetrically[36]. The third method measured the liberation of protons as the hydrolysis proceeded, and could be used most easily at or near pH 8.0, when hydrolysis of one mole of ATP generated 1 mole of protons[37].

Engelhardt and Ljubimova[38] and Banga[39] have shown that myosin-ATPase is greatly affected by externally added divalent metal ions, and Fig. 2 shows the relationship between the ionic radius and the ATPase activity. The activity is highest at an ionic radius of about 1Å, and decreases at smaller or greater radii[40]. Two mechanisms can be advanced for the case when the enzymic activity is solely dependent upon the metal ion. Either binding of the enzyme to the metal ion activates the enzyme, or else complex formation between the metal ion and the substrate activates the substrate. In either case, at sufficiently high substrate concentrations the relationship between the hydrolysis rate and the divalent metal ion concentration would be given by $v = V / \left(1 + \dfrac{K_{Me}}{[Me]}\right)$, where V is the rate when the concentration of Me is high and K_{Me} is the dissociation constant of the metal ion-enzyme complex in the first case or of the metal ion-substrate complex in the second case. Comparison of the value of K_{Me} found from kinetic measurements with the dissociation constants of the metal ion-enzyme and metal ion-substrate complexes measured directly can be used to determine which of the two mechanisms is the more plausible. Another method to differentiate these two mechanisms is to express the velocity in terms of free concentrations of the divalent cation, Me^{2+}, and ATP. If it is found that the apparent Michaelis constant (K_m) for ATP_{free} at one concentration of Me^{2+}_{free} is equal to the apparent K_m for Me^{2+}_{free} at the same concentration of ATP_{free}, then the mechanism of the reaction involves the addition of Me-ATP to the enzyme[41].

The K_{Me} value found from our work[42] using Ca^{2+} agrees with the dissociation constant of ATP with Ca^{2+} measured by Smith and Alberty[11]. Hence Ca^{2+} seems to activate the hydrolysis by the second mechanism (cf. Ref. 43). However, increasing the Ca^{2+} concentration to very high levels causes an inhibition of the ATPase activity which can be related to the Ca^{2+} concentration by a first order dissociation curve. This suggests that at high concentrations the Ca^{2+} complexes not only with the ATP but also with the myosin, thus suppressing the hydrolysis, since the ATP cannot now bind to the active site. The inhibition of the ATPase by Mg^{2+} also depends upon the Mg^{2+} concentration as a sum of two first order dissociation curves[40]. However, Mühlrad et al.[44] and Offer[45] have shown that since Mg^{2+} strongly inhibits the ATPase activity of myosin, the activity is strongly inhibited even by Mg^{2+} which contaminates usually a reaction medium for myosin-ATPase. Furthermore, Kiely and Martonosi[46] showed that the binding of PP_i, a competitive inhibitor of ATPase, to myosin is dependent on the binding of Mg^{2+} to myosin, of which the binding constant is about 10^7 M^{-1}. Thus, myosin-ATPase is inhibited by a concentration of Mg^{2+} which is too low to form an appreciable amount of the complex Mg-ATP, and this inhibition appears to be due to the tight binding of Mg^{2+} with myosin.

3. The Michaelis Complex Formed by the ATPase

Myosin-ATPase exhibits very low substrate specificity. As will be described in chapter 9, it hydrolyses all triphosphates. The dependence of the reaction rate, v, on the substrate concentration, [S], has been clearly shown by us[42] and by Ouellet et al.[47] to follow the Michaelis-Menten equation, at least at ATP concentrations higher than several μM (cf. section 7). Thus phenomenologically there are at least two steps:

$$E + S \underset{k_{-1}}{\overset{k_{+1}}{\rightleftharpoons}} ES \xrightarrow{k_{+2}} E + P.^*$$

The steady-state rate is

$$v = V_{max} / \left(1 + \frac{K_m}{[S]}\right) = k_{+2}\varepsilon / \left(1 + \frac{k_{-1} + k_{+2}}{k_{+1}[S]}\right).$$

E and S denote myosin and ATP respectively, and P is the reaction product. The Lineweaver-Burk plot of $1/v$ versus $1/[S]$ is linear. $1/k_{+2}$ is obtained from its intercept, and $(k_{-1} + k_{+2})/k_{+1}$ from its slope.

Nanninga and Mommaerts[48-50] also calculated k_{+1} from the initial rate of decrease in concentration of free ATP after rapid mixing with myosin. They followed the reaction by measuring the intensity of luminescence of luciferin-luciferase, which is proportional to the concentration of free ATP in the system. k_{+2} was also derived from the reaction rate at high

---

* ES in this scheme contains, of course, various complexes formed from E and S. Furthermore, E might be in a state or chemical form different from the original enzyme. These two points will be treated later by analysing the pre-steady-state of ATPase.

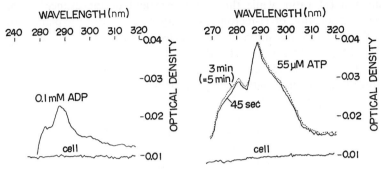

Fig. 3. Difference spectra of ATP and ADP with H-meromyosin. 0.5M KCl, 1mM MgCl$_2$, 2mg/ml H-MM, 10mM Tris-HCl buffer at pH 7.0. 0.4mM PEP and 20μg/ml pyruvate kinase were added together with ATP to maintain a constant ATP concentration. Time shown is the interval between the mixing of H-MM with ATP and taking the measurement.

ATP concentrations, and combination of the two rate constants with K_m permitted calculation of k_{-1}. In order that the luciferin-luciferase method could be used, these measurements were performed in the presence of NaCl, MgCl$_2$ and high concentrations of NaHAsO$_4$. If the results obtained under these special conditions follow the general ATPase mechanism, then it seems that k_{-1} is considerably smaller than k_{+2}.

As described in chapter 2, when myosin is mixed with sufficient ATP, its helical content, as measured by its optical rotatory dispersion, changes several per cent, and this has been attributed to the formation of an ES complex[51,52]. Recently Morita[53-55] observed changes in the absorption of H-meromyosin (H-MM) around 280nm upon the addition of ADP or ATP (Fig. 3). The effect of ADP can be ascribed to its binding to H-mero-

TABLE I. Dissociation Constant for the Binding of H-Meromyosin with ATP and ADP

Substrate	Method	Conditions	Dissociation constant (μM)
ATP	Optical	0.6M KCl, 1mM MgCl$_2$, pH 7.5, 25°C	4
	Kinetic	4-6mM KCl, 1mM MgCl$_2$, pH 7.5, 25°C	3
		0.5M KCl, 1mM MgCl$_2$, pH 8.2, 17°C	3.8
		0.5M KCl, 2mM MgCl$_2$, pH 7.5, 25°C	2.7
		0.5M KCl, 2.5mM MgCl$_2$, pH 7.8, 0°C	1.0
ADP	Optical	0.5M KCl, 2mM MgCl$_2$, pH 7.5, 23°C	14
	Kinetic	0.5M KCl, 2mM MgCl$_2$, pH 7.5, 25°C	16

myosin in a 1:1 ratio, while the appearance and disappearance of the change in absorption of H-meromyosin with ATP could be explained by formation and decomposition of a Michaelis complex. We[56] have measured the dissociation constants for the interaction of ATP and ADP with H-MM from the dependence of the difference spectra on ATP and ADP concentration. The dissociation constant for ATP was almost equal to the Michaelis constant for the ATPase, while that for ADP was equal to the competitive inhibition constant of ADP in the ATPase reaction at steady-state (Table I). The difference spectra are thus caused by the formation of an ES complex. However the mechanism of hydrolysis is much more complicated than the simple scheme shown above, and there are two distinct ES complexes involved. In section 6 we shall present a more detailed mechanism and discuss further which of these two complexes is responsible for the changes in absorption.

4. The Effect of pH on Myosin-ATPase

Kinetic studies of the effect of pH on an enzymic reaction have often played an important role in the elucidation of its mechanism. Since each dissociable group in a protein molecule has a characteristic dissociation constant, and since there are usually several such groups present at the active site of an enzyme, the effect of pH can provide information about the identity of the groups present at the active site. The reaction of ATP with myosin-ATPase follows Michaelis-Menten kinetics, and as pointed out by Dixon and Webb[57], in order to understand the effect of the medium on an enzyme one must measure the variation of both V_{max} and K_m.

The hydrolysis by myosin of inosine triphosphate, ITP (in which the 6-NH$_2$ of ATP is replaced by 6-OH), is relatively uncomplicated and so will be described first[58]. Figure 4 shows the pH-dependence of both V_{max} and K_m in the presence of sufficiently high concentrations of Ca^{2+}. V_{max} is constant at high acidities but increases with increasing pH to a maximum at pH 9.0. At about pH 7.0 the activity is half the maximum value. The variation of K_m with pH corresponds proportionately to that of V_{max}, although at high acidities K_m is somewhat larger than would be expected from this relationship. These observations are easily interpreted by assuming that k_{+2} is much greater than k_{-1}, as was shown by the kinetic measurements described previously, and that k_{+1} is independent of pH. The higher value of K_m at low pH than would be expected from its proportionality to V_{max} (Fig. 4) is thought to indicate that under these conditions the asso-

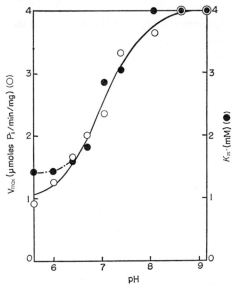

Fig. 4. pH-Dependence of V_{max} and K_m for the hydrolysis of ITP by myosin. 0.6M KCl, 7mM CaCl$_2$, 20°C. The line is represented by $V_{max}=1.0+3.0/(1+[H^+]/10^{-7.0})$. ○, V_{max}; ●, K_m.

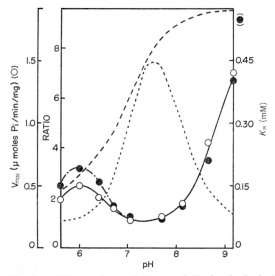

Fig. 5. pH-Dependence of V_{max} and K_m for the hydrolysis of ATP by myosin. Conditions as for Fig. 4. ---- and ······ show the pH-dependencies of k_{+2} and $(1+K)$ (see text). ○, V_{max}; ●, K_m.

ciation between Ca^{2+} and ITP is suppressed, increasing the K_m correspondingly and necessitating a greater concentration of ITP.

When ATP is the substrate, the pH-dependencies of V_{max} and K_m, shown in Fig. 5, demonstrate some similarity to those for the reaction of ITP, in that V_{max} and K_m change proportionately over a broad pH range. Again, at high acidities K_m is somewhat greater than would be expected from the proportionality relationship, since the substrate complexes less easily with Ca^{2+}. However, the curves of V_{max} and K_m versus pH are more complex than with ITP, revealing a minimum at pH 7.5–8.0 and a maximum at pH 6.0. If one assumes a simple pH-activity curve like that for the hydrolysis of ITP, as shown by the broken line, and then plots the ratio of this to the observed values for ATP, one obtains a well-defined bell shaped curve with a maximum at about pH 7.5–7.8. This can be explained plausibly by assuming that the enzyme forms two complexes with the substrate, one (E_1S) reacting further to give the product, the other (E_2S) being inactive[40,58].

$$E+S \underset{k_{-1}}{\overset{k_{+1}}{\rightleftharpoons}} E_1S \overset{k_{+2}}{\longrightarrow} E+P,$$
$$(1) \quad \updownarrow K \ (3)$$
$$E_2S \quad (2)$$

$$v = \frac{V_{max}}{1 + \dfrac{K_m}{[S]}},$$

$$V_{max} = k_{+2}\varepsilon/(1+K), \quad K_m = (k_{+2}+k_{-1})/[k_{+1}(1+K)],$$

where ε is the total concentration of enzyme active sites and K is the equilibrium constant between the two enzyme-substrate complexes. If $k_{+2} \gg k_{-1}$ and k_{+1} is independent of pH, then k_{+2} simply increases with pH for the hydrolyses of both ITP and ATP. In the ATPase case, the equilibrium $E_1S \rightleftharpoons E_2S$ lies towards E_2S, near neutrality, but shifts towards E_1S under alkaline or acid conditions. In contrast, with the ITPase there is almost no E_2S formed over the entire pH range. This mechanism adequately explains the observed data. We shall describe the properties of the two complexes E_1S and E_2S in more detail later, after a discussion of the pre-steady-state of hydrolysis of ATP by myosin.

If this mechanism is correct, the temperature dependence of the reaction should be more complicated than if the enzyme formed only one Michaelis complex, since it must include the temperature dependence of the equilibrium $E_1S \rightleftharpoons E_2S$[58]. However, very little E_2S is formed during the hydroly-

TABLE II. The Effect of Chemical Modification of Myosin-ATPase

Modifying reagent	Group which is modified	Mg^{2+}	ATPase Ca^{2+}	EDTA	pH-Activity dependence	Reference
IAA	Cys	Activation	Activation	Inactivation	Having a maximum and minimum near neutrality	60
NEM	Cys	Activation	Activation	Inactivation	Disappearance of maximum and minimum near neutrality	62, 63
PCMB	Cys	Activation	Activation	Inactivation	Disappearance of maximum and minimum near neutrality	62, 63
Diazobenzene-p-sulphonic acid	Cys	—	No change	Inactivition	Having a maximum and minimum near neutrality	64
2,2'-Dihydroxy-6,6'-dinaphthyl disulphide	Cys	—	Activation	—	Disappearance of maximum and minimum near neutrality	65
2-Bromoacetamide-4-nitrophenol	Cys	—	Activation	Inactivation	Disappearance of maximum and minimum near neutrality	65a
EDTA	Removal of Mg^{2+}	—	—	—	Disappearance of maximum and minimum near neutrality	66
2,4-Dinitrophenol	—	—	Activation	—	—	67, 68
TBS	Lys	Activation	Inactivation	No effect	Disappearance of maximum and minimum near neutrality	70
1-Dimethylamino-naphthalene-5-sulphonyl chloride	Lys	—	Inactivation (Ca^{2+} alone) Activation ($Ca^{2+}+K^+$)	—	Disappearance of maximum and minimum near neutrality	71
DHT	Tyr	Inactivation	Inactivation	Inactivation	—	72
Photo-oxidation	His	—	Inactivation	—	—	73
Dioxane	Solvent effect	Activation	Activation	Inactivation	Disappearance of maximum and minimum near neutrality	74

sis of ATP at pH 6.0, and as would be expected, the temperature dependencies of V_{max} and K_m under these conditions are normal Arrhenius plots. However, at pH 7.05, when there is appreciable formation of E_2S, there is a significant break in the temperature dependencies of V_{max} and K_m at 10°C, and the values of V_{max} and K_m found at lower temperatures are much smaller than would be predicted from those measured at higher temperatures. This may mean that even at pH 7.05, E_1S is formed more easily at lower temperatures, and this is supported by the absence of either a maximum or a minimum near neutrality in the pH-ATPase activity curve at low temperatures. The temperature dependence of V_{max} for the ITPase was measured by Levy et al.[59] and is different above and below 15°C. In this case the values of K_m are proportional to those of V_{max} over the entire concentration range used, unlike the situation with the ATPase. From this temperature dependence of V_{max}, these authors postulated a change in the secondary structure of myosin caused by binding of the substrate.

Kinetic studies are essential in the elucidation of reaction mechanisms, but the conclusions from such studies about the structure of the active site are inherently indirect. Chemical modification gives more direct information about the structure of the active site, and the results of such chemical modification of myosin are summarised in Table II. For the purpose of this discussion, we will give a brief summary of several chemical modifications of myosin-ATPase, and these will be described in detail in chapter 4. The activity of myosin-ATPase increases significantly in the presence of Mg^{2+} or Ca^{2+} after carboxamidomethylation by monoiodoacetamide (IAA), but it decreases in the presence of EDTA[60]. The pH-activity curve of the carboxamidomethyl myosin is not different from that of unmodified myosin, showing both a maximum and a minimum near neutrality. This result can be most easily explained by assuming that IAA activates reaction step (2) in the reaction sequence given above. On the other hand, high concentrations of PCMB or -SH reagents such as N-ethylmaleimide inhibit the myosin-ATPase[61,62], while low concentrations of these agents substantially promote its activity[63]. Friess[66] showed that the ATPase activity is also promoted by EDTA which chelates with Mg^{2+}, a powerful inhibitor of ATPase, as described in section 2. The ATPase activity is also activated by 2,4-dinitrophenol[67,68]. The pH-activity curve of myosin-ATPase activated by the above reagents is much simpler, and shows neither maximum nor minimum near neutrality. These phenomena can be rationalised if we suppose that the production of one of the two myosin-ATP complexes, E_2S, which is normally easily formed at pH

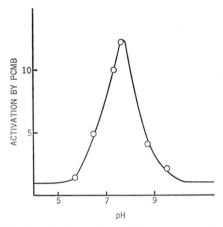

Fig. 6. pH-Dependence of the maximum degree of activation of myosin-ATPase by PCMB. Binding of PCMB at 0.8M KCl, pH 6.7, 20°C, 30 min. Assay of the ATPase at 0.6M KCl, 7mM Ca^{2+}, 1mM ATP, 20°C.

7.5, is suppressed by EDTA and the SH reagents. Of particular interest is the relationship between pH and the activation of the ATPase by PCMB. We measured the activity at various pH's after incubation of myosin with PCMB at a constant pH[69], as shown in Fig. 6. As would be expected if reaction with PCMB prevents the formation of E_2S, the pH-dependence of activation by PCMB was almost identical to that of $(1+K)$ shown in Fig. 5.

Trinitrobenzenesulphonate (TBS) modifies myosin-ATPase in a different fashion. Kubo et al.[70] discovered that the activity of myosin-ATPase is increased considerably in the presence of Mg^{2+} when the ε-amino group of one mole of lysine residue per mole of myosin combines specifically and rapidly with TBS. As will be explained later, this activation by TBS is due to promotion of the decomposition of a myosin-phosphate-ADP complex upon trinitrophenylation.

5. The Pre-steady-state Condition of the Hydrolysis of ATP by Myosin

As described above, the investigation of the myosin-ATPase reaction at steady-state has elucidated several features of the reaction mechanism. However, it is difficult to study directly the reaction intermediates at steady-state. In order to obtain information important in the elucidation of the molecular mechanism of muscle contraction, it was thought necessary to investigate the reaction intermediates by studying the myosin-ATPase under pre-steady-state conditions. As a result of this work it was proposed that myosin hydrolyses ATP by two routes—the simple hydrolysis described in sections 3 and 4, and a second path involving the phosphorylation of myosin by ATP[75-88].

The phosphorylation of myosin by ATP has been repeatedly postulated since its initial proposal by Kalckar[30] in 1941, but previous searches for the phosphorylated myosin have produced ambiguous results. Hence, we found it necessary to determine whether phosphorylated myosin is actually produced, and to elucidate its mode of formation and decomposition, in order to understand the mechanism of muscle contraction. As a result of our work, it appears that the reaction between myosin (E) and ATP (S) can be explained by the mechanism shown in Fig. 7. As will be described in detail in this chapter, myosin has two kinds of ATPase active sites, one for simple hydrolysis, (site 1) and the other for hydrolysis *via* a phos-

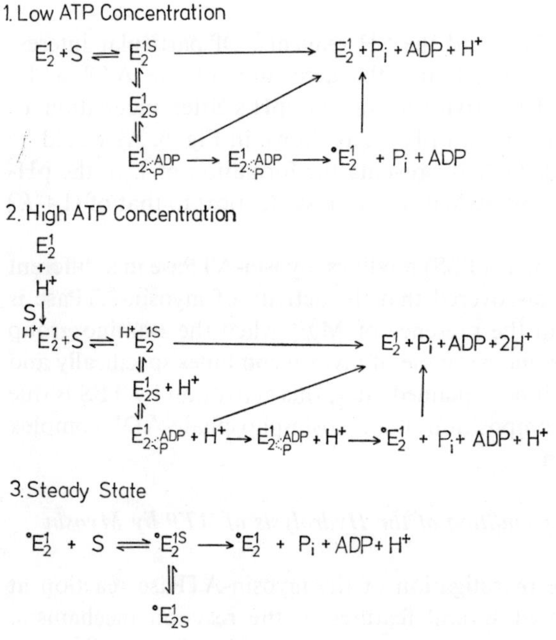

1. Low ATP Concentration

$$E_2^1 + S \rightleftharpoons E_2^{1S} \longrightarrow E_2^1 + P_i + ADP + H^+$$
$$\Updownarrow$$
$$E_{2S}^1$$
$$\Updownarrow$$
$$E_{2 \sim P}^1 \cdot ADP \longrightarrow E_{2 \sim P}^1 \cdot ADP \longrightarrow {}^\circ E_2^1 + P_i + ADP$$

2. High ATP Concentration

$$E_2^1$$
$$+$$
$$H^+$$
$$S\downarrow$$
$$H^+E_2^1 + S \rightleftharpoons H^+E_2^{1S} \longrightarrow E_2^1 + P_i + ADP + 2H^+$$
$$\Updownarrow$$
$$E_{2S}^1 + H^+$$
$$\Updownarrow$$
$$E_{2 \sim P}^1 \cdot ADP + H^+ \longrightarrow E_{2 \sim P}^1 \cdot ADP + H^+ \longrightarrow {}^\circ E_2^1 + P_i + ADP + H^+$$

3. Steady State

$$^\circ E_2^1 + S \rightleftharpoons {}^\circ E_2^{1S} \longrightarrow {}^\circ E_2^1 + P_i + ADP + H^+$$
$$\Updownarrow$$
$$^\circ E_{2S}^1$$

Fig. 7. Proposed scheme for the reaction of myosin with ATP. For explanation see text.

phorylated intermediate (site 2). Results from the chemical modification of myosin (chapter 4) and from other studies indicate that these sites are present in the proportion of about 1 mole per mole of myosin of molecular weight, 4.8×10^5 (see chapter 2). In Fig. 7 myosin is designated by E_2^1, and $E_2{}^{1S}$ and $E_{2S}{}^1$ denote intermediates involving sites 1 and 2 respectively. $E_2^1{\cdot}_{P}^{ADP}$ is an intermediate capable of exchanging a phosphate group with ATP, and $E_2^1{:}_{\cdot P}^{ADP}$ and $°E_2^1$ are intermediates incapable of exchanging a phosphate group with ATP. Henceforth $E_2^1{\cdot}_{P}^{ADP}$ and $E_2^1{:}_{\cdot P}^{ADP}$ will be called phosphorylated myosin and myosin-phosphate-ADP complex, respectively; collectively they will be called reactive myosin-

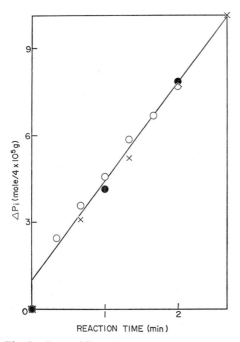

Fig. 8. Rate of liberation of TCA-labile P_i after adding ATP to various concentrations of myosin. 53 μM ATP, 1.08M KCl, 5mM $MgCl_2$ and 20mM Tris-HCl buffer at pH 7.5 and 25°C. Concentrations of myosin: ○, 0.5; ×, 0.2; ●, 0.1mg/ml.

phosphate-ADP complex ($E_2^1{}_{\text{ADP} \atop \text{P}}$). These two intermediates are labile with respect to TCA.

The starting point in our investigations was the sudden initial fast liberation of free P_i from the myosin-ATP system shown by quenching the reaction with TCA[75,89]. As described later, this 'initial burst' amounted to one mole of P_i per mole of myosin over a broad range of experimental conditions, as long as the Mg^{2+} concentration was 1mM or greater[76]. As the Mg^{2+} concentration was decreased below 1mM, the size of the initial burst of P_i increased[76]. We shall first discuss the stoichiometric release of P_i when the Mg^{2+} concentration is 1mM or more.

6. The Reaction of Myosin-ATPase with ATP in the Presence of High Concentrations of Mg^{2+}

Kinetic measurements of the reaction of myosin with ATP, performed by quenching with TCA, show that in the presence of Mg^{2+} there is a very rapid initial liberation of P_i (Fig. 8), followed by a slower liberation at steady-state. Hydrolysis of γ-^{32}P-ATP showed that the P_i liberated in the initial burst comes from the terminal phosphate group of the ATP. At Mg^{2+} concentrations of 1mM or more the size of the initial burst increased with increasing ATP up to one mole of ATP per 4–5×10^5g of myosin, and then remained constant. As shown in Fig. 8, it is also proportional to the myosin concentration over a broad range of KCl concentration, temperature, and pH[76].

The most likely molecular weight of myosin is 4.8×10^5 (chapter 2), but in chapter 4 it is shown that there is no general agreement as to the number of active sites per molecule of myosin-ATPase. However, our results indicate that there is 1 mole of both sites 1 and 2 per 4–5×10^5g of myosin[60, 70,72,90,91], as will be discussed in chapter 4. Hence we conclude that the initial burst of P_i amounts to 1 mole per mole of active sites of myosin. A similar conclusion holds for H-meromyosin and for myosin obtained from cardiac muscle[79]. However, the initial burst for subfragment-1 amounts to about only 0.5–0.6 moles of P_i per mole, and we shall discuss the significance of this difference with regard to the submolecular structure of S-1 in chapter 4, section 2.*

* In many of our experiments the initial burst of P_i amounted to 1.1, 1.1 and 0.55 moles/mole for myosin (molecular weight 4.8×10^5), H-meromyosin (3.4×10^5) and S-1 (1.2×10^5) respectively. The fact that the measured values are somewhat higher than the theoretical values of 1.0 and 0.5 is due to an unavoidable contamination by the 'extra-burst' described in section 8.

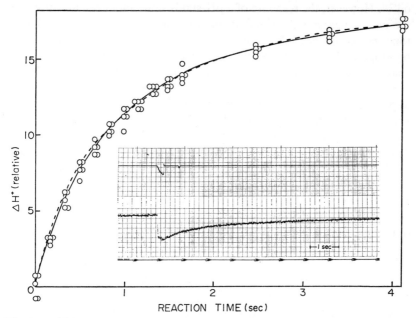

Fig. 9. Initial rapid change in [H⁺] after adding ATP to myosin. 1.0M KCl, 2mM MgCl$_2$, 2mg/ml myosin, 5μM ATP and 52μM cresol red at pH 8.2 and 21°C. The solid line is drawn through the means of 5 experimental measurements (○) of the initial liberation of protons. The broken line represents the mean of 5 measurements of the change in optical density at 293nm. (The inserted figure shows on the lower trace the initial rapid H⁺ liberation, and on the upper trace the flow-rate.)

There is no initial burst of P_i when the SH groups of myosin are modified with PCMB[76], or when Mg^{2+} is removed from the system by the addition of EDTA[89]. Several mechanisms could account for the stoichiometric initial burst. One is that since both Mg^{2+} and the SH groups of the protein are required, the formation of an intermediate phosphorylated on the carboxyl group of the myosin occurs very rapidly, and this, on decomposition by added TCA, liberates the excess P_i[75].

At pH 8.0 the hydrolysis of 1 mole of ATP liberates 1 mole of protons[37] but phosphorylation of the carboxyl group should not be accompanied by any generation of protons. Therefore the liberation of protons upon addition of a large amount of ATP to myosin at pH 8.0 was measured with a pH-stat. The initial burst of P_i was observed, but the amount of protons liberated increased linearly with time, and there was no initial fast period[75]. However, the response of a pH-stat is slow, and it can measure only the net change, not the simultaneous generation and absorption of protons. Hence we have determined the initial change in H^+ concentration by stopped flow, using cresol red as a pH indicator and following the change in absorption of 570nm[77,88]. Figure 9 shows the increase in H^+ concentration immediately after the addition of ATP to myosin, and it appears that 1 mole of protons per mole of myosin is produced at a rate similar to the initial burst of P_i-liberation (the relationship between the rapid initial generation of protons and P_i will be described on p. 87). These results do not mean that addition of ATP to myosin is accompanied solely by liberation of protons during the initial period. We have concluded from the succeeding measurements that there is also a very fast absorption of 1 mole of protons per mole of myosin, but that this occurs so rapidly that it is completed within the mixing period of the stopped flow apparatus. Hence as mentioned above, there is no overall change in acidity measured by the pH-stat during the initial phase. The rate of liberation of protons over the complete hydrolysis of ATP to $ADP + P_i$ when one mole of ATP is added to one mole of myosin at pH 8.0 is several hundred times slower than the rapid initial discharge described above (see section 7). Therefore unless we assume that there is also a very rapid initial absorption of protons, immeasurable even by stopped flow, prior to the initial rapid discharge, the liberation of 2 moles of protons per mole of ATP would be indicated by the pH-stat. When the solutions of ATP and enzyme to be reacted are prepared at the same pH, the initial H^+ concentration is equal to the value obtained by extrapolation to zero time of the generation of protons at steady-state. This shows that on mixing myosin with ATP, there is a very rapid absorption of 1 mole of protons per mole of myosin, followed by an equivalent rapid discharge, leaving no net change in the concentration of H^+ during the initial period of the reaction. These results seemed to support the assumption that the initial reaction of myosin with ATP probably involves phosphorylation of a group, such as carboxyl, with unit negative charge. Our recent experiments[83] described later (p. 90) have, however, suggested that, when several μM of ATP is added to myosin as in the above experiments, the initial very rapid absorption of protons is attributable to a conformational change of myosin induced by ATP, and the rapid liberation is accompanied by the formation of an intermediate situated just before phosphorylated myosin in the reaction sequence:

$$E_2^1$$
$$+$$
$$H^+$$
$$S\downarrow$$
$$^{H+}E_2^1 + S \rightleftharpoons {}^{H+}E_2^{1S}$$
$$\Updownarrow$$
$$E_{2S}^1 + H^+.$$

In this scheme, $^{H+}E_2^1$ indicates myosin of which the conformation has been changed by a high concentration (several μM) of ATP (S). ATP is assumed to bind first to the simple hydrolysis site 1 and to be transferred to the phosphorylation site 2, since this mechanism is consistent not only with that of myosin-ATPase at steady-state (section 4) but also with those of transport ATPase (chapters 11 and 12) (also cf. p. 90).

As described in section 3, Morita[54] and we[56] have reported that the change in the UV absorption spectrum of H-meromyosin with ATP can be explained by the formation of a Michaelis intermediate (ES). However it was not clear which of the two Michaelis complexes E_1S and E_2S postulated from the pH-dependency of the steady-state rate was responsible for the change. Therefore we have compared the change in the difference spectrum at 293nm with time to the time dependence of the rapid generation of protons (after the first very rapid absorption of protons) when ATP is added to myosin in a molar ratio of 1 : 1. The two phenomena correlated reasonably well[88] (Fig. 9). If we accept the above mechanism for the initial change in [H$^+$] when several μM ATP is added to myosin, these results indicate that the observed change in the UV difference spectrum is not due to the formation of E_2^{1S} from myosin and ATP but to the formation of E_{2S}^1 from E_2^{1S}, i.e. $^{H+}E_2^{1S} \rightleftharpoons E_{2S}^1 + H^+$. The fact that this change in optical density in the UV region occurs under conditions where no initial burst of P_i is observed does not conflict with our argument[55,88].

We have also measured the rate of liberation of $^{32}P_i$ during the initial burst using γ-^{32}P-ATP simultaneously with the liberation of protons[82]. At low temperatures, the rapid release of protons was faster than the formation of TCA-unstable $^{32}P_i$, but the rates were similar at room temperature[82,88]. Hence a simplified mechanism for the initial period of the reaction between myosin and ATP could be

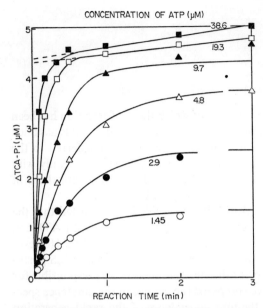

Fig. 10. Rate of liberation of P_i after adding various concentrations of ATP to myosin. 2.8M KCl, 10mM $MgCl_2$, 2mg/ml myosin and 20mM Tris-HCl at pH 7.5 and 0°C. Concentrations of ATP: ○, 1.45; ●, 2.9; △, 4.8; ▲, 9.7; □, 19.3; ■, 38.6 μM.

$$\begin{array}{c} E_2^1 \\ + \\ H^+ \\ S \downarrow \\ {}^{H^+}E_2^1 + S \rightleftharpoons {}^{H^+}E_2^{1S} \\ \Updownarrow \\ E_{2S}^1 + H^+ \\ \text{(dif. sp.)} \\ \Updownarrow \\ E_2{}^1_{\substack{ADP \\ P}} + H^+. \end{array}$$

(initial burst of P_i-liberation)

The dependence of the initial reaction on ATP concentration was measured as follows. Figure 10 shows the rate of liberation of TCA-unstable P_i after addition of various concentrations of ATP to myosin. At concentration ratios of ATP to myosin less than 1, the time, $\tau_{1/2}$, necessary to form half the maximum amount of $E_2{}^1_{\substack{ADP \\ P}}$ at steady-state was almost inde-

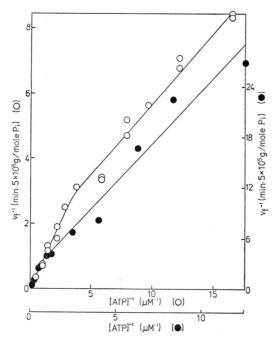

Fig. 11. Double reciprocal plot of the initial rate of formation of myosin-phosphate-ADP complex (v_f) against ATP concentration. 0.03mg/ml myosin, 2.5mM $MgCl_2$, pH 7.8, 0°C. ○, 0.5M KCl; ●, 1.5M KCl.

pendent of the concentration of ATP added. However when the ATP concentration is greater than the stoichiometric amount, the value of $\tau_{1/2}$ decreased with increase in the ATP concentration. The rapid liberation of protons also showed a similar dependence on the concentration of ATP[82].

The experiments shown in Fig. 10 were performed in the presence of several μM myosin. Hence the ATP concentration decreased greatly during the reaction due to binding of ATP to myosin, and we could not determine the kinetic constants of $E_2^1{}_P^{ADP}$-formation. Therefore, we[83] have measured the dependence of the initial rate of $E_2^1{}_P^{ADP}$-formation, v_f, on the ATP concentration at a very low concentration of myosin, 0.07μM (Fig. 11). In the range of low concentrations of ATP, the Michaelis constant of $E_2^1{}_P^{ADP}$-formation, K_f, was 0.3μM, and the maximum rate, V_f, was

0.75 min^{-1} in 0.5M KCl and pH 7.8 and 0°C. Furthermore, in the range of high ATP concentrations v_f increased with increase in the ATP concentration more than expected from the Michaelis equation. However, we could not determine the values of K_f and V_f for formation of $E_2^1{}_{\substack{ADP\\P}}$ in the range of high concentrations of ATP. On the other hand, we measured these values for formation of $E_{2S}{}^1$ from the dependence of the rate in change in UV spectra on the concentration of ATP[83a]. The values of K_f and V_f for formation of $E_{2S}{}^1$ in the range of high concentrations of ATP in 0.2M KCl at pH 7.8 and 4°C were 0.2mM and 30 sec^{-1}, as shown in Fig. 12. These results show clearly that $E_{2S}{}^1$ is formed *via* a myosin-ATP complex ($E_2{}^{1S}$), and that ATP at high concentrations induces a conformational change in myosin ($E_2{}^1 \rightarrow {}^{H^+}E_2{}^1$)* and the Michaelis constant and the maximum rate of conversion of ${}^{H^+}E_2{}^1$ to $E_2^1{}_{\substack{ADP\\P}} + H^+$ at high concentrations of ATP are about 700 and 1,200 times** as high as those of conversion of $E_2{}^1$ to $E_2^1{}_{\substack{ADP\\P}}$ at low concentrations of ATP (Fig. 7), although the relation between the first very rapid absorption of protons and the conformational change suggested above remains to be demonstrated. The physiological function of the acceleration of $E_2^1{}_{\substack{ADP\\P}}$-formation by MgATP will be discussed in chapters 6 and 10. The same type of acceleration of the formation of phosphorylated intermediates is also observed in transport ATPases, as will be described in chapters 11 and 12.

According to this mechanism, the effects of temperature on the liberation of H$^+$ and TCA-P$_i$ mentioned above seem to suggest that at room temperatures and in the presence of low concentrations of ATP the concentration of $E_{2S}{}^1$ is very small, while at low temperatures the concentration of $E_{2S}{}^1$ increases. As will be discussed in chapter 10, section 5, this might be one of the causes that at low temperatures the 'clearing response' of actomyosin occurs even at low concentrations of ATP.

---

* For simplicity, the conformational change in myosin is not indicated explicitly in the reaction scheme given in this chapter.
** This value was obtained by assuming that the maximum rate of $E_2^1{}_{\substack{ADP\\P}}$-formation at high ATP concentrations was 15 sec^{-1}, since, as stated above, at low temperatures the rate for formation of $E_{2S}{}^1$ was higher than (about 2 times) that of formation of $E_2^1{}_{\substack{ADP\\P}}$.

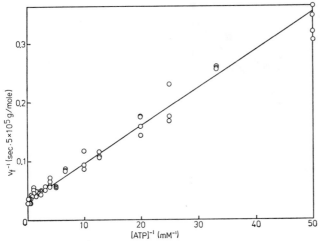

Fig. 12. Double reciprocal plot of the initial rate of formation of E_2S^1 deduced from change in OD against ATP concentration. 3mg/ml H-meromyosin, 0.2M KCl, 50mM Tris- maleate, pH 7.8, 4°C. The MgCl$_2$ concentration is always higher than that of ATP by 2.5mM.

7. The Properties of the Reactive Myosin-Phosphate-ADP Complex

The TCA-labile reactive myosin-phosphate-ADP complex ($E_2{}^1{}_P^{ADP}$) was proposed above as an intermediate in the reaction of myosin with ATP. Its formation was studied by the experiments described in the previous section and the investigation of its properties will now be discussed. The formation of this complex is not accompanied by a net change in the concentration of H^+[77], and in consequence the complex was assumed to be formed by phosphorylation of the carboxyl group of myosin. If this labile complex is a phosphorylated myosin, it might be possible to trap it with a nucleophile, and we have attempted to do this with p-nitrothiophenol (NTP). NTP combines with myosin only in the presence of Mg^{2+} and ATP[75,92]. Hence myosin was p-nitrothiophenylated in a medium containing 1mM or greater concentrations of Mg^{2+} and a large amount of ATP, giving NTP (1)-myosin.* The initial burst of P_i liberated from

* Myosin which is nitrothiophenylated under these conditions has different properties from myosin nitrothiophenylated in the presence of low Mg^{2+} concentrations, and hence the present form is called NTP (1)-myosin.

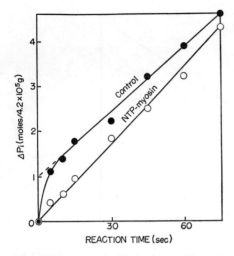

Fig. 13. Rate of liberation of TCA-labile P_i after adding ATP to NTP(1)-myosin and control unmodified myosin. 1.0M KCl, 20mM $MgCl_2$, 0.64mM ATP and 20mM Tris-maleate at pH 7.0 and 25°C. ●, 4mg/ml control myosin; ○, 4mg/ml NTP(1)-myosin (0.8 moles NTP/4×10^5g).

ATP by NTP (1)-myosin decreases in proportion to the degree of nitrothiophenylation, and is completely eliminated when 1 mole of NTP is bound per mole of myosin (Fig. 13). However, there is no change in the maximum velocity, V_{max}, of the reaction at steady-state, or in its dependence on pH[85].

In order to elucidate the structure of the site of bonding of myosin with NTP, we have digested NTP (1)-myosin with Nagarse and Pronase to remove the NTP-peptide. The pH-stability curve of this peptide indicated the presence of an acyl-thiol bond[85], and its further digestion with aminopeptidase and prolidase gave an NTP-amino acid. After correction for losses in the isolation procedure, it appears that this contained 1 mole of glutamic acid per mole of NTP[85].

However, these results do not show conclusively that the phosphate group is covalently bound to myosin in the reactive myosin-phosphate-ADP complex ($E_2^1{}_P^{ADP}$). Sartorelli et al.[93] found that the amount of ^{18}O incorporated into P_i upon the decomposition of the reactive myosin-phosphate-ADP complex in $H_2{}^{18}O$ at neutral pH is 16% or less. Thus our results could also be explained by assuming that a myosin-monomeric

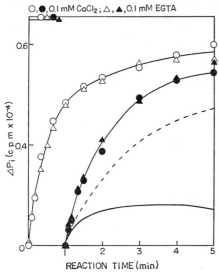

Fig. 14. Exchange of a phosphate group between the 'reactive myosin-phosphate-ADP complex' and γ-^{32}P-labelled ATP. 2.78M KCl, 10mM MgCl$_2$, 2mg/ml myosin and 20 mM Tris-HCl at pH 7.5 and 0°C. ●, ▲: 0.22μM γ-^{32}P-ATP was added 1 min after adding 5μM of 'cold' ATP to 2mg/ml myosin. ○, △: control reaction, started by adding 5μM of γ-^{32}P-ATP to 2mg of myosin/ml. ○, ●: 0.1mM CaCl$_2$, △, ▲: 0.1mM EGTA. ----: curve calculated from the control experiment on the assumption that none of the reactive myosin-phosphate-ADP complex exchanges its phosphate group with ATP; ———: difference between the experimental and calculated values. The activity of the γ-^{32}P-ATP used in a given experiment is indicated at the upper left of the figure.

metaphosphate-ADP complex is formed *via* phosphorylated myosin by the reaction of myosin with ATP[94]. Therefore we have investigated whether there is exchange of a phosphate group between ATP and $E_2^1{}_\text{P}^\text{ADP}$ [80]. As a control experiment, we measured the amount of ^{32}P$_i$ liberated when γ-^{32}P-ATP was added to myosin in a 1:1 molar ratio and the reaction then quenched after a specific time with TCA. Non-radioactive ATP was then reacted with myosin at a molar ratio of 1:1, and after the $E_2^1{}_\text{P}^\text{ADP}$ complex had been allowed to form for a fixed time a very small amount of γ-^{32}P-ATP with a very high specific activity was added. The reaction was then quenched with TCA after a further period to find the rate of liberation of ^{32}P$_i$. The reactive myosin-phosphate-ADP complex ($E_2^1{}_\text{P}^\text{ADP}$) is decomposed by TCA[80], while ATP is not. Thus the amount of $^{32}P_i$ which would be liberated if there were no exchange of a phosphate group can be calculated easily from the control experiment. As shown in Fig. 14, the differences between the calculated and observed values can be attributed to exchange of a phosphate group between the TCA-labile intermediates and ATP. These results show that under usual experimental conditions, almost all the reactive myosin-phosphate-ADP complex ($E_2{}^1{}_{\text{ADP}}^{\text{P}}$) does not undergo exchange of a phosphate group with ATP[80], that even in the presence of 2.8M KCl and 10mM Mg^{2+}, 10–20% of $E_2{}^1{}_{\text{ADP}}^{\text{P}}$ undergoes the exchange, and that the exchange is not affected by trace amounts of Ca^{2+}, which controls the myosin-actin-regulatory protein-ATP system under normal physiological conditions (see chapter 10). This means that there are at least two reactive myosin-phosphate-ADP complex intermediates ($E_2{}^1{}_{\text{ADP}}^{\text{P}}$), one capable of exchange of a phosphate group with ATP, the other not. The formation of the first complex suggests that ATP phosphorylates a carboxyl group of myosin, which binds a phosphoryl group covalently, since its formation results in no net absorption or liberation of H^{+}[77]. Since only one mole of glutamic acid residue is nitrothiophenylated per mole of myosin under the conditions of the stoichiometric burst, yet this completely inhibits the initial burst of P_i, it was concluded that the carboxyl group of this specific glutamic acid residue is phosphorylated. Therefore the first complex is phosphoryl myosin ($E_2{}^1{\!\!\diagdown\!}_{\text{P}}^{\cdots\text{ADP}}$), while the second is the myosin-phosphate-ADP complex ($E_2{}^1{\!:\!\cdot\!\cdot\!}_{\cdot\cdot\text{P}}^{\cdots\text{ADP}}$). Acyl-phosphates are generally assumed to be hydrolysed *via* the metaphosphate[94], and the formation of $E_2{}^1{\!:\!\cdot\!\cdot\!}_{\cdot\cdot\text{P}}^{\cdots\text{ADP}}$ is not accompanied by liberation of protons. Hence, as suggested by Sartorelli *et al.*[93], $E_2{}^1{\!:\!\cdot\!\cdot\!}_{\cdot\cdot\text{P}}^{\cdots\text{ADP}}$ may be a myosin-metaphosphate-ADP complex.

$$E_2{}^1 + S \rightleftharpoons E_2{}^1S$$
$$\updownarrow$$
$$E_2S^1$$
$$\updownarrow$$
$$E_2{}^1{\!\!\diagdown\!}_{\text{P}}^{\cdots\text{ADP}} \longrightarrow E_2{}^1{\!:\!\cdot\!\cdot\!}_{\cdot\cdot\text{P}}^{\cdots\text{ADP}}$$

(phosphoryl myosin)   (myosin-phosphate-ADP complex)

$$\downarrow +HS-\langle\rangle-NO_2$$

$$E_2^1-Glu-\overset{O}{\underset{\|}{C}}-S-\langle\rangle-NO_2.$$

(NTP(1)-myosin)

There was no observable exchange when a large amount of 'cold' ATP was added during the formation of the reactive myosin-phosphate-ADP complex after addition of a stoichiometric amount of $\gamma\text{-}^{32}P$-ATP to the myosin in the presence of 2.8M KCl. Hence the reaction $E_2^1{\cdot}{\cdot}{\cdot}\underset{P}{ADP} \rightarrow E_2^1{:}{\cdot}{\cdot}\underset{P}{ADP}$ is markedly promoted by the presence of a large amount of ATP. The physiological significance of this observation will be described in chapters 6 and 10. Thus phosphorylation of myosin has been suggested by rather indirect evidence. Furthermore, phosphoryl myosin occurs transiently during the initial phase of the reaction, and its amount is so small that it can be neglected in kinetic analyses under usual conditions, as mentioned above. However, to emphasise the similarity of the reaction mechanisms of myosin-ATPase and transport ATPases which is described in chapters 11 and 12 in this monograph, we designate conventionally the route via E_{2S}^1, $E_2^1{\cdot}{\cdot}{\cdot}\underset{P}{ADP}$ and $E_2^1{:}{\cdot}{\cdot}\underset{P}{ADP}$ as the one via phosphoryl myosin.

We have investigated the relationship between the rate of decomposition of the reactive myosin-phosphate-ADP complex and the ATPase activity at steady-state. Myosin was added to ATP in a 1:1 molar ratio, followed by a pyruvate kinase system after different time intervals. The rate of formation of free ADP from the complex was then measured by following the liberation of pyruvate[85], and found to follow first order kinetics, with a rate constant of the same magnitude as that for the maximum rate of the ATPase at steady-state in the presence of high concentrations of ATP (cf. Fig. 18). The rate of decay of the UV absorption after the addition of a 1 : 1 molar ratio of ATP to myosin was also approximately equal to the rate of liberation of ADP[86]. Thus, the UV spectrum of myosin changes as the result of formation of E_{2S}^1 and $E_2^1\underset{P}{ADP}$, and the OD of E_{2S}^1 and $E_2^1\underset{P}{ADP}$ are almost equal at least at 293nm, while upon release of ADP from the latter complex the original spectrum returns. Furthermore, we[83a] have recently measured the rate of liberation of ADP from $E_2^1\underset{P}{ADP}$ from the rate of dialysis[94a] across a Millipore membrane. We added ^{3}H-ATP to myosin at a molar ratio of 1:1, under the conditions where the binding of ADP to myosin was rather strong, and measured the rate of liberation of ^{3}H-ADP from the complex after adding a high concentration of non-radioactive ADP. It was found that 17 sec after the addition of ^{3}H-ATP to myosin the rate of liberation of ADP from $E_2{}^1_{\mathrm{P}}{}_{\mathrm{ADP}}$ was of the same order of magnitude as that for V_{max} of the ATPase at the steady-state, while 10 min after the addition of ^{3}H-ATP the rate of ADP-liberation from the myosin-ADP complex was too high to be measured by the method used. Furthermore, the rate of ADP-liberation from $E_2{}^1_{\mathrm{P}}{}_{\mathrm{ADP}}$ after adding non-radioactive ATP was equal to that after adding non-radioactive ADP.

Taylor et al.[95] have recently shown that the rate constant for liberation of P_i from $E_2{}^1_{\mathrm{P}}{}_{\mathrm{ADP}}$ is similar to or even higher than that for liberation of ADP. This result has been confirmed by us[83a] by measuring the rate of liberation of P_i from $E_2{}^1_{\mathrm{P}}{}_{\mathrm{ADP}}$ from the rate of dialysis[94a] of P_i across a Millipore membrane. Thus the rate constants of the release of free P_i after adding ATP to myosin at a molar ratio 1:1 were estimated by flow dialysis under various conditions. As shown in Fig. 15, the rate of liberation of P_i from $E_2{}^1_{\mathrm{P}}{}_{\mathrm{ADP}}$ was slightly higher than that of the ATPase reaction at steady-state in the high ATP concentration range. Furthermore, the former rate was unaffected by adding a large amount of ATP.

The above results suggest that the reactive myosin-phosphate-ADP complex is an intermediate in the steady-state hydrolysis of ATP by myosin under usual conditions. However, further experiments refute this idea, and we have therefore assumed that myosin has a changed conformation ($°E_2{}^1$), after ADP and P_i are liberated from $E_2{}^1_{\mathrm{P}}{}_{\mathrm{ADP}}$, and that $°E_2{}^1$ shows no initial burst of P_i-liberation on adding ATP again, but shows usual activity for the simple hydrolysis of ATP. Thus, the following simplified scheme can be written for the myosin-ATP reaction.*

---

* It has not been elucidated which step in the initial reaction of myosin with ATP accompanies the conformational change, $E_2{}^1 \to °E_2{}^1$. It is probable that the conformational change occurs accompanied by the conversion of phosphoryl myosin to myosin-phosphate-ADP complex.

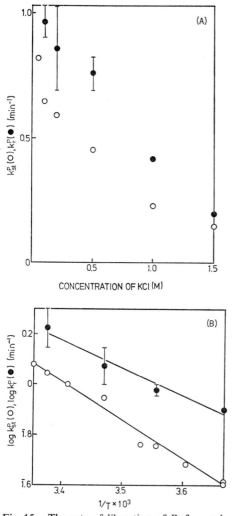

Fig. 15. The rate of liberation of P_i from $E_2^1{}_{ADP}^P$. The rate constant of liberation of P_i from $E_2^1{}_{ADP}^P$ ($k_r{}^p$) (●) was measured by flow dialysis, after adding 5μM ATP to 2.4 mg/ml myosin. The rate constant of the ATPase reaction at steady-state ($k_{st}{}^p$) (○) was measured in the presence of 20μM ATP and 0.05mg/ml myosin. Therefore, $k_{st}{}^p$ corresponds to V_{max} in the high ATP concentration range in Fig. 18. The value of $k_{st}{}^p$ was calculated by assuming that the weight of myosin per mole of the active site of ATPase is 4.8×10^5g. 2.5mM $MgCl_2$, 75mM Tris-HCl at pH 7.8. A: Dependence of the rate constants at 0°C on KCl concentration. B: Dependence of the rate constants in 0.5M KCl on temperature.

$$E_2{}^1 + S \underset{}{\overset{k_s}{\rightleftharpoons}} E_2{}^{1S} \longrightarrow E_2{}^1 + P_i + ADP$$
$$\Updownarrow \qquad\qquad \Big\uparrow k_d$$
$$E_{2S}{}^1$$
$$\Updownarrow$$
$$E_2{}^1_{\underset{P}{ADP}} \longrightarrow {}^\circ E_2{}^1 + P_i + ADP$$

(reactive myosin-phosphate-ADP complex)

$${}^\circ E_2{}^1 + S \overset{k_s}{\rightleftharpoons} {}^\circ E_2{}^{1S} \longrightarrow {}^\circ E_2{}^1 + ADP + P_i,$$
$$\Updownarrow$$
$${}^\circ E_{2S}{}^1$$

where k_s and k_d are the apparent rate constants for the simple hydrolysis of $E_2{}^{1S}$ and for the return of ${}^\circ E_{2S}{}^1$ to the original $E_2{}^1$ respectively. Therefore we have attempted to measure the rate of the return of ${}^\circ E_{2S}{}^1$ to $E_2{}^1$. According to this mechanism and that shown in Fig. 7, the formation of $E_2{}^{1S}$ is very rapid when ATP is added to myosin in a 1:1 molar ratio. The equilibria $E_2{}^{1S} \rightleftharpoons E_{2S}{}^1 \rightleftharpoons E_2{}^1_{\underset{P}{ADP}}$ are also rapid, and are shifted to the right, while the simple hydrolysis of $E_2{}^{1S}$ is slow relative to the equilibria, so that the amount of the reactive myosin-phosphate-ADP complex ($E_2{}^1_{\underset{P}{ADP}}$) produced quickly reaches the level of the ATP added, since it is rapidly formed by a stoichiometric reaction of myosin with ATP. As already described, the liberation of protons does not occur in the formation of the complex, suggesting that the rate of liberation of H+ equals the rate of return of the changed conformation of the enzyme (${}^\circ E_2{}^1 \rightarrow E_2{}^1 + H^+$). The rate of ${}^\circ E_2{}^1 \rightarrow E_2{}^1$ thus obtained was about one-tenth of the maximum rate of the ATPase in the simple hydrolysis of ${}^\circ E_2{}^{1S}$ at 1M KCl, 2mM MgCl$_2$, pH 7.5 and at 27°C, *i.e.*, $k_d \approx 1/10 \ k_s$[78]. For reference, Fig. 16 gives a summary of the rates of formation of the various products when myosin is added to ATP in a 1:1 molar ratio.

The possibility that the protons are liberated by a reaction other than ${}^\circ E_2{}^1 \rightarrow E_2{}^1$ cannot be excluded. Therefore the following experiments were performed under conditions such that amount of exchange of phosphate groups between $E_2{}^1_{\underset{P}{ADP}}$ and ATP was negligible. 'Cold' ATP was first added to myosin in a 1:1 molar ratio, followed by γ-^{32}P-ATP also in a molar ratio of 1:1 after various time intervals, and the time-course of liberation of TCA-unstable ^{32}P$_i$ was then measured (Fig. 17). The initial reaction of myosin was almost complete when the γ-^{32}P-ATP was added

REACTION MECHANISM OF MYOSIN-ATPase

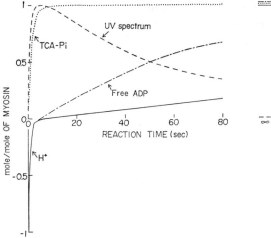

Fig. 16. Schematic view of the course of the pre-steady-state reaction of myosin with ATP. ATP was added to myosin at a molar ratio of 1:1 at high concentrations of KCl and MgCl$_2$. ———, H$^+$; ······, P$_i$ after quenching with TCA; - - -, UV difference spectrum; — · —, free ADP.

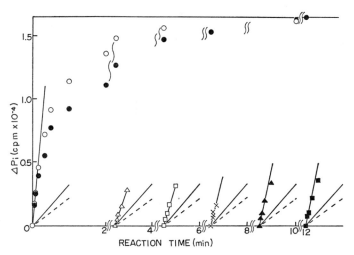

Fig. 17. Rate of liberation of TCA-labile ^{32}P$_i$ after addition of γ-^{32}P-ATP to myosin (molar ratio, 1:1) at various times after addition of non-radioactive ATP to myosin (molar ratio, 1:1). 1.08M KCl, 5mM MgCl$_2$, 2mg/ml myosin and 20mM Tris-HCl at pH 7.5 and 0°C. ○: 5μM γ-^{32}P-ATP; ●: 5μM γ-^{32}P-ATP plus 5μM ADP. △, □, ×, ▲, ■: 5μM of AT^{32}P was added to the reaction mixture at various times after addition of 5μM of non-labelled ATP.———: simple ATP-hydrolysis when the value of K_m is assumed to be much less than 5μM;- - -: simple ATP-hydrolysis when the value of K_m is assumed to be 2.5μM.

so that the amount of E_2^1 formed by return of the changed conformation of $°E_2^1$ after any time interval is proportional to the initial rate of liberation of the TCA-unstable $^{32}P_i$. The rate constant of $°E_2^1 \rightarrow E_2^1$ at 1.1M KCl, 5 mM $MgCl_2$, pH 7.5 and 0°C, as given from the results of Fig. 16, is 0.04 min^{-1}, which is about one-fifth of the maximum rate of simple hydrolysis of E_2^{1S} at steady-state[80].

If ATP is hydrolysed by myosin *via* two routes, we may expect that the rate, v, at steady-state can be expressed as the sum of two rates which have different maximum values (V_{max}) and dependencies on the ATP concentration. We have measured the rates, v, in the presence of ATP concentrations from 0.1 to $2\mu M$ and a very low concentration of myosin[83]. As shown in Fig. 18, the Lineweaver-Burk plot in 0.5M KCl and 2.5mM $MgCl_2$ at pH 7.8 and 0°C consisted of two lines. The value of K_m in the range of high ATP concentration was $1.0\mu M$, while that in the low concentration range was too low to be accurately determined. The value of V_{max} in the low concentration range was about one quarter of that in the high concentration range.* As mentioned on p. 89 (*cf*. Fig. 11), the

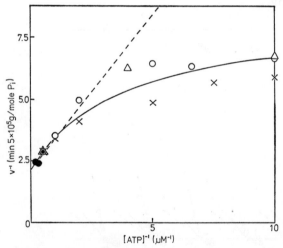

Fig. 18. Lineweaver-Burk plot of myosin-ATPase over a broad range of ATP concentrations. 0.5M KCl, 2.5mM $MgCl_2$, 0.005mg/ml myosin and 50mM Tris-maleate at pH 7.8 and 0°C. The line is represented by $v(min^{-1})=0.12+0.35/\left(1+\dfrac{K}{[S]}\right)$, when $K=1\mu M$. Four different preparations of myosin (○, △, ×, ●) were used.

* In the experiments on the ATPase at steady-state described in the preceding section and also in pp. 95–98, only the rates in the high ATP concentration range were measured.

Michaelis constant for $E_2^1{}_{\substack{ADP\\P}}$-formation, K_f, was $0.3\mu M$. The value of K_m for ATP-hydrolysis via $E_2^1{}_{\substack{ADP\\P}}$ and $°E_2{}^1$ must be much lower than 0.3 μM, since the formation of $E_2^1{}_{\substack{ADP\\P}}$ was much faster than its decomposition into $E_2{}^1+P_1+ADP$. The rate-determining step of ATP-hydrolysis via $E_2^1{}_{\substack{ADP\\P}}$ is conversion of $°E_2{}^1$ to $E_2{}^1$, and the rate constant of this step is about 1/5 of that of liberation of ADP and P_i from $E_2^1{}_{\substack{ADP\\P}}$. In other words, the rate constant of the rate-determining step of the reaction sequence deduced from transient kinetics ($E_2{}^1+S \rightleftharpoons E_2{}^{1S} \rightleftharpoons E_{2S}{}^1 \rightleftharpoons E_2^1{}_{\substack{ADP\\P}} \rightarrow °E_2{}^1+P_1+ADP \rightarrow E_2{}^1+P_i+ADP$) is essentially equal to the value of V_{max} in the steady-state in a low concentration range of ATP, since V_{max} at low ATP concentrations is about 1/4 of V_{max} at high ATP concentrations, while the latter value is almost equal to the rate of liberation of ADP and P_i from $E_2^1{}_{\substack{ADP\\P}}$. Furthermore, the fact shown by the luciferin-luciferase method[48] and UV spectroscopy[56] that myosin contains almost no bound nucleotide when the concentration of ATP is less than $0.2\mu M$ is consistent with the mechanism that the rate-determining step is $°E_2{}^1 \rightarrow E_2{}^1$, as mentioned above, because according to the mechanism the most stable intermediate, $°E_2{}^1$, contains no bound nucleotide.

Thus, it is almost certain that the ATPase reaction in the range of low concentrations of ATP occurs via $E_2^1{}_{\substack{ADP\\P}}$ and $°E_2{}^1$. Then the problem is how the steady-state ATPase in the range of high concentrations of ATP can be explained. Several different mechanisms may be considered. One possible mechanism is acceleration of the conversion of $°E_2{}^1$ to $E_2{}^1$ by ATP at high concentrations, since liberation of ADP and P_i from $E_2^1{}_{\substack{ADP\\P}}$ was unaffected by adding a large amount of ATP (p. 96). However, it seems to be rather improbable that ATP which induces the conformational change of myosin from $E_2{}^1$ to $°E_2{}^1$ also accelerates the reverse reaction at high concentrations. Furthermore, the value of K_m of ATPase in the steady-state in the range of high concentrations of ATP was much higher than the value calculated by the assumption of acceleration of $°E_2{}^1 \rightarrow E_2{}^1$ by ATP itself.[83a]

On the other hand, we proposed that the mechanism of the ATPase reaction in the steady-state at high concentrations of ATP is different from that at a low concentration range, and that the main route at high ATP concentrations is simple hydrolysis of ATP catalysed by $°E_2^1$ or E_2^1, as shown is the reaction sequence on p. 82. When the initial burst was considerably reduced, as when the samples were stored for a long time at room temperature, the rate of reaction of ATPase at steady-state hardly changed. Also, when the degree of binding of *p*-nitrothiophenol was varied from 0 to 1 mole per mole of myosin in NTP (1)-myosin, the amount of the initial burst decreased from 1 to 0 mole per mole of myosin but there was no change in the steady-state rate and its pH-dependence[81,85]. There was no resemblance of the pH-dependence of the rate at steady-state to that of the rate of formation of the myosin-phosphate-ADP complex[82]. As will be shown in Fig. 2 of chapter 4 (*cf*. section 6), not only the ATPase activity in the steady-state but also the amount of initial burst per mole of H-meromyosin retained completely in two moles of S-1 prepared by degestion of H-meromyosin with trypsin. However, Yagi *et al.*[95a] have recently com-

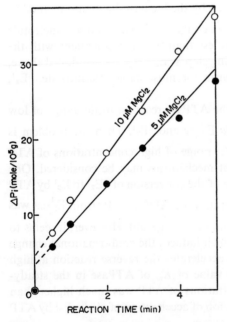

Fig. 19. Rate of liberation of P_i from the reaction of myosin with ATP at low $MgCl_2$ concentrations. 0.6M KCl, 2.1mg/ml myosin, 1mM ATP and 20mM Tris-maleate buffer at pH 7.0 and 25°C. Concentrations of $MgCl_2$: ○, 10; ●, 5μM.

pared kinetic parameters of various kinds of S-1, and found that the ATPase activities in the steady-state were almost the same, while the amount of initial burst decreased by drastic proteolysis. For example, the amount of initial burst of S-1 prepared by digestion of usual S-1 with Nagarse was only about 0.1 mole/mole. Thus the amount of initial burst of P_i-liberation could be reduced markedly by some kinds of proteolysis, without impairing the ATPase activity in the steady-state. Furthermore, Bárány[95b] have recently shown that in living muscle myosin contains ATP but not ADP at the relaxed state, *i.e.*, in the state, where only the myosin-type of ATPase occurs. These results indicate that the myosin-phosphate-ADP complex is not an intermediate in the decomposition of ATP at steady-state in the high ATP concentration range, which must therefore proceed by a different route, *i.e. via* a simple myosin-ATP complex.

8. *The Reaction of Myosin-ATPase with ATP in the Presence of Low Concentrations of Mg^{2+}*

As described in the previous section, the amount of the initial burst of TCA-labile P_i generated by the reaction of myosin with ATP in the presence of high concentrations (greater than 1mM) of Mg^{2+}, is about 1 mole per mole of myosin. However, about 5-20 moles of TCA-labile P_i are rapidly liberated per mole of myosin when the concentration of Mg^{2+} is reduced to only several μM[76,85] (Fig. 19). We have named this phenomenon the 'extra-burst.' The ratio of the 'extra-burst' obtained per mole of myosin, H-MM, and S-1 was 1:1:0.5, as with the stoichiometric burst. We have also measured the liberation of pyruvate from a pyruvate kinase system in order to determine the initial rapid liberation of free ADP accompanying this 'extra-burst', and found it to be one mole less per mole of myosin than the amount of the 'extra-burst' of P_i[85]. The 'extra-burst' was also accompanied by rapid liberation of protons, unlike the initial stoichiometric burst[85]. These two results suggest that ATP is decomposed to free ADP and P_i during the 'extra-burst,' in contrast to the initial stoichiometric burst, and at the termination of the 'extra-burst,' myosin is bound to ADP and TCA-labile P_i.

The following experiments were performed to clarify further the relationship between the initial stoichiometric burst and the 'extra-burst.' The initial stoichiometric burst from the addition of ATP to myosin in the presence of a high concentration of Mg^{2+}, was allowed to reach comple-

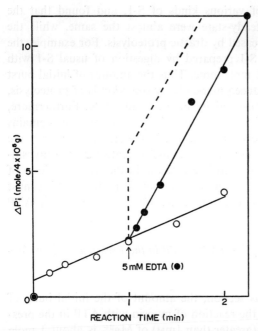

Fig. 20. Effect of addition of EDTA on the rate of liberation of P_i in 2mM $MgCl_2$ solution. 0.6M KCl, 1mM ATP, 4mg/ml myosin and 20mM Tris-maleate buffer at pH 7.0 and 25°C. ○: 2mM $MgCl_2$. ●: 5mM EDTA was added 1 min after the start of the reaction. The arrow indicates the time of addition of 5mM EDTA. - - -: assumed rate of liberation of P_i if the 'extra-burst' occurred with 2mM $MgCl_2$ and 5mM EDTA.

tion. EDTA was then added to reduce the Mg^{2+} concentration, and as expected, the steady-state rate increased, but the ATP still decomposed at a constant rate and no 'extra-burst' was observed (Fig. 20). In contrast, an 'extra-burst' was observed when ATP was added to myosin in the presence of Mg^{2+} and EDTA at the same concentrations used for the above experiment. No stoichiometric burst was observed upon addition of a high concentration of Mg^{2+} after completion of the 'extra-burst' at a low concentration of Mg^{2+}, but a new burst occurred when more ATP was added after completion of ATP decomposition and return of $°E_2^1$ to E_2^1 in a system which had already undergone the initial burst once[75]. These results could be easily explained if phosphoryl myosin ($E_2^1 {\cdot \cdot \cdot ADP \atop \sim P}$) were very unstable and could decompose into enzyme, ADP, P_i and H^+, as well as to $E_2^1 {\cdot \cdot \cdot ADP \atop \cdot \cdot \cdot P}$:

$$E_2^1 + S \rightleftharpoons E_2^{1S} \longrightarrow E_2^1 + P_i + ADP$$
$$\updownarrow$$
$$E_{2S}^1$$
$$\updownarrow$$
$$E_2^1{\cdots}^{ADP}_P \longrightarrow E_2^1{\cdots}^{ADP}_P \longrightarrow °E_2^1 + P_i + ADP$$
(initial stoichiometric burst)

$$°E_2^1 + S \rightleftharpoons °E_2^{1S} \longrightarrow °E_2^1 + P_i + ADP.$$
$$\updownarrow \quad \text{(simple hydrolysis)}$$
$$°E_{2S}^1$$

As already described, in the presence of high concentrations of Mg^{2+}, $E_2^1{\cdots}^{ADP}_P$ is rapidly converted to the more inert $E_2^1{\cdots}^{ADP}_P$, and according to the above mechanism the rate constant for this conversion is much greater than that for the direct decomposition of $E_2^1{\cdots}^{ADP}_P$. On the other hand, when the Mg^{2+} concentration is low, the rate constant for this conversion is thought to be of similar magnitude to that for the direct decomposition. Thus until all of the phosphoryl myosin ($E_2^1{\cdots}^{ADP}_P$) is completely changed to the more inert myosin-phosphate-ADP complex ($E_2^1{\cdots}^{ADP}_P$) and free enzyme with changed conformation ($°E_2^1$), the hydrolysis of ATP continues to proceed through the first intermediate, and this direct decomposition of $E_2^1{\cdots}^{ADP}_P$ is observable as the 'extra-burst.'

We[88] have modified myosin with NTP under conditions necessary for the 'extra-burst,' *i.e.*, at 0.6M KCl, 0.5mM ATP, 10µM $MgCl_2$, pH 6.5 and room temperature. The NTP (2)-myosin* has about 2 moles of NTP bound to each mole of myosin, but there is little effect on the steady-state rate of the ATPase, or on the initial stoichiometric burst. However as shown in Fig. 21, the 'extra-burst' was completely eliminated. Thus alteration of the conditions for modification of myosin with NTP can cause rather specific changes in the steps of the reaction of myosin with ATP, and the significance of each reaction step to the mechanism of muscle contraction has now been successfully elucidated (chapter 7).

In our proposed mechanism, the most stable intermediate is $°E_2^1$, and this species must be present at steady-state when a large amount of ATP

---

* We shall designate as NTP (2)-myosin the modified myosin obtained by nitrothiophenylation of myosin in the presence of ATP and a low concentration of $MgCl_2$.

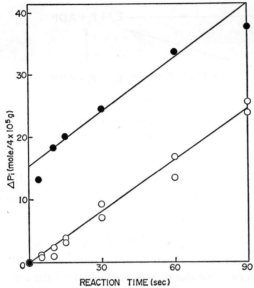

Fig. 21. Effect of *p*-nitrothiophenylation (2) of myosin on the rate of liberation of P_i from the reaction of myosin with ATP at low $MgCl_2$ concentrations. 0.6M KCl, 10 μM $MgCl_2$, 1.0mg/ml myosin, 0.5mM ATP and 20mM Tris-maleate buffer at pH 7.0 and 25°C. ○: NTP(2)-myosin (2.37 moles of NTP/4×10^5g of myosin), prepared by incubation of myosin with NTP in the presence of 2mM ATP, 0.6M KCl, 10μM $MgCl_2$ and 50mM Tris-maleate buffer at pH 6.5 and 25°C for 2 hr; ●: control myosin prepared by incubation with NTP under the same conditions except that ATP was omitted.

is added to myosin. Hence the reaction scheme at steady-state predicted from our analysis of the pre-steady-state is:

$$°E_2^1 + S \rightleftharpoons °E_2^{1S} \longrightarrow °E_2^1 + P_i + ADP .$$
$$\Updownarrow$$
$$°E_{2S}^1$$

Thus the myosin-ATP complexes, E_1S and E_2S, predicted from the pH dependence at steady-state in section 4, can be written as $°E_2^{1S}$ and $°E_{2S}^1$. This mechanism not only explains the results of chemical modification but also the agreement between the dissociation constants of ATP and ADP with myosin obtained from kinetic measurements and from the UV difference spectra shown in Table I.

In conclusion, we shall briefly describe Taylor and his co-workers' recent kinetic studies[95,96-98] on the pre-steady-state of myosin-ATPase. Their results essentially confirm ours, but they have proposed a simpler reaction mechanism, since they have not carried out comparative studies of the many other properties of the myosin-ATP reaction described herein; for example, the exchange of a phosphate group between the reactive myosin-phosphate-ADP complex and ATP, the effect of *p*-nitrothiophenylation of myosin, the 'extra-burst' of P_i-liberation at low Mg^{2+} concentrations, the dependences on ATP concentration of the rates of formation of several intermediates in a range of low ATP concentrations or the change in the UV spectrum with time. They assumed that the rate-determining step of myosin-ATPase is the liberation of ADP from the enzyme. However, a simple complex of myosin with ADP cannot be the most stable intermediate, since the UV spectrum of myosin in the presence of ATP is different from that in the presence of ADP (*cf.* Fig. 3), and since an initial burst of P_i-liberation of normal magnitude (1 mole/mole myosin) is observed when ATP is added to myosin in the presence of ADP, the concentration of which is much higher than K_{ADP} for competitive inhibition[83a], not only at 0°C but also at 20°C. Furthermore, we[83a] have recently shown by a rapid-flow dialysis method that the rate of liberation of ADP from the myosin-ADP complex is much larger than that from the reaction intermediate, *i.e.*, myosin-phosphate-ADP complex (p. 96). Trentham *et al.*[98a] also showed that the rate of liberation of thio-ADP from the simple H-meromyosin-thio-ADP complex was higher than that from the reaction intermediate of thio-ATPase, *i.e.*, the H-meromyosin-P-thio-ADP

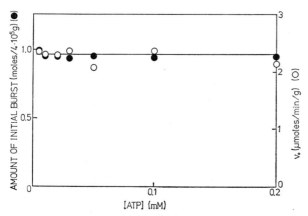

Fig. 22. Dependence on ATP concentration of amount of initial burst and ATPase activity at steady-state. 1mg/ml myosin, 0.5M KCl, 0.1M Tris-HCl, 10mM $MgCl_2$, pH 8.0, 20°C. ●, amount of initial burst; ○, rate of ATPase at steady-state.

complex. Taylor and his colleagues explained several complicated kinetic properties of the myosin-ATP system by assuming that the myosin molecule contains two ATPase active sites interacting with each other, and that the amount of the reaction intermediate increases from 1 to 2 moles per mole of myosin as the ATP concentration is increased to several tens of μM. However, as shown in Fig. 22, both the amount of the initial burst and the rate of ATP-hydrolysis at steady-state were independent of the ATP concentration from 5μM to 0.2mM under the same conditions as those employed by Taylor and his colleagues. For reference, Table III shows how predictions from our mechanism are consistent with the observed results, while those from Taylor's new mechanism are inconsistent.

TABLE III. Comparison of Predictions from Taylor's Mechanism with Those of Ours

	Predicted from the mechanism proposed by		Observed
	Taylor et al.	Tonomura et al.	
K_m, low ATP range (μM)	0.05	$\gg 0.3$	≥ 0.1
v_o, high ATP range: v_o, low ATP range	2:1	$\sim$5:1	4:1
ATP conc. for transition (μM)	$\sim$50	Unpredictable	$\sim$0.5
Nucleotide binding related to ATPase (mole/mole Myo):			
Low ATP range ($\leq 0.2\mu$M)	1 (ADP)	0	0
High ATP range ($\geq 3\mu$M)	2 (ADP)	1 (ATP)	1 (ATP)
v_o, high ATP range: initial burst	Constant	Independent, var.	Independent, var.

Another problem to be discussed in connection with the molecular mechanism of the ATPase is the exchange reaction catalysed by myosin. To show the existence of phosphorylated myosin in the myosin-ATP system, many researchers, especially Koshland et al.[99], investigated the ATP-ADP exchange reaction catalysed by myosin, but always with negative results. Recently, we reinvestigated the exchange reaction in the presence of Mg^{2+} and D_2O to minimise the ATP-decomposition[100]. No exchange between ATP and ADP was observed. On the other hand, Levy, Koshland, Yount and their co-workers[101-105] showed that in the presence of Mg^{2+} there is incorporation of more than 1 mole of ^{18}O per mole of P_i generated from the myosin-ATP system in $H_2^{18}O$. The similarity of the conditions for the initial burst and ^{18}O-incorporation[79] suggests that phosphorylation of the carboxyl group occurs at site 2, and ^{18}O-exchange takes place according to:

$$-\overset{O}{\underset{|}{\underset{O^-}{C}}}-O-\overset{O}{\underset{|}{P}}-OH \underset{-H^+}{\overset{+H^+}{\rightleftharpoons}} -\overset{O}{\underset{|}{\underset{O^-}{C}}}-O-\overset{O}{\underset{|}{P}}-O^+ H_2 \underset{+H_2O}{\overset{-H_2O}{\rightleftharpoons}} -\overset{O}{\underset{|}{C}}-O^+=P\overset{\diagup O}{\diagdown O^-}$$

Thus, it may be assumed that even in the steady-state there exists a rapid equilibrium reaction between $°E_{2S}{}^1$ and $°E_2{}^1\cdots_{ADP}$ although the concentration of the latter compound is negligibly small, as mentioned above. The results that myosin catalyses the incorporation of ^{18}O into P_i, but neither the incorporation of ^{18}O into ATP[101–105] nor the exchange reaction between ATP and ADP[99,100] suggest that the rate constant of the step $E_2{}^{1S} \rightarrow E_2{}^1 + P_i + ADP$ is much larger than that of the dissociation of $E_2{}^{1S}$ into $E_2{}^1 + S$ (see section 3).

Another type of oxygen exchange reaction catalysed by contractile protein is the one occurring between water and P_i of the medium (medium exchange[106]). The medium exchange reaction is catalysed by myosin in the presence of ATP and by actomyosin and glycerol-treated fibre bundles in the absence or presence of ATP[107,108]. The relevance of the phosphate-oxygen exchange reaction to the process of muscle contraction was discussed by Dempsy et al.[107] and R. W. Benson et al.[108]

In this chapter we have discussed the mechanism of reaction of myosin with ATP. It is also necessary to know how F-actin interacts with the myosin-ATP system before the molecular mechanism of muscle contraction can be elucidated. This is discussed in detail in chapter 7, based on the reaction mechanism for the myosin-ATP system developed in this chapter.

REFERENCES

1 K. Lohmann, *Biochem. Z.*, **233**, 260 (1929).
2 K. Lohmann and P. Schester, *Biochem. Z.*, **272**, 24 (1934).
3 C. H. Fiske and Y. SubbaRow, *Science*, **70**, 381 (1929).
4 J. Baddiley, A. M. Michelom and A. R. Todd, *J. Chem. Soc.*, 582 (1949).
5 O. Kennard, N. W. Issacs, J. C. Coppola, A. J. Kirby, S. Warren, W. D. S. Motherwell, D. G. Watson, D. L. Wampler, D. H. Cheney, A. C. Larson, K. A. Kerr and L. R. D. Sanseverino, *Nature*, **225**, 333 (1970).
6 F. Lipmann, *Advan. Enzymol.*, **1**, 99 (1941).
7 R. J. Podolsky and M. F. Morales, *J. Biol. Chem.*, **218**, 945 (1956).
8 C. Kitzinger and T. Benzinger, *Z. Naturforsch.*, **10b**, 375 (1955).
9 T. Benzinger, C. Kitzinger, R. Hems and K. Burton, *Biochem. J.*, **71**, 400 (1959).

10   R. C. Phillips, P. George and R. J. Rutman, *J. Biol. Chem.*, **244**, 3330 (1969).
10a J. Rosing and E. C. Slater, *Biochim. Biophys. Acta*, **267**, 275 (1972).
11   R. M. Smith and R. A. Alberty, *J. Amer. Chem. Soc.*, **78**, 2376 (1956).
12   R. A. Alberty, R. M. Smith and R. M. Bock, *J. Biol. Chem.*, **193**, 425 (1951).
13   R. C. Phillips, P. George and R. J. Rutman, *J. Amer. Chem. Soc.*, **88**, 2631 (1966).
14   R. C. Phillips, P. George and R. J. Rutman, *Biochemistry*, **2**, 501 (1963).
15   A. E. Martell and G. Schwarzenbach, *Helv. Chim. Acta*, **39**, 653 (1956).
16   E. Walaas, *Acta Chem. Scand.*, **12**, 528 (1958).
17   K. Burton, *Biochem. J.*, **71**, 388 (1959).
18   R. M. Bock, in "The Enzymes," 2nd ed., ed. by P. D. Boyer, H. Lardy and K. Myrbäck, Academic Press, New York, Vol. 2, p. 3 (1960).
19   H. Brintzinger and S. Fallab, *Helv. Chim. Acta*, **43**, 43 (1960).
20   L. B. Nanninga, *Biochim. Biophys. Acta*, **54**, 330 (1961).
21   M. M. Taqui Khan and A. E. Martell, *J. Phys. Chem.*, **66**, 10 (1962).
22   W. J. O'Sullivan and D. D. Perrin, *Biochemistry*, **3**, 18 (1964).
23   M. M. Taqui Khan and A. E. Martell, *J. Amer. Chem. Soc.*, **88**, 668 (1966).
24   R. A. Alberty, *J. Biol. Chem.*, **244**, 3290 (1969).
25   H. Diebler, M. Eigen and G. G. Hammes, *Z. Naturforsch.*, **15b**, 554 (1960).
26   M. Eigen and G. G. Hammes, *Advan. Enzymol.*, **25**, 1 (1963).
27   M. Eigen and G. G. Hammes, *J. Amer. Chem. Soc.*, **82**, 5951 (1960).
28   M. Eigen and G. G. Hammes, *J. Amer. Chem. Soc.*, **83**, 2786 (1961).
29   B. Pullman and A. Pullman, *Radiation Res., Suppl.*, **2**, 160 (1960).
30   H. Kalckar, *Chem. Rev.*, **28**, 71 (1941).
31   T. Hill and M. Morales, *Arch. Biochem.*, **29**, 450 (1950).
32   T. Hill and M. Morales, *J. Amer. Chem. Soc.*, **73**, 1656 (1951).
33   P. George, R. J. Witonsky, M. Trechtman, C. Wu, W. Dorwart, L. Richman, W. Richman, F. Shurayh and B. Lentz, *Biochim. Biophys. Acta*, **223**, 1 (1970).
34   D. E. Koshland Jr., Z. Budenstein and A. Kowalsky, *J. Biol. Chem.*, **211**, 279 (1954).
35   J. B. Martin and D. M. Doty, *Anal. Chem.*, **21**, 965 (1949).
36   A. M. Reynard, L. F. Hass, D. D. Jacobsen and P. D. Boyer, *J. Biol. Chem.*, **236**, 2277 (1961).
37   I. Green and W. F. H. M. Mommaerts, *J. Biol. Chem.*, **202**, 541 (1953).
38   W. A. Engelhardt and M. N. Ljubimova, *Nature*, **144**, 668 (1939).
39   I. Banga, *Stud. Inst. Med. Chem. Univ. Szeged*, **1**, 27 (1941).
40   T. Nihei and Y. Tonomura, *J. Biochem.*, **46**, 305 (1959).
41   M. Dixon and E. C. Webb, "Enzymes," chapter IX, 2nd ed., Longmans, Green & Co., London (1964).
42   S. Watanabe, Y. Tonomura and H. Shiokawa, *J. Biochem.*, **40**, 387 (1953).
43   E. A. Sugden and T. Nihei, *Biochem. J.*, **113**, 821 (1969).

44 A. Mühlrad, F. Fábián and N. A. Biró, *Biochim. Biophys. Acta*, **89**, 136 (1964).
45 G. W. Offer, *Biochim. Biophys. Acta*, **89**, 566 (1964).
46 B. Kiely and A. Martonosi, *J. Biol. Chem.*, **243**, 2273 (1968).
47 L. Ouellet, K. J. Laidler and M. F. Morales, *Arch. Biochem. Biophys.*, **39**, 37 (1952).
48 L. B. Nanninga and W. F. H. M. Mommaerts, *Proc. Natl. Acad. Sci. U. S.*, **46**, 1155 (1960).
49 L. B. Nanninga and W. F. H. M. Mommaerts, *Proc. Natl. Acad. Sci. U. S.*, **46**, 1166 (1960).
50 L. B. Nanninga, *Biochim. Biophys. Acta*, **60**, 112 (1962).
51 Y. Tonomura, K. Sekiya, K. Imamura and T. Tokiwa, *Biochim. Biophys. Acta*, **69**, 305 (1963).
52 K. Sekiya, S. Mii, K. Takeuchi and Y. Tonomura, *J. Biochem.*, **59**, 584 (1966).
53 F. Morita and K. Yagi, *Biochem. Biophys. Res. Commun.*, **22**, 297 (1966).
54 F. Morita, *J. Biol. Chem.*, **242**, 4501 (1967).
55 F. Morita, *Biochim. Biophys. Acta*, **172**, 319 (1969).
56 K. Sekiya and Y. Tonomura, *J. Biochem.*, **61**, 787 (1967).
57 M. Dixon and E. C. Webb, "Enzymes," chapter IV, section C, 2nd ed., Longmans, Green & Co., London (1964).
58 N. Azuma and Y. Tonomura, *Biochim. Biophys. Acta*, **73**, 499 (1963).
59 H. M. Levy, N. Sharon, E. M. Rayan and D. E. Koshland, Jr., *Biochim. Biophys. Acta*, **56**, 118 (1962).
60 M. Ohe, B. K. Seon, K. Titani and Y. Tonomura, *J. Biochem.*, **67**, 513 (1970).
61 E. R. G. Barron and T. P. Singer, *Science*, **97**, 356 (1943).
62 K. Bailey and S. V. Perry, *Biochim. Biophys. Acta*, **1**, 506 (1947).
63 W. W. Kielley, H. M. Kalckar and L. B. Bradly, *J. Biol. Chem.*, **219**, 95 (1956).
64 T. Yamashita, I. Kabasawa and T. Sekine, *J. Biochem.*, **63**, 608 (1968).
65 M. Onodera and K. Yagi, *J. Biochem.*, **66**, 751 (1969).
65a K. Uchida, K. Tanaka and T. Hiratsuka, *Biochim. Biophys. Acta*, **256**, 132 (1972).
66 E. T. Friess, *Arch. Biochem. Biophys.*, **51**, 17 (1954).
67 G. D. Greville and D. M. Needham, *Biochim. Biophys. Acta*, **16**, 284 (1955).
68 J. B. Chappell and S. V. Perry, *Biochim. Biophys. Acta*, **16**, 285 (1955).
69 Y. Tonomura and K. Furuya, *J. Biochem.*, **48**, 899 (1960).
70 S. Kubo, S. Tokura and Y. Tonomura, *J. Biol. Chem.*, **235**, 2835 (1960).
71 H. Takashina, *Biochim. Biophys. Acta*, **200**, 319 (1970).
72 T. Shimada, *J. Biochem.*, **67**, 185 (1970).
73 K. Sekiya, S. Mii and Y. Tonomura, *J. Biochem.*, **57**, 192 (1965).
74 Y. Tonomura, S. Tokura, K. Sekiya and K. Imamura, *Arch. Biochem. Biophys.*, **95**, 229 (1961).

75 Y. Tonomura, S. Kitagawa and J. Yoshimura, *J. Biol. Chem.*, **237**, 3660 (1962).
76 T. Kanazawa and Y. Tonomura, *J. Biochem.*, **57**, 604 (1965).
77 T. Tokiwa and Y. Tonomura, *J. Biochem.*, **57**, 616 (1965).
78 K. Imamura, T. Kanazawa, M. Tada and Y. Tonomura, *J. Biochem.*, **57**, 627 (1965).
79 K. Imamura, M. Tada and Y. Tonomura, *J. Biochem.*, **59**, 280 (1966).
80 H. Nakamura and Y. Tonomura, *J. Biochem.*, **63**, 279 (1968).
81 H. Tokuyama and Y. Tonomura, *J. Biochem.*, **62**, 456 (1967).
82 H. Onishi, H. Nakamura and Y. Tonomura, *J. Biochem.*, **63**, 739 (1968).
83 A. Inoue, K. Sekiya-Shibata and Y. Tonomura, *J. Biochem.*, **71**, 115 (1972).
83a A. Inoue and Y. Tonomura, *J. Biochem.*, in press.
84 H. Onishi, H. Nakamura and Y. Tonomura, *J. Biochem.*, **64**, 769 (1968).
85 N. Kinoshita, S. Kubo, H. Onishi and Y. Tonomura, *J. Biochem.*, **65**, 285 (1969).
86 N. Kinoshita, T. Kanazawa, H. Onishi and Y. Tonomura, *J. Biochem.*, **65**, 567 (1969).
87 Y. Tonomura and T. Kanazawa, *J. Biol. Chem.*, **240**, PC4110 (1965).
88 Y. Tonomura, H. Nakamura, N. Kinoshita, H. Onishi and M. Shigekawa, *J. Biochem.*, **66**, 599 (1969).
89 Y. Tonomura and S. Kitagawa, *Biochim. Biophys. Acta*, **40**, 135 (1960).
90 S. Kubo, H. Tokuyama and Y. Tonomura, *Biochim. Biophys. Acta*, **100**, 459 (1965).
91 H. Tokuyama, S. Kubo and Y. Tonomura, *Biochem. Z.*, **345**, 57 (1966).
92 S. Kitagawa, K.-K. Chiang and Y. Tonomura, *Biochim. Biophys. Acta*, **82**, 83 (1964).
93 L. Sartorelli, H. J. Fromm, R. W. Benson and P. D. Boyer, *Biochemistry*, **5**, 2877 (1966).
94 G. DiSabato and W. P. Jencks, *J. Amer. Chem. Soc.*, **83**, 4400 (1961).
94a S. P. Colowick and F. C. Womack, *J. Biol. Chem.*, **244**, 774 (1969).
95 E. W. Taylor, R. W. Lymn and G. Moll, *Biochemistry*, **9**, 2984 (1970).
95a K. Yagi, Y. Yazawa, F. Ohtani and Y. Okamoto, presented at a Japan-U.S. Seminar, Tokyo (1972).
95b M. Bárány, personal communication.
96 B. Finlayson and E. W. Taylor, *Biochemistry*, **8**, 802 (1969).
97 B. Finlayson, R. W. Lymn and E. W. Taylor, *Biochemistry*, **8**, 811 (1969).
98 R. W. Lymn and E. W. Taylor, *Biochemistry*, **9**, 2975 (1970).
98a D. R. Trentham, R. G. Bardsley, J. F. Eccleston and A. G. Weeds, *Biochem. J.*, **126**, 635 (1972).
99 D. E. Kohland, Jr., Z. Budenstein and A. Kowalsky, *J. Biol. Chem.*, **211**, 279 (1954).

100 H. Ikezawa, K. Ikezawa, Y. Tonomura and M. F. Morales, *J. Biochem.*, **69**, 901 (1971).
101 H. M. Levy and D. E. Koshland, Jr., *J. Amer. Chem. Soc.*, **80**, 3164 (1958).
102 H. M. Levy and D. E. Koshland, Jr., *J. Biol. Chem.*, **234**, 1102 (1959).
103 H. M. Levy, N. Sharon, E. Lindeman and D. E. Koshland, Jr., *J. Biol. Chem.*, **235**, 2628 (1960).
104 R. G. Yount and D. E. Koshland, Jr., *J. Biol. Chem.*, **238**, 1708 (1963).
105 J. R. Swanson and R. G. Yount, *Biochem. Z.*, **345**, 395 (1966).
106 M. E. Dempsey and P. D. Boyer, *J. Biol. Chem.*, **236**, PC6 (1961).
107 M. E. Dempsey, P. D. Boyer and E. S. Benson, *J. Biol. Chem.*, **238**, 2708 (1963).
108 R. W. Benson, M. E. Dempsey and E. S. Benson, *J. Biol. Chem.*, **242**, 1612 (1967).

4
THE CHEMICAL STRUCTURES OF THE ACTIVE SITES OF MYOSIN*

The biological functions of myosin are ATPase activity, formation of filaments by molecular aggregation (chapters 2 and 3) and binding with actin (chapter 6). Since myosin is a key contractile protein, the chemical structures of the active centres for these reactions must be known in order to understand the molecular mechanism of muscle contraction.

There are several ways to investigate the chemical structures of the active centres of myosin, including the kinetic methods described in the previous chapter. The present chapter describes studies of experimental chemical modification of myosin. These studies can provide direct information about the structures of the active sites for the various functions of myosin.

* Contributor: Takamichi Shimada

TABLE I. The Amino Acid Composition of Myosin and Its Subfragments (moles/10^5g)

	Myosin[2]	L-MM[2]	L-MM fr. 1[2]	H-MM[2]	Subfragment-1[3]	Subfragment-2[4]
Cys/2	8.8	(7.3)	4.0	7.4	10	6.4
Asp	85	82	83	82	91	87
Thr	44	37	33	44	50	43
Ser	39	38	34	39	43	35
Glu	157	198	210	137	126	242
Pro	22	7.3	0	32	30	0
Gly	40	22	18	50	55	19
Ala	78	79	81	73	67	87
Val	43	41	38	48	51	24
Met	23	15	19	26	29	26
Ileu	42	41	39	44	52	34
Leu	81	92	96	73	73	99
Tyr	20	10	9	21	31	2.6
Phe	29	6.1	4	36	46	8.6
His	16	20	21	14	15	9
Lys	92	90	94	86	71	121
Arg	43	57	60	34	32	37
(NH_3)	(92)	(107)	(108)	(90)	(93)	—
Total	863	842	843	846	872	881

1. The Chemical Structure of the Myosin Molecule

Before describing the active centres, we shall discuss briefly the structure of the myosin molecule. The amino acid composition was first reported by Kominz et al.[1], but there have been several more recent studies[2-4]. Table I shows typical amino acid compositions of myosin and the subfragments described in chapter 2. The following conclusions can be drawn about the distribution of amino acid residues in the myosin molecule: (1) Many residues with charged side chains are present, and are fairly uniformly distributed throughout the molecule. (2) Those residues which cannot form part of a helix structure (particularly proline) are abundant in subfragment-1, which is derived from the head portion of the molecule. These two facts are important determinants of the shape and function of the myosin molecule, especially (2), which indicates that it is difficult to form an α-helix structure in the head portion. (3) Many investigators have found that the number of SH groups in myosin, which are titratable with PCMB, is approximately equal to the half cystine content found by amino acid analysis. Myosin probably, therefore, contains no S-S bridge.

Myosin contains three unusual amino acids: methylhistidine and methyl-lysines. The presence of 3-methylhistidine in myosin was first reported by Asatoor and Armstrong[5] and Johnson et al.[6] It was later shown that the heavy chains of myosin from white fibres contain two residues of 3-methylhistidine per molecule[7,8], but red and cardiac myosins are devoid of 3-methylhistidine[8]. This is the first clear demonstration of a difference in chemical structure between the heavy chains of myosin of different origins. Adult rabbit skeletal myosin has also been shown to contain both ε-N-monomethyllysine and ε-N-trimethyllysine[9-11a]. These two unusual amino acids occur in an approximate 1 : 2 : 1 ratio with 3-methylhistidine respectively. Evidence for *in vitro* methylation of histidine and lysine residues in myosin and actin (see chapter 5, section 1) by S-adenosyl-L-methionine has been reported by several workers[11b-11e].

The identity of the terminal amino acid residue of myosin was first studied by Bailey[12]. He[12] and Gaetzens et al.[13] concluded that there is no N-terminal amino acid. On the other hand, Kielley et al.[14] reported that there is one mole of N-terminal histidine per 2.0×10^5 g of myosin. Offer[15] obtained one mole of an N-acetyl peptide per 2.5×10^5 g of myosin. The C-terminal amino acid was studied by Locker[16], who found that 5.0×10^5 g of myosin contained 1.7 moles of C-terminal isoleucine. As described in chapter 2, myosin is composed of two heavy chains and probably four light chains, and Trotta et al.[17] have shown that the C-terminal amino acids of the light chains are isoleucine, while those of the heavy chains are undetectable.

Recently, Kimura and Kielley[18] and Weeds and Hartley[19] investigated in detail the peptides near the SH groups in myosin. They found that although the myosin molecule contains 44 SH groups, only 22 different SH peptides result from tryptic and chymotryptic digestion. Hence, at least in the vicinity of the SH groups, myosin contains two identical polypeptide chains. White skeletal muscle myosin from adult rabbits contains about 2 moles of 3-methylhistidine per 5×10^5g, as described above. It has recently been shown that 3-methylhistidine is located in the heavy chain of S-1[7,9,10,20-22], and that all the 3-methylhistidine is found only in one unique peptide, which can be isolated from the peptides obtained by cleavage of S-1 with cyanogen bromide or digestion of S-1 with trypsin[23]. These results agree with the double headed structure for myosin suggested by the electron microscopic work of Slayter and Lowey[24]. More recently, the peptides containing histidine were isolated from tryptic digests of cardiac myosin and were found to be homologous to the peptide containing 3-methylhistidine from skeletal muscle myosin[24a].

2. The Number of Active Centres in Myosin

In the vicinity of the SH groups and 3-methylhistidine, the chemical structures of the two heavy chains—the main constituent of myosin—are identical. We shall now describe the measurement of the number of molecules of ATP, ADP or PP_i which can bind to the myosin molecule. The binding of substrate analogues was first studied by equilibrium dialysis to determine the number of active sites in myosin[25]. Since this method requires long measurement times during which ATP itself would be hydrolysed, PP_i, a competitive inhibitor of the ATPase, was used instead. Myosin solution was contained in a cellophane membrane and immersed in a solution containing $^{32}PP_i$. After time sufficient to reach equilibrium the PP_i concentration of the external liquid was estimated from its radioactivity. If the activity coefficient of the PP_i is assumed to be unity, the concentration of PP_i within the membrane equals the sum of free PP_i and that bound to myosin, while the free PP_i concentrations inside and outside the membrane should be equal. Figure 1 shows the amounts of PP_i bound to myosin at various free PP_i concentrations measured by this tech-

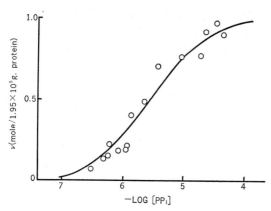

Fig. 1. The binding of myosin with PP_i. 0.6M KCl, 0.3mM $MgCl_2$, pH 7.5, 5°C. The curve is represented by $\nu = \dfrac{10^{5.9}[PP_i](1+2\times 10^{5.2}[PP_i])}{1+10^{5.9}[PP_i](1+10^{5.2}[PP_i])}$.

nique. One mole of PP_i is bound per $2\text{--}2.5 \times 10^5$g of myosin, and since the most likely molecular weight of myosin is 4.8×10^5 (chapter 2) 2 moles of PP_i bind to 1 mole of myosin, but with different dissociation constants. It is of interest that only 1 mole of PP_i binds per mole of myosin in actomyosin, showing that the complete dissociation of actomyosin is not accompanied by this binding (see chapter 6).

One of the latest methods for measuring the binding between proteins and low molecular weight materials, developed by Hummel and Dreyer[25a], involves Sephadex G-25 gel filtration. This technique is rapid, yet very precise, and was used by Morita[26,27] to study the binding of H-MM with ADP in order to compare the values for this with the values obtained from changes in the absorption of H-MM at 288nm (see chapter 3, section 3). She showed that the maximum number of ADP molecules binding to H-MM, measured by gel filtration, is 2 per molecule of H-MM. In the presence of Mg^{2+} at low ionic strength[27], and especially in the presence of Mn^{2+}[27a], the binding constants of 2 moles of ADP with 1 mole of H-MM were very different from each other. By comparing the binding of ADP to H-MM with changes in the UV absorption of H-MM induced by ADP, Morita and her co-workers[27a] showed conclusively that only 1 of the 2 moles of ADP, that which binds more strongly, induces the difference spectrum. Iino et al.[28] observed that there are 2 moles of tryptophane residue per mole of H-meromyosin which bind characteristically with 2-

hydroxy-5-nitrobenzyl bromide. The shoulder near 293nm in the UV spectrum, which changes when H-meromyosin binds with ATP, is not altered when only 1 mole of tryptophane reacts, but decreases linearly when the modification increases from 1 to 2 moles. More recently, Iino et al.[28a] have shown that only one of these two tryptophane residues is located in S-1, and Morita[28b] has shown that 0.5 moles of ADP binding with S-1 induces the maximum change in the UV spectrum of S-1.

The binding of myosin with ATP can be determined, first by measuring the free ATP concentration under conditions in which any ADP formed is immediately reconverted to ATP by phosphokinase, and then by subtracting the free ATP from the total ATP which had been added. Nanninga and Mommaerts[29] measured the concentration of free ATP by the luciferin-luciferase method, and concluded that 1 mole of ATP binds to 1 mole of myosin. As described above, the UV absorption changes when ATP or ADP is added to H-MM. Sekiya and Tonomura[30] and Morita[26] studied the extent of binding as a function of ATP or ADP concentration under conditions in which binding is strong, and concluded that the complexing of 1 mole of ATP or ADP per mole of H-MM causes the change in absorption. Measurements of the dependence of the rate of myosin-ATPase upon ATP concentration showed that one mole of ATP binds to myosin, assuming the relation $\frac{v}{V_{max}} = \frac{[ES]}{\varepsilon}$ and $[\Sigma S] = [S] + [ES]$, where ε is the total concentration of active sites, [S] the concentration of free ATP, and $[\Sigma S]$ the total ATP concentration[31]. Kinetic studies also indicate that there is one site for simple hydrolysis and one site for phosphorylation per molecule of myosin (chapter 3). As described later, the maximum change in enzyme activity occurs when 1 mole of specific residues per mole of myosin is chemically modified with reagents such as p-nitrothiophenol (NTP), trinitrobenzenesulphonate (TBS), diazonium-1H tetrazole (DHT) and monoiodoacetamide (IAA).

Thus, various lines of evidence obtained by us suggest strongly that the myosin molecule contains two ATP-binding sites but that their physiological functions are not equivalent to each other. However, our view is not generally accepted by investigators in this field, since many results reported by other workers, particularly those in the United States, can be explained most easily by assuming that the myosin molecule has two identical active sites.

The previous results of Gergely, Martonosi et al.[32,33] conflicted with ours, since they found that only 1 mole of PP_i was bound per mole of myosin, but they have now obtained new data agreeing with ours[25], except that under their conditions the two dissociation constants were approximately equal[34,35]. Young[36] and Lowey and Luck[37] reported that myosin and H-MM bind with ADP at two sites which have identical association constants. The dissociation constants for binding of ATP to myosin A and B were also determined by Bowen and Evans[38], by rapid filtration using an ATP-regenerating system. The existence of two sets of binding sites was revealed, and the combined weights of ATP with the sites of higher association constant on myosin A and B were reported to be 200,000 and 250,000g per mole of ATP.

The kinetic analysis of the ATPase in the presence of excess Mg^{2+}, performed by Eisenberg and Moos[39], showed that there are two sites for ATP per myosin or H-MM molecule, but only one site per S-1 molecule. The titration of myosin with the spin label, N-2,2,6,6-tetramethyl piperidine nitroxide iodoacetamide, reported by Seidel et al.[40], indicated the presence of two rapidly reacting thiol groups per molecule. Labelling of these groups increased the Ca^{2+}-activated ATPase activity (see section 3). The addition of ATP, ITP, ADP or PP_i increased the mobility of spin labels bound to the thiol groups, the maximum changes in EPR spectra being obtained in the presence of 2 moles of PP_i per mole of enzyme. Furthermore, Murphy and Morales[41] measured the change in the absorption spectrum of 6-mercapto 9-β-D ribofuranosyl purine 5′-triphosphate induced by its binding to myosin, and concluded that 1.8 moles bind to one mole of myosin with the same association constant. The view contrary to ours, that there are two identical active sites in the myosin molecule, is thus supported by many experimental results. The fact that 2 moles of ADP or PP_i bind to one mole of myosin with identical dissociation constants under fixed conditions, support but do not prove the presence of two identical active sites. It seems to be questionable to deduce the number of binding sites from kinetic analysis of myosin-ATPase, since its mechanism is rather complicated, as described in chapter 3. Furthermore, the results obtained by chemical modification are not conclusive. For example, Seidel et al.[40] have not shown by chemical analysis whether only one unique thiol group is titrated by the spin label in their studies. Moreover, the results suggesting two seemingly identical sites can be expected from the substructure of myosin, which contains two similar heavy chains and two sets of light chains with similar chemical structures, as mentioned in chapter 2, section 5.

Under these circumstances, we cannot decide conclusively whether the two substrate binding sites on the myosin molecule are identical or not. However, the model with two non-identical active sites is supported not only by our results on the binding of ATP, ADP or PP_i with myosin described above, but also by the molecular structure of myosin (chapter 2, section 5), the kinetic properties of myosin-ATPase at the pre-steady-state (chapter 3, section 6) and the mechanism of the reaction of actomyosin with ATP (chapter 6). Therefore, in this monograph we will analyse the various experimental results on the basis of the non-identical, two-active-site model of myosin.

If we accept this model, the following problem is apparent: although the main constituents of the myosin molecule are two heavy chains with almost identical chemical structures, there is but one site directly related to each function. The modification of myosin with IAA, which is described in the following section[42], is pertinent to this. In the presence of urea, IAA reacts with 2 moles of the cysteine in Ileu·Cys-SH·Arg per mole of myosin. In spite of this, the maximum alteration of the ATPase is attained when only one of the two cysteine residues in the intact myosin molecule is bound to IAA. Comparison of the ATPase activities of H-MM and S-1 obtained from carboxamidomethyl myosin with those from a control unmodified myosin showed that the ATPase site attacked by IAA amounted

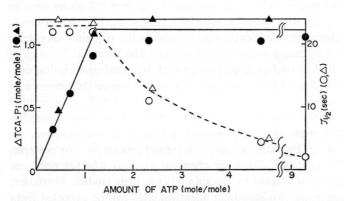

Fig. 2. The dependence on ATP concentration of the amouht of the stoichiometric burst and $\tau_{1/2}$ for myosin and subfragment-1. 2.8M KCl, 10mM $MgCl_2$, pH 7.5, 0°C, 4.2μM protein. ○, ●: myosin (molecular weight taken as 4.8×10^5); △, ▲: S-1 (taken as two fragments of molecular weight of 1.2×10^5). ●, ▲, amount of initial stoichiometric burst of P_i; ○, △: $\tau_{1/2}$ of stoichiometric burst.

to 1 mole per mole of myosin. One possibility, already discussed in chapter 2, section 6, is that only one of the two heavy chains is capable of forming an active site on interaction with the light chain essential to ATPase activity. Furthermore, as shown in Fig. 2, the amount of the stoichiometric burst of liberation of P_i and its rate per $2 \times 1.2 \times 10^5$g of S-1 were identical to those per 4.8×10^5g of myosin. This indicates that, with respect to the phosphorylation reaction, the two S-1 in one myosin molecule are independent of each other, and that only one S-1 of the two has the active site for phosphorylation.

The relationship between the submolecular structure of the myosin molecule and its function is important in considering the role of myosin during contraction. We shall return to this question in relation to the binding of myosin with F-actin in chapter 6, and its physiological significance will be discussed in chapter 13.

3. Chemical Modification of Myosin-ATPase

Before we discuss the chemical modification of myosin-ATPase itself we should consider what this technique can reveal concerning the active sites of enzymes in general. In order for a particular chemical modification* to help elucidate the structure of the active site, it must cause stoichiometric inactivation, and there must have been specific protection by the substrate or by competitive inhibitors against the inactivation. Even so, the possibility remains that the particular modification occurs elsewhere than at the catalytic site and that specific protecting agents induce a change in conformation at a site other than the catalytic site. However, the affinity labelling method[43] is free from such ambiguity of interpretation. So far the work performed on the chemical modification of myosin-ATPase is incomplete, and the foregoing points have not always been taken into account. Hence we must sound a *caveat* that there is still substantial ambiguity in the interpretation of the experimental results.

In chapter 3 it was shown that there are separate pathways involving different active sites for the simple hydrolysis of ATP by myosin and for the formation of phosphoryl myosin. We shall now discuss the changes in enzymic activity caused by chemical modification (a) with respect to the formation and decomposition of phosphoryl myosin, for example with NTP and TBS, and (b) with respect to the simple hydrolysis, for example

* See recent excellent reviews *42a–42d* on the chemical modification of amino acid residues.

with IAA, N-ethyl maleimide (NEM), DHT, photo-oxidation and TBS.

A) The effect of chemical modification on the formation of phosphoryl myosin

Under conditions in which phosphoryl myosin is formed, *i.e.* in the presence of high concentrations of Mg^{2+} and ATP, 1 mole of NTP binds to each mole of myosin[44,45], probably by nucleophilic attack by NTP on phosphoryl myosin. The *p*-nitrothiophenylation scarcely affects the simple hydrolysis of ATP by myosin; only the formation of phosphoryl myosin, as measured by the 'initial burst' of P_i, is suppressed[45,46]. In order to determine the amino acid residue bound to NTP, NTP (1)-myosin was exhaustively fragmented with proteolytic enzymes. One mole of NTP-glutamic acid, and no other NTP-amino acid, was obtained per mole of NTP, indicating that the binding site of NTP with myosin is a glutamic acid residue[45,47].

We[48] have *p*-nitrothiophenylated myosin at low concentrations of Mg^{2+} and ATP, under which conditions the 'extra-burst' of P_i is observed (chapter 3). The NTP (2)-myosin contained about 2 moles of NTP per mole of myosin and there was little change in the ATPase activity at steady-state or the stoichiometric initial burst of P_i, although the 'extra-burst' was completely eliminated.

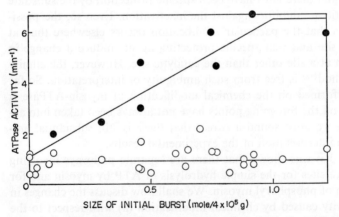

Fig. 3. Relation between the amount of P_i liberated in the initial burst before trinitrophenylation and the ATPase activity at steady-state before and after trinitrophenylation. The amount of initial burst of P_i was changed by *p*-nitrothiophenylation (1) of myosin. ○: ATPase activity at steady-state before trinitrophenylation; ●: ATPase activity at steady-state of trinitrophenyl myosin (1 mole TNP/mole of myosin).

Kubo et al.[49] showed that about 2 moles of lysine ε-amino groups per mole of myosin are rapidly trinitrophenylated by 2,4,6-trinitrobenzenesulphonate (TBS) at neutral pH and low temperatures. The TBS is shown to be bound near the active site by the suppression of its specific binding on addition of PP_i and ATP and by changing the secondary structure near the active site with LiBr (see chapter 2), which markedly alters the binding rate[50]. Recently, Fábián and Mühlrad[51,52] observed that there was almost no change in ATPase activity upon reaction with TBS in the presence of ATP, thereby supporting the above conclusion. The Mg^{2+}-ATPase activity is significantly promoted by trinitrophenylation. As mentioned above, p-nitrothiophenylation (1) of myosin decreases the initial burst but induces no change in the ATPase activity at steady-state. In contrast, the Mg^{2+}-ATPase activity of trinitrophenyl myosin decreases as the size of the initial burst decreases upon p-nitrothiophenylation. When the initial burst becomes zero, the activation by TBS ceases (Fig. 3). A reasonable explanation is that the decomposition of phosphoryl myosin is accelerated by trinitrophenylation[53]. As described later, F-actin has a similar influence, but since the binding sites of TBS and F-actin with myosin are different, both effects are thought to be allosteric. This hypothesis is supported by studies of the nucleotides in actin, described in chapter 7. Both K_m and V_{max} for the simple hydrolysis of ATP by the Ca^{2+}-activated myosin-ATPase at steady-state are decreased by trinitrophenylation, and this may be due to molecular association between the adenine group of ATP and the trinitrophenyl ring, which stabilises the Michaelis complex. The TNP-H-MM obtained by tryptic digestion of trinitrophenyl myosin contained about 90% of the TNP-lysine of the original myosin. Furthermore, the subfragment-1 (MW, 1.2×10^5) obtained by tryptic digestion of TNP-H-MM contained 0.6 moles* of lysine residues which react rapidly and specifically with TBS to the extent of 2 moles per mole of myosin or H-MM[54]. The simplest explanation of all these results is that only 1 of the 2 moles of lysine per mole of myosin which rapidly bind with TBS is contained in the active site, and that this lysine is present in only one of the two S-1 contained in myosin. To determine the chemical structure in the region of the trinitrophenylated lysine residue, TNP-myosin (or H-MM or S-1) was digested with Nagarse or Pronase to isolate the TNP-peptide [54,55]. Edman degradation and fragmentation with hydrazine showed that 1 of the 2 moles of lysine per mole of myosin which bind rapidly with

* The results of Tokuyama et al.[54] were recalculated using a molecular weight for S-1 of 1.2×10^5.

TBS is contained in the peptide Asp(NH$_2$)·Pro·Pro·Lys. Since this peptide is present in S-1, it is probably near the active site.

Recently, Takashina[56] also reported the presence of a lysine residue near the active site of the ATPase. He showed that one mole of 1-dimethyl-aminonaphthalene-5-sulphonyl chloride reacts selectively with 1 mole of lysine residues per mole of heavy meromyosin to form a sulphonamide linkage, and the ATPase activity in the presence of 5mM CaCl$_2$ or 0.5M KCl disappears completely. The labelling of H-MM with the reagent was suppressed by the addition of ATP. However, the chemical structure around the binding site of this reagent has not yet been clarified.

B) Chemical modification and simple hydrolysis

The physiological function of the simple hydrolysis of ATP by myosin described in chapter 3 is not yet clearly understood, but chemical modification has cast considerable light on the nature of the groups present at the active site for this reaction. The chemical modification of the SH group has been particularly studied, and the roles of the histidine and tyrosine residues have also been largely determined.

In 1944 Singer and Barron[57] reported the inactivation of myosin-ATPase by an organic mercury reagent, but it was Kielley and Bradley[58] in 1956 who attracted interest to the chemical modification of the SH group when they discovered that a low concentration of PCMB activated the ATPase, while a high concentration inactivated it. This led eventually to the discovery that two kinds of SH group are present in the active sites. Reagents inducing similar changes in enzyme activity include 2,4-dinitrophenol (DNP)[59], cysteine ethyl-ester[60], disulphides[61-64], 2,3-dimercaptopropanol arsenite[65], carbonylcyanide p-chlorophenylhydrazone[66], diazobenzene p-sulphonic acid[67], and 2,4-dinitrofluorobenzene[68]. Many of these compounds react with SH groups, but it is difficult to elucidate the amino acid sequence in the vicinity of the SH groups actually responsible for the activity, because other SH groups are also modified.

Sekine et al.[69,70] attempted to modify myosin with NEM. By reacting with ^{14}C-NEM, they demonstrated, from the SH peptide map of the digest of NEM-myosin with trypsin, that NEM combines specifically with 1 mole of SH groups (which they called S$_1$) per 2×10^5g of myosin. In this case, the Ca^{2+}-promoted ATPase was activated, and the minimum, near neutrality, of its pH-activity curve disappeared, while the EDTA-ATPase was inhibited. Yamashita et al.[71] showed that the nitrophenol derivative of maleimide (DDPM) also causes an activation similar to that of NEM and the production of a tripeptide containing $DDPS_1$ with an amino acid sequence Ileu·Cys-SH·Arg. More recently, Seidel et al.[72] have spin-labelled the S_1 thiol groups of H-MM with a paramagnetic analogue of iodoacetamide[73] and studied the effects of tryptic digestion on the EPR spectrum and ATPase activity. These studies indicated that the S_1 thiol groups are present in subfragment-1. On reacting S_1-modified myosin with NEM in the presence of ATP, Yamashita et al.[74] found that the remaining SH groups (S_2) are attacked rapidly, and that the enhancement of the Ca^{2+}-ATPase activity obtained on S_1 modification disappears. Using DDPM, the SH peptide containing S_2 was separated and an amino acid sequence, Cys-SH·Asp·Gly, determined. Sekine[75] postulated that the activation of myosin-ATPase described above is due to conformational changes induced by modification at the active site. This implies that the allosteric effects on the ATPase are similar, since not only SH reagents but also EDTA (cf. chapter 3, section 4), DNP and dioxane (cf. chapter 2, section 8) change the pH dependency and increase the K_m of myosin-ATPase in the same way as S_1 modification. The high reactivity of S_2 with NEM in the presence of ATP is ascribed to a local conformational change in the presence of ATP which increases the reactivity of an otherwise unreactive SH group. On the other hand, when myosin is photo-oxidized, heated or treated with PCMB or IAA in the presence of ATP, these groups are protected from inactivation. As pointed out by Stracher and Dreizen[76], these results could suggest a different interpretation from that of Sekine concerning the significance of the role of the thiol groups.

Myosin-ATPase shows a maximum activity at pH 6.0 and a minimum at pH 7.5, as described in chapter 3. Modification with PCMB, NEM or EDTA simplifies the pH-activity curve by removing the maximum and minimum. Sekine explained this change as an allosteric effect, but it can also be easily explained kinetically, by assuming that at steady-state myosin forms both active (E_1S) and inactive (E_2S) complexes with ATP, and that the production of the inactive complex occurs mostly near pH 7.5 but is inhibited by the various reagents[77], as described in chapter 3, section 4. However, it remains to be determined whether this inhibition is caused by specific chemical modification at site 2 or by an allosteric effect of chemical modification at other sites, as advocated by Sekine.

With reference to the work of Sekine and his colleagues, it is important to determine why there is 1 mole of active sites per mole of myosin, even though NEM and other reagents combine with 2 moles of specific SH groups, and whether the change in the pH-activity curve is a necessary sequel to the modification of the SH groups. To clarify these points, Ohe et al.[42] modified myosin with monoiodoacetamide (IAA), and found that the Ca^{2+}-ATPase and Mg^{2+}-ATPase activities increased about ten fold over those of control, unmodified myosin, while the EDTA-ATPase activity decreased to about 20% of its initial value. In contrast, the pH-activity curve was unaltered from that of the control myosin. Myosin carboxamidomethylated (CM) with IAA was then digested with trypsin to obtain a peptide containing CM-cysteine, which is related to the ATPase activity. It was found that the changes described above occur proportionately to the degree of carboxamidomethylation of the specific SH group, and that a maximum of 1 mole of IAA is bound to each mole of myosin (Fig. 4). The structure near the CM-cysteine was Ileu·Cys-SH·Arg, which is identical to the S_1 proposed by Sekine and his colleagues. Hence modification of 1 mole of SH groups per mole of myosin with IAA activated

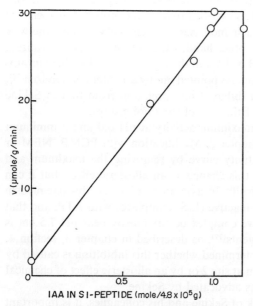

Fig. 4. Relation between the ATPase activity of carboxamidomethyl myosin and the amount of IAA incorporated into the main fraction (S_1-peptide). The ATPase activity was measured in 1M KCl, 20mM $MgCl_2$, 0.5mM ATP and 20mM Tris-maleate at pH 7.0 and 25°C.

the ATPase but did not affect the pH-activity curve. In spite of these results, measurements of the incorporation of IAA into the S_1 peptide upon reaction of myosin with IAA in urea indicated that 2 moles combined to 1 mole of myosin. This can be understood if myosin molecules have an active site on only one of their two heads, or if both heads contain sites but only one site can be active because of mutual steric hindrance. However, the EDTA-ATPase of CM-S-1, a subfragment from CM-myosin, is only 20% as active as the control. This indicates that the first suggestion is correct, since in S-1 there is no interaction between two heads, and in the

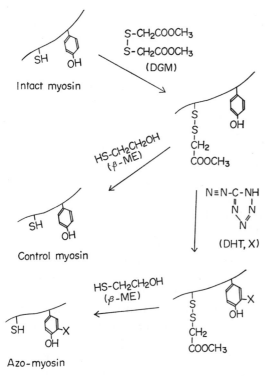

Fig. 5. Procedures for the preparation of azo-myosin. DGM=dithiodiglycolic acid dimethyl ether, βME=β-mercaptoethanol, DHT=diazonium-1H-tetrazole,

$$X = -N = N - C - NH$$
$$\qquad\quad \| \quad |$$
$$\qquad\quad N\ \ N$$
$$\qquad\quad\ \ \backslash\ //$$
$$\qquad\qquad\ \ N$$

second case 50% of the S-1 would be intact, so that the EDTA-ATPase in CM-S-1 would be 60% of that of the control[78].

Chapter 3 showed how the addition of ATP to H-MM enhances the UV absorption similarly to the red shift observed upon changing the medium surrounding the tyrosine residue[26,27]. The interaction of synthetic ATP analogues with actomyosin[79] described in chapter 9 suggests that there is donation of π electrons from the adenine base to a benzene nucleus in the side chain of myosin. Chemical modification by diazonium-1H-tetrazole (DHT) was attempted[80] in order to clarify further the role of the tyrosine residue in myosin. Since the SH group of myosin also reacts with DHT, it was protected by conversion to the -S-S- group with dithiodiglycolic acid dimethyl ester (DGM), and after the treatment with DHT, was regenerated with β-mercaptoethanol or dithiothreitol (Fig. 5). The Mg^{2+}-ATPase and Ca^{2+}-ATPase activities of myosin modified in this way with DHT were reduced proportionately to the formation of monoazotyrosine, and completely disappeared when 1 mole of monoazotyrosine was produced per mole of myosin. There was substantially less inactivation when the modification with DHT was performed in the presence of Mg-ATP. The tyrosine residue which combines with DHT is contained in a heavy chain of H-MM. These results indicate that a specific tyrosine residue in the H-MM of myosin is involved in the ATPase reaction.

Morales and Hotta[81] suggested that there is a histidine residue in the ATPase active site of myosin, since the ITPase activity of myosin increases with the pH, the activity at pH 6.8 is one-half of its maximum value observed at alkaline pH (chapter 3, section 4), and there is zinc ion inhibition[82]. In support of this, Stracher[83] showed that myosin-ATPase is inactivated by photo-oxidation, but this evidence is inconclusive, since there is also simultaneous photo-oxidation of the SH group as well as that of the histidine residue. Therefore Sekiya et al.[84] photo-oxidized myosin with a protected SH group and subsequently regenerated the SH. The ATPase activity disappeared when 25% of the histidine residues present in the original myosin were destroyed, suggesting that histidine occurs near the active centre. However, since there was a concurrent large change in helical content, the role of the histidine residue remains uncertain. More recently, Hegyi and Mühlrad[85] reported that when all the histidyl residues of H-meromyosin were blocked by diethylpyrocarbonate, about 10% of the initial activity in the presence of EDTA was maintained, while the activity in the presence of Mg^{2+} was higher than that of the untreated control. It seems, therefore, highly probable that the histidyl residues are not constituents of the myosin centre responsible for enzymatic activity, but their blocking will affect myosin-ATPase through changes in the conformation.

To conclude this section we shall briefly discuss the relationship between the active site of myosin-ATPase and the subunit structure of the myosin molecule. We have already mentioned that modification of 1 mole of specific SH groups per mole with IAA changed the ATPase activity dramatically. Reaction of myosin with ^{14}C-IAA, followed by determination of the ^{14}C-distribution in the light and heavy chains, showed that the specific SH group was in a heavy chain[86]. On the other hand, Murphy and Morales[41] reported that affinity labelling occurs when myosin is treated under alkaline conditions with an ATP substrate analogue — 6-mercaptopurine ribose TP — which binds to the light chain.* Hence it can be concluded tentatively that the active site is present in a region of the molecule containing both light and heavy chains.

4. The Effect of Chemical Modification on the Ability of Myosin to Bind with Actin

The binding of actin with myosin will be described in detail in chapter 6, but we shall now outline how this reaction is affected by chemical modification of myosin.

Bailey[87] first observed that the SH group of myosin is involved in binding with actin. Recently, comparison has been made between the changes in binding ability and ATPase activity upon chemical modification of myosin, particularly with SH reagents. Stracher[88] has investigated the reactivity of the various thiol groups by treating myosin with disulphides which can exchange with -SH. He found that (1) only 1 mole of SH groups per 2.0×10^5g of myosin did not react with bis-β-carboxyletherdisulphide; (2) in the presence of ATP, 2 moles of SH remained inert and (3) in the presence of actin, 3 moles of SH groups were inert. Only the modified myosin obtained from (3) did bind with actin. The myosin from (1) and (2) did not bind with actin, while that from (1) and (3) had lost its ATPase activity. Since the modified myosin could bind with actin even though its ATPase activity had been completely eliminated, the active sites for the two functions must be different.

* Unfortunately, they isolated the light chains by raising the pH of the solution of myosin labelled with the analogue to 11, and it is highly probable that the analogue is transferred from sulphhydryl groups of the heavy chain to those of the light chains by the exchange reaction.

5. The Effect of Chemical Modification on the Formation of Myosin Filaments

Myosin becomes soluble at low ionic strength after most of its lysine residues have been modified by succinylation[89] or poly-DL-alanylation[90]. This change in solubility is attributed to an alteration in the number of ionised groups in the molecule. In order to locate the specific sites which are important in the formation of myosin filaments, Shimada[80] reacted L-MM fr. 1 — a subfragment which forms filaments similar to myosin — with DHT, after having first protected the thiol groups by the method described above. The solubility of L-MM fr. 1 at low ionic strengths increased proportionately to the degree of formation of monoazotyrosine, and it became completely soluble when one mole of monoazotyrosine per mole of L-MM fr. 1 was produced. The ability to form filaments was simultaneously lost. However, lysine residues are also markedly affected by DHT, and so it is not certain that any specific tyrosine residue in L-MM fr. 1 is essential to the site of filament formation in myosin.

6. The Chemical Structure of the Active Site

This chapter has described the effect of chemical modification on the biological functions of myosin and particularly on ATPase activity. Two kinds of SH group are related to the ATPase function, and the amino acid sequences near these SH groups responsible for activation and suppression of the ATPase activity are Ileu·Cys-SH·Arg and Cys-SH·Asp·Gly respectively. The promotion of decomposition of phosphoryl myosin, induced by the modification of myosin with TBS, indicates that the amino acid sequence near the lysine residue which reacts specifically with TBS is Asp(NH_2)·Pro·Pro·Lys. The effect of modification by TBS on the simple hydrolysis of ATP in the presence of Ca^{2+} may be due to stabilisation of the Michaelis complex by molecular association between the adenine and the trinitrophenyl groups (see section 3).

Tonomura and his co-workers[79,91] have studied the interaction of ATP analogues with actomyosin (chapter 9), and have concluded that the binding of myosin with nucleoside phosphates requires that there be an N or O (not S) in the 6 position of the substrate base, that a certain distance be maintained between the triphosphate and the base by a ribose moiety, that O be present at the 3' position of the sugar and that the phosphate moiety be a triphosphate. The binding of the terminal O atom of the triphosphate chain with myosin is also suggested from the results that the dissociation constant of binding of adenylylmethylene DP to H-meromyosin is much higher than that of adenylylimido DP and the Michaelis constant of ATPase[92]. Another problem to be resolved before a model of the active site can be constructed is the mode of binding of divalent cations with ATP. Szent-Györgyi[93] previously assumed that in a solution containing only Mg^{2+} and ATP, Mg^{2+} interacts simultaneously with ring N atoms and triphosphate O atoms, while Hotta et al.[94] have obtained UV difference spectra which support this suggestion. However, later studies have invalidated this conclusion[95-97]. In particular, Happe and Morales[98] have obtained the nuclear magnetic resonance spectrum of ^{15}N-substituted ATP, and concluded that Mg^{2+} does not interact with the N atoms, while Zn^{2+} does. Rimai et al.[99-101] investigated the interaction of ATP with divalent cations by Raman spectroscopy, and found that Ca^{2+} and Mg^{2+} bind only to the phosphate group, while Mn^{2+} and Zn^{2+} bind to both the phosphate and the adenine base. Cohn and Hughes[102] investigated the structure of the complexes of ADP and ATP with divalent metal ions by

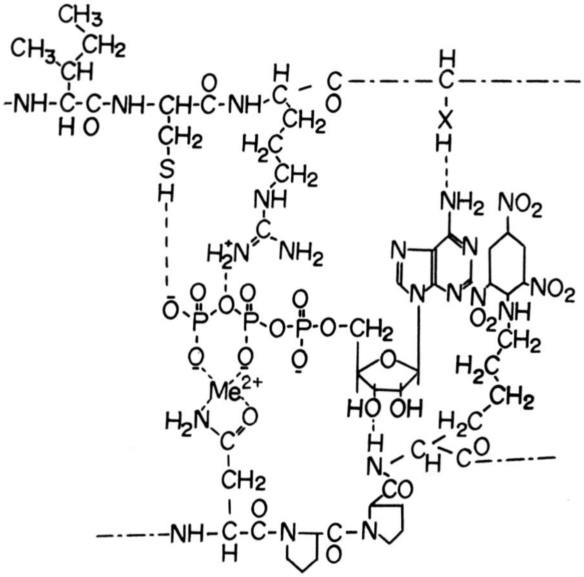

Fig. 6. A schematic model of the binding of ATP with trinitrophenyl myosin. For explanation see text.

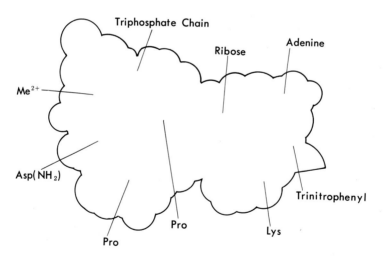

Fig. 7. A molecular model of the binding of ATP with Asp(NH$_2$)·Pro·Pro·TNP-Lys chain.

studying the nuclear magnetic resonance spectra of the hydrogen and phosphorus nuclei of the nucleotides. The chemical shifts of the phosphorus nuclei in the presence of Mg^{2+} and Ca^{2+} have indicated that these metals form complexes with the β- and γ-phosphate groups of ATP and with the α- and β-groups of ADP.

On the assumption that the sequence containing the SH group is Ileu·Cys-SH·Arg, and that the sequence Asp(NH$_2$)·Pro·Pro·Lys also occurs near the active site, a structural model of ATP bound to trinitrophenyl myosin was formulated, based on the Pauling-Corey rules for the structure of a peptide chain[91] (Figs. 6 and 7). However, many problems remain to be solved about the nature of the active centre of myosin: (1) the physiological function of simple hydrolysis is not clear, and a tyrosine residue, as well as an SH group and a lysine residue, appears to be involved; (2) the ATPase activity changes on chemical modification, but it is difficult to decide whether this is induced by alteration of the active site itself or whether it is an allosteric effect; (3) little is known of the structure of the physiologically more important phosphoryl myosin, except that the carboxyl group of a specific glutamic acid residue is phosphorylated; the amino acid sequence about the glutamic acid is not known; (4) the active sites for the binding of actin to myosin and for the formation of myosin filaments have not yet been determined. Thus the model described above is only a preliminary step in the elucidation of the structure of all the active centres of myosin. We hope that future advances will allow the mechanism of action of myosin-ATPase to be understood in terms of the organic chemistry involved.

REFERENCES

1 D. R. Kominz, A. Hough, P. Symonds and K. Laki, *Arch. Biochem. Biophys.*, **50**, 148 (1954).
2 S. Lowey and C. Cohen, *J. Mol. Biol.*, **4**, 293 (1962).
3 J. M. Jones and S. V. Perry, *Biochem. J.*, **100**, 120 (1966).
4 S. Lowey, H. S. Sleyter, A. G. Weeds and H. Barker, *J. Mol. Biol.*, **42**, 1 (1969).
5 A. M. Asatoor and M. D. Armstrong, *Biochem. Biophys. Res. Commun.*, **26**, 168 (1967).
6 P. Johnson, C. I. Harris and S. V. Perry, *Biochem. J.*, **105**, 361 (1967).
7 I. P. Trayer, C. I. Harris and S. V. Perry, *Nature*, **217**, 452 (1968).

8 W. M. Kuehl and R. S. Adelstein, *Biochem. Biophys. Res. Commun.*, **35**, 956 (1968).
9 G. Huszar and M. Elzinga, *Nature*, **223**, 834 (1969).
10 W. M. Kuehl and R. S. Adelstein, *Biochem. Biophys. Res. Commun.*, **37**, 59 (1969).
11 M. F. Hardy, C. I. Harris, S. V. Perry and D. Stone, *Biochem. J.*, **120**, 653 (1970).
11a G. Haszar, *J. Biol. Chem.*, **247**, 4057 (1972).
11b M. F. Hardy and S. V. Perry, *Nature*, **223**, 300 (1969).
11c M. F. Hardy, C. I. Harris, S. V. Perry and D. Stone, *Biochem. J.*, **117**, 44p (1970).
11d K. Laki and E. Wilson, *Physiol. Chem. Phys.*, **1**, 536 (1969).
11e B. Krzysik, J. P. Vergnes and I. R. McManes, *Arch. Biochem. Biophys.*, **146**, 34 (1971).
12 K. Bailey, *Biochem. J.*, **49**, 23 (1951).
13 E. Gaetzens, H. S. Cheung and M. Bárány, *Biochim. Biophys. Acta*, **93**, 188 (1964).
14 W. W. Kielley, M. Kimura and J. P. Cooke, *Abstr. 6th Int. Congr.Biochem.*, 634 (1964).
15 G. W. Offer, *Biochim. Biophys. Acta*, **90**, 193 (1964).
16 R. H. Locker, *Biochim. Biophys. Acta*, **14**, 533 (1954).
17 P. P. Trotta, P. Dreizen and A. Stracher, *Proc. Natl. Acad. Sci. U. S.*, **61**, 659 (1968).
18 M. Kimura and W. W. Kielley, *Biochem. Z.*, **345**, 188 (1966).
19 A. G. Weeds and B. S. Hartley, *Biochem. J.*, **107**, 531 (1968).
20 M. Elzinga, *Biochemistry*, **9**, 1365 (1970).
21 M. Elzinga, *Biochemistry*, **10**, 224 (1971).
22 R. S. Adelstein and W. M. Kuehl, *Biochemistry*, **9**, 1355 (1970).
23 G. Huszar and M. Elzinga, *Biochemistry*, **10**, 229 (1971).
24 H. S. Slayter and S. Lowey, *Proc. Natl. Acad. Sci. U. S.*, **58**, 1611 (1967).
24a G. Huszar and M. Elzinga, *J. Biol. Chem.*, **247**, 745 (1972).
25 Y. Tonomura and F. Morita, *J. Biochem.*, **46**, 1367 (1959).
25a J. P. Hummel and W. J. Dreyer, *Biochim. Biophys. Acta*, **63**, 530 (1962).
26 F. Morita, *J. Biol. Chem.*, **242**, 4501 (1967).
27 F. Morita, *J. Biochem.*, **69**, 517 (1971).
27a M. Yazawa, F. Morita and K. Yagi, presented at the 45th meeting of the Biochemical Society of Japan, Tokyo (1972).
28 H. Iino, F. Morita and K. Yagi, presented at the 43rd meeting of the Biochemical Society of Japan, Tokyo (1970).
28a H. Iino and F. Morita, presented at the 11th meeting of the Biophysical Society of Japan, Kyoto (1972).

28b F. Morita, presented at a seminar at the Institute for Protein Research, Osaka University, Osaka (1971).
29 L. B. Nanninga and W. F. H. M. Mommaerts, *Proc. Natl. Acad. Sci. U. S.*, **46**, 1155 (1960).
30 K. Sekiya and Y. Tonomura, *J. Biochem.*, **61**, 787 (1967).
31 K. Imamura, M. Tada and Y. Tonomura, *J. Biochem.*, **59**, 280 (1966).
32 J. Gergely, A. Martonosi and M. A. Gouvea, *in* "Sulfur in Proteins," ed. by R. Benesch, R. E. Benesch, P. D. Boyer, I. M. Klotz, W. R. Middlebrock, A. G. Szent-Györgyi and D. R. Schwarz, Academic Press, New York, p. 297 (1959).
33 A. Martonosi and H. Meyer, *J. Biol. Chem.*, **239**, 640 (1964).
34 K. M. Nauss, S. Kitagawa and J. Gergely, *J. Biol. Chem.*, **244**, 755 (1969).
35 B. Kiely and A. Martonosi, *J. Biol. Chem.*, **243**, 2273 (1968).
36 D. M. Young, *J. Biol. Chem.*, **242**, 2790 (1967).
37 S. Lowey and S. M. Luck, *Biochemistry*, **8**, 3195 (1969).
38 W. J. Bowen and T. C. Evans, *Eur. J. Biochem.*, **5**, 507 (1968).
39 E. Eisenberg and C. Moos, *Biochemistry*, **9**, 4106 (1970).
40 J. C. Seidel, M. Chopek and J. Gergely, *Biochemistry*, **9**, 3265 (1970).
41 A. J. Murphy and M. F. Morales, *Biochemistry*, **9**, 1528 (1970).
42 M. Ohe, B. K. Seon, K. Titani and Y. Tonomura, *J. Biochem.*, **67**, 513 (1970).
42a C. H. W. Hirs, *in* "Methods in Enzymology," Vol. 11, sections VIII-X (1967).
42b G. R. Stark, *Advan. Protein Chem.*, **24**, 261 (1970).
42c B. L. Vallee and J. F. Riordan, *Annu. Rev. Biochem.*, **38**, 733 (1969).
42d A. N. Glazer, *Annu. Rev. Biochem.*, **39**, 101 (1970).
43 S. J. Singer, *Advan. Protein Chem.*, **22**, 1 (1967).
44 K. Imamura, T. Kanazawa, M. Tada and Y. Tonomura, *J. Biochem.*, **57**, 627 (1965).
45 N. Kinoshita, S. Kubo, H. Onishi and Y. Tonomura, *J. Biochem.*, **65**, 285 (1969).
46 Y. Tonomura and T. Kanazawa, *J. Biol. Chem.*, **240**, PC4110 (1965).
47 S. Kubo, N. Kinoshita and Y. Tonomura, *J. Biochem.*, **60**, 476 (1966).
48 Y. Tonomura, H. Nakamura, N. Kinoshita, H. Onishi and M. Shigekawa, *J. Biochem.*, **66**, 599 (1969).
49 S. Kubo, S. Tokura and Y. Tonomura, *J. Biol. Chem.*, **235**, 2835 (1960).
50 Y. Tonomura, J. Yoshimura and T. Ohnishi, *Biochim. Biophys. Acta*, **70**, 698 (1963).
51 F. Fábián and A. Mühlrad, *Biochim. Biophys. Acta*, **162**, 596 (1968).
52 A. Mühlrad and F. Fábián, *Biochim. Biophys. Acta*, **216**, 422 (1970).
53 H. Tokuyama and Y. Tonomura, *J. Biochem.*, **62**, 456 (1967).
54 H. Tokuyama, S. Kubo and Y. Tonomura, *Biochem. Z.*, **345**, 57 (1966).

55 S. Kubo, H. Tokuyama and Y. Tonomura, *Biochim. Biophys. Acta*, **100**, 459 (1965).
56 H. Takashina, *Biochim. Biophys. Acta*, **200**, 319 (1970).
57 T. P. Singer and E. S. G. Barron, *Proc. Soc. Exp. Biol. Med.*, **56**, 120 (1944).
58 W. W. Kielley and L. B. Bradley, *J. Biol. Chem.*, **218**, 653 (1956).
59 J. B. Chappell and S. V. Perry, *Biochim. Biophys. Acta*, **16**, 285 (1955).
60 J. J. Blum, *Arch. Biochem. Biophys.*, **97**, 309 (1962).
61 M. Bárány, *in* "Sulfur in Proteins," ed. by R. Benesch, R. E. Benesch, P. D. Boyer, I. M. Klotz, W. R. Middlebrock, A. G. Szent-Györgyi and D. R. Schwarz, Academic Press, New York, p. 317 (1959).
62 E. Gaetjens, T. Therattil-Antony and M. Bárány, *Biochim. Biophys. Acta*, **86**, 554 (1964).
63 D. Hartshorne and M. Morales, *Biochemistry*, **4**, 18 (1965).
64 M. Onodera and K. Yagi, *J. Biochem.*, **66**, 751 (1969).
65 A. L. Fluharty and D. R. Sanadi, *Arch. Biochem. Biophys.*, **97**, 164 (1962).
66 J. J. Blum and D. R. Sanadi, *J. Biol. Chem.*, **239**, 452 (1964).
67 T. Yamashita, I. Kabasawa and T. Sekine, *J. Biochem.*, **63**, 608 (1968).
68 G. Bailin and M. Bárány, *Biochim. Biophys. Acta*, **168**, 282 (1968).
69 T. Sekine, L. M. Bernett and W. W. Kielley, *J. Biol. Chem.*, **237**, 2769 (1962).
70 T. Sekine and W. W. Kielley, *Biochim. Biophys. Acta*, **81**, 336 (1964).
71 T. Yamashita, Y. Soma, S. Kobayashi, T. Sekine, K. Titani and K. Narita, *J. Biochem.*, **55**, 576 (1964).
72 J. S. Seidel, M. Chopek and J. Gergely, *Arch. Biochem. Biophys.*, **142**, 223 (1971).
73 J. S. Seidel, M. Chopek and J. Gergely, *Biochemistry*, **9**, 3265 (1970).
74 T. Yamashita, Y. Soma, S. Kobayashi and T. Sekine, *J. Biochem.*, **57**, 460 (1964).
75 T. Sekine, *in* "Molecular Biology of Muscular Contraction," ed. by S. Ebashi, F. Oosawa, T. Sekine and Y. Tonomura, Igaku Shoin, Tokyo, p. 33 (1965).
76 A. Stracher and P. Dreizen, *Curr. Topics Bioenergetics*, **1**, 153 (1966).
77 Y. Tonomura, *J. Res. Inst. Catalysis, Hokkaido Univ.*, **16**, 323 (1968).
78 Y. Hayashi and Y. Tonomura, *J. Biochem.*, **68**, 665 (1970).
79 Y. Tonomura, K. Imamura, M. Ikehara, H. Uno and F. Harada, *J. Biochem.*, **61**, 460 (1967).
80 T. Shimada, *J. Biochem.*, **67**, 185 (1970).
81 M. F. Morales and K. Hotta, *J. Biol. Chem.*, **235**, 1979 (1960).
82 K. Hotta and S. Kojima, *J. Biochem.*, **55**, 486 (1964).
83 A. Stracher, *J. Biol. Chem.*, **240**, PC958 (1965).
84 K. Sekiya, S. Mii and Y. Tonomura, *J. Biochem.*, **57**, 192 (1965).
85 G. Hegyi and A. Mühlrad, *Acta Biochim. Biophys., Acad. Sci. Hung.*, **3**, 425 (1968).

86 Y. Hayashi, *J. Biochem.*, **72**, 83 (1972).
87 K. Bailey, *in* "The Proteins," ed. by H. Neurath and K. Bailey, Academic Press, New York, Vol. 2, pt. B, p. 951 (1954).
88 A. Stracher, *J. Biol. Chem.*, **239**, 1118 (1964).
89 H. Oppenheimer, K. Bárány, G. Hamoir and J. Fenton, *Arch. Biochem. Biophys.*, **120**, 108 (1967).
90 I. S. Edelman, E. Hoffer, S. Bauminger and M. Sela, *Arch. Biochem. Biophys.*, **123**, 211 (1968).
91 Y. Tonomura, S. Kubo and K. Imamura, *in* "Molecular Biology of Muscular Contraction," ed. by S. Ebashi, F. Oosawa, T. Sekine and Y. Tonomura, Igaku Shoin, Tokyo, p. 11 (1965).
92 R. G. Yount, D. Ojala and D. Babock, *Biochemistry*, **10**, 2490 (1971).
93 A. Szent-Györgyi, *in* "Enzymes, Units of Biological Structure and Function," ed. by O. H. Gaebler, Academic Press, New York, p. 393 (1956).
94 K. Hotta, J. Brahms and M. Morales, *J. Amer. Chem. Soc.*, **83**, 997 (1961).
95 P. W. Schneider, H. Brintzinger and H. Erlenmyer, *Helv. Chim. Acta*, **47**, 992 (1964).
96 M. Cohn and T. R. Hughes, *J. Biol. Chem.*, **237**, 176 (1962).
97 G. G. Hammes, G. E. Maciel and J. S. Waugh, *J. Amer. Chem. Soc.*, **83**, 2394 (1961).
98 J. A. Happe and M. Morales, *J. Amer. Chem. Soc.*, **88**, 2077 (1966).
99 L. Rimai, M. E. Heyde and E. B. Carew, *Biochem. Biophys. Res. Commun.*, **38**, 231 (1970).
100 L. Rimai and M. E. Heyde, *Biochem. Biophys. Res. Commun.*, **41**, 313 (1970).
101 M. E. Heyde and L. Rimai, *Biochemistry*, **10**, 1121 (1971).
102 M. Cohn and T. R. Hughes, *J. Biol. Chem.*, **237**, 176 (1962).

5
THE STRUCTURE AND FUNCTION OF ACTIN*

Chapters 2 to 4 have described the structure and functions of myosin. This chapter describes the properties of actin which, together with myosin, constitutes the major part of the structural protein of muscle. Actin was first isolated from muscle stroma by Straub[1,2]. It accounts for 20–25% of myofibril protein and is the main component of the thin filaments[3]. Straub's method of isolation from stroma produces globular G-actin, but on the addition of salts this aggregates to fibrous F-actin[1]. One mole of G-actin is bound to 1 mole of ATP. On transformation to F-actin the ATP is dephosphorylated, so that F-actin contains bound ADP[4]. The basic properties of actin were reported by Straub and Feuer in 1950[4], and the reversible G⇌F transformation accompanied by the dephosphorylation of ATP was intensively studied, since it was thought that it might be a key reaction in muscle contraction. However present evidence suggests that it

* Contributor: Kazuko Shibata-Sekiya

does not participate directly in contraction (chapter 7). The G⇌F transformation upon slight alteration of the medium has been known for a long time, and was one of the first reactions basic to the organized fine structures of living things to be understood at a molecular level. It is, for instance, relevant to the formation of fibrous polymers by self-assembly of globular monomers, such as in flagellae and spindles. It is also a possible model for the general interaction of proteins.

1. G-Actin

A typical method for the extraction and purification of G-actin is that of Mommaerts[5]. G-Actin contained in the aqueous extract of muscle powder treated with acetone is converted to F-actin by adding salt. It may then be precipitated by ultracentrifugation and dissolved in ATP solution in order to reconvert it to G-actin. Further purification can be obtained by repeating the process. However Drabikowski and Gergely[6] and Laki et al.[7] later showed that actin prepared in this way by extraction at room temperature is contaminated with 10–25% of tropomyosin and that most of the tropomyosin can be removed by extracting at 0°C instead. The actin used in our work was usually prepared in the latter manner, but it was sometimes purified further by partial polymerisation and fractionation with ammonium sulphate[8]. Recently, Rees and Young[9] showed that even these specimens are contaminated with some inert, non-polymerisable dimer which is not bound to a nucleotide, but this can be partly removed during the extraction if ATP, Ca^{2+} and reducing agents are added, and can be completely removed by gel filtration. G-Actin thus prepared was shown by a variety of methods to be of high purity.

Reported estimates of the molecular weight of G-actin vary from 50,000 to 110,000[10-12], but as a result of Mommaerts' measurements[5], many workers adopted the use of the value of 60,000. However, Johnson et al.[13] recently determined, from the 3-methylhistidine content, that the minimum molecular weight of rabbit actin is 47,600. Also, Adelstein et al.[14] recently found a molecular weight of 47,000 for G-actin, while Rees and Young[9] determined a value of 46,000 for their purified actin, both measurements being made by the sedimentation equilibrium method. The molecular weight of G-actin purified by reversible polymerisation and gel filtration was estimated by Tsuboi[15] to be 45,000 by analytical gel filtration, and it contained an average of 1 mole of bound nucleotide per 45,000g protein. Sakakibara and Yagi[16] obtained a value of 43,000 by the light-scattering method. The molecular weight now universally accepted for G-actin is 46,000–47,000, but since there is some question of the purity of actin specimens produced by conventional methods, in this chapter we have used the molecular weights found for the samples employed in our studies, which ranged from 57,000 to 61,000.

TABLE I. Amino Acid Composition of Actin

Amino acid	No. of amino acid residues for molecular weight of 4.7×10^4		
	(18)	(19)	(20)
Lys	24.3	23.6	22.0
His	8.4	8.76	8.62
3-Methyl His	—	0.95	1.10
Arg	20.8	21.4	20.4
Cys/2	4.3	4.75–5.23	—
Asp	38.7	39.5	38.4
Thr	31.2	29.5	28.2
Ser	26.6	24.6	25.9
Glu	45.5	46.1	44.7
Pro	20.8	20.7	20.4
Gly	30.8	31.9	29.8
Ala	33.4	33.6	32.9
Val	20.6	20.2	21.9
Met	18.0	16.4	17.2
Ileu	30.3	28.7	31.4
Leu	28.5	30.5	29.0
Tyr	17.7	17.2	17.2
Phe	13.0	12.9	13.3
Try	4.2	5.0	—

The molecular weight of G-actin is unaltered in guanidine-urea solution, which indicates that it contains only one polypeptide chain[9,17]. Its amino acid composition[18–20] (Table I) shows that it contains an unusually large proportion of acidic and basic residues compared to a normal protein. This is probably related to the important role of electrostatic forces in many reactions of actin. 4.7–5.2 moles of SH groups per 4.7×10^4g of G-actin are titratable with PCMB[21], and 4.8–5.0 moles per 4.4×10^4g are titratable with 2,3-dicarboxyl-4-iodoacetamide azobenzene[22]. These values approximate to the amount of half cystine in the molecule. Like myosin, actin contains no S-S bridge. Johnson et al.[13] and Asatoor and Armstrong[20] have shown that 3-methylhistidine is a normal component of actin. Locker[23] identified the C-terminal group as Phe·Ileu·His by the carboxypeptidase method and by hydrazinolysis. The N-terminal group has been assumed to be acetylated, since there is no reaction with dinitrofluorobenzene, and Alving and Laki[24] and Gaetjens and Bárány[25] both reported it to be N-acetyl-Asp. Elzinga[19] and Adelstein and Kuehl[26] are at present determining the amino acid sequence of the peptide obtained by cleavage of actin with cyanogen bromide. So far Elzinga[27] has reported that all of the 3-methylhistidine (1 mole/mole of actin) is found in only 1 of the 17 cyanogen bromide peptides of actin, and also Collins et al.[28] have shown that the N-terminal peptide of actin has the sequence: Ac-Asp·Glu·Thr·Glu·Asp·Thr·Ala·Leu·Val·Cys·Asp·Asp·Gly·Ser·Gly·Leu·Val·Lys·Ala·Gly·Phe·Gly·Ala·Asp·Asp·Ala·Pro·Arg·Ala·Val·Phe·Pro·Ser·Ileu·Val·Gly·Arg·Pro·Arg·His·Gln·Gly·Val·Met. Now that this much is known, elucidation of the complete primary structure should follow soon.

With respect to its secondary structure, optical rotatory dispersion measurements of a sample of G-actin containing no tropomyosin indicate a helical content of 29%[24,29]. Murphy[29a] has concluded from the circular dichroic spectra that the contents of α-helix, β-structure and random coil are 26, 26 and 48%, respectively, while Nagy and Strzelecka-Golaszewska[29b] have concluded that α-helix is the main optically active conformation and amounts to 30%, but the contribution of β-structure may not exceed 10%.

2. The Binding of G-Actin with ATP

One mole of G-actin binds with about 1 mole of ATP[4], but the ATP rapidly exchanges with ATP in the solvent[30]. When G-actin is polymerised with salt, the ADP produced by dephosphorylation of ATP is not liberated, but forms F-actin-ADP. This ADP is difficult to exchange with a nucleotide in the solvent. Martonosi et al.[31,32] measured the exchange of ^{14}C-ATP bound to G-actin with a nucleotide in the surrounding medium, and found a reactivity order ATP>ITP>UTP≈ADP>GTP>CTP, which agrees with the order of the association constants for complexes of G-actin with nucleotides, which were measured afterwards by Iyengar and Weber[33]. There is almost no exchange of ATP in G-actin with deoxy-ATP, deoxy-GTP, deoxy-CTP, ribose, or adenosine[32]. These results suggest that not only the base and the phosphate chain, but also the ribose-OH group are important in the binding of actin with a nucleotide. By comparing the UV absorption spectra of G-actin-ATP, G-actin-ADP and

G-actin-ITP, West[34] suggested a conformational change in actin close to its tyrosine and tryptophane residues caused by its binding with nucleotides.

The mode of binding of G-actin to ATP was largely elucidated using actin modified with EDTA and SH reagents. Section 3 describes the chemical modification of actin.

Removal of the nucleotides from actin does not affect its function. Mommaerts[35] dialysed F-actin for a long time in 0.1M KCl and reduced the amount of nucleotide by over 80% without causing depolymerisation. Kasai et al.[36] treated G-actin with Dowex in 50% sucrose and 1.5M urea containing EDTA in order to remove ATP. The nucleotide-free G-actin almost completely polymerised to F-actin upon the addition of cations. Hayashi and Rosenbluth[37,38] depolymerised F-actin-ADP in water containing no salt in order to obtain G-actin-ADP, which undergoes a reversible $G \rightleftharpoons F$ transformation when the temperature is raised from 0° to 29°C in the presence of Mg^{2+}. When this G-actin-ADP was incubated with a myosin filament, tension could be generated upon addition of ATP[38]. Bárány et al.[39] used ultrasonic vibrations to convert F-actin-ADP in the presence of AMP to F-actin-AMP, and then removed 90% more of the bound nucleotide with charcoal. This actin specimen depolymerised reversibly, similarly to F-actin-ADP. It also complexed with myosin, promoted the Mg^{2+}-activated CTPase and deoxy-ATPase of myosin at low ionic strength, and superprecipitated with CTP and deoxy-ATP[39]. The physiological significance of the nucleotide contained in actin will be discussed again in chapter 7, mainly on the basis of results obtained using ADP-free F-actin prepared by the methods of Mommaerts[35] and of Kasai et al.[36] However, it seems that the nucleotide contained in actin, and its dephosphorylation, is not directly related to the $G \rightleftharpoons F$ transformation or to the reaction of the intermediates in myosin-ATPase with actin.

3. Chemical Modification of Actin

Since Kuschinsky and Turba[40] first observed that SH reagents inhibit the polymerisation of actin, there has been considerable work on the chemical modification of actin and its effect on function. The binding of G-actin to ATP has been studied by Gergely et al.[41,42], Martonosi et al.[31,43], Bárány et al.[44-46], Katz and Mommaerts[47], Strohman and Samorodin[48], Grubhofer and Weber[49] and by us[21,50], and almost identical results were obtained by each research group. The addition of EDTA to G-actin suppresses its polymerisation, liberates bound ATP and Ca^{2+}, and alters its

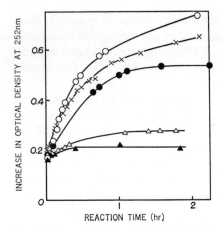

Fig. 1. Rate of binding of PCMB to G-actin. The amount of PCMB bound was determined from the increase in absorption at 252nm. Room temperature, pH 7.9, amounts of PCMB (moles/6.1×10^4 g actin): ○, 15.0; ×, 10.0; ●, 5.0; △, 2.5; ▲, 2.0.

conformation[41], but addition of EDTA to F-actin induces very little depolymerisation[50].

SH reagents have also been used to study the complexing of actin with ATP[21]. The five SH groups in the G-actin molecule comprise two or more types with different properties. We have measured the rate of increase of absorption at 252nm after the addition of various concentrations of PCMB[51] to actin (Fig. 1) and found that only 2 moles of SH per 6.1×10^4 g of actin bound rapidly with PCMB, while subsequently there was a slower reaction.* With F-actin also, only 2 moles of SH groups bound per 6.1×10^4g, while the remainder reacted even more slowly than with G-actin. NEM likewise reacted rapidly with only 2 moles of SH groups per 6.1×10^4 g of actin, but the remaining SH groups were inert. Figure 2 shows clearly that the polymerisation of G-actin is not suppressed until it is bound by 2 moles of PCMB, and binding more PCMB suppresses polymerisation. The ATP complexed to G-actin is removed when PCMB binds the less reactive SH groups, and the amount of bound Ca^{2+} decreases from 1.5 moles to 0.5 moles per $6.1 \times 10^4 g$[21].

Thus, Ca^{2+} is removed from actin when PCMB binds the SH groups of G-actin. Ca^{2+} and ATP are also removed from G-actin by EDTA, as men-

---

* Recently, Justly and Fasold[22], and also Martonosi[52] have presented conclusive evidence that the content of SH groups fast reactive with SH reagents is 1 mole/4.4×10^4g. Our conclusion that the content is 2 moles/6.1×10^4g may be due to contamination of our actin preparation by other proteins.

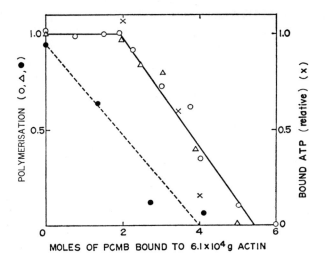

Fig. 2. Inhibition of polymerisation and removal of bound ATP by treatment with PCMB. PCMB and actin were incubated at room temperature and pH 8.3. The ability to polymerise was determined from the increase in viscosity after addition of 0.1M KCl to actin at pH 7.0. ● indicates preliminary treatment with 63 moles of NEM/6.1×10^4 g actin. △ and ○ refer to different actin specimens. ×, amount of ATP bound to actin.

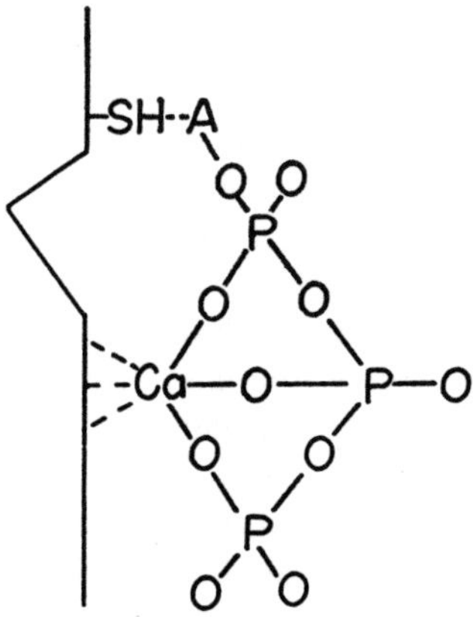

Fig. 3. Tentative model for the binding of ATP to G-actin. (After Strohman and Samorodin, redrawn from Ref. 48).

tioned above. The action of EDTA and PCMB on G-actin can be explained reasonably by Strohman and Samorodin's[48] model of the binding of G-actin with ATP (Fig. 3). Adenosine binds to an SH group, while the triphosphate group is thought to chelate a Ca^{2+} ion strongly bound to the G-actin. Hence ATP will be removed whether PCMB binds with the SH or EDTA removes the Ca^{2+}. Ca^{2+}, even in high concentration, has little effect on the rate of binding of PCMB to G-actin, but in contrast, addition of ATP suppresses the binding of about 1 mole of PCMB per mole of G-actin[21]. West[34a] has recently shown that the binding constant of cation-free actin for ATP and ADP is approximately one thousandth that of the constant for ATP in the presence of bound cation, while binding of ATP is appreciably higher than of ADP when cation is present, as mentioned above. All these phenomena can be explained by the model (Fig. 3).

However, this simple model cannot account for several other experimental results. First, the conformation of G-actin changes markedly upon reaction with EDTA and PCMB[21,41] and secondly, conformational changes after binding 1–2 moles of PCMB per mole of G-actin do not affect the $G \rightleftharpoons F$ transformation, while they facilitate the removal of Ca^{2+} from G-actin by treatment with resin[21]. Hence the liberation of Ca^{2+} by PCMB may not be due directly to the removal of the triphosphate chain of ATP, as Strohman and Samorodin's model would suggest, but may be better represented by a change in the secondary structure of G-actin, accompanied by the removal of ATP upon the binding of PCMB. Thus the nucleotide in actin is thought to help maintain the secondary and tertiary structures of the molecule.

Chemical modification of other functional groups such as histidine, tyrosine, tryptophane and lysine[53–57] is summarised in Table II. Martonosi and Gouvea[53] photo-oxidized G-actin in the presence of methylene blue, and found that oxidation of 2 moles of imidazole per mole suppressed polymerisation, while further oxidation gradually inhibited the ability to react with myosin. They also found that diazotization of about 1 mole of tyrosine per mole of actin suppressed both polymerisation and activation of myosin-ATPase in the presence of Mg^{2+}. Recently Mühlrad et al.[55] modified about 5 moles of tyrosine, about 2 moles of tryptophane and a small number of SH groups with tetranitromethane, and this sharply decreased the ability of actin to polymerise and also decreased the sensitivity of actomyosin to ATP.* However, activation of the myosin-ATPase in the

* The ATP-sensitivity is defined as $\dfrac{\log \eta_{rel} - \log \eta_{rel \cdot ATP}}{\log \eta_{rel \cdot ATP}} \times 100$ where η_{rel} and $\eta_{rel \cdot ATP}$ are the relative viscosities before and after the addition of ATP, respectively[58].

TABLE II. Chemical Modification of Actin

Modification	Modified group		Effects on biological activity				Authors
	Identity	mole/5–6×10^4 g	Polymerisation	Binding with myosin	Activation of myosin-ATPase Mg^{2+}, low I	Superprecipitation	
Photo-oxidation (M.B. O$_2$ hν)	His	2	Decrease	—	Constant	—	Martonosi and Gouvea[53]
	His	2	Decrease	Decrease	Decrease	Decrease	Mühlrad et al.[55]
Carbethoxylation	His	3	Decrease	Decrease	Decrease	—	Mühlrad et al.[54]
Diazotization (Diazosulphanilic acid)	Tyr	1	Decrease	—	Decrease	—	Martonosi and Gouvea[53]
Nitration (Tetranitromethane)	Tyr Try (SH)	5 2	Decrease	Decrease	Slight decrease	Constant	Mühlrad et al.[55]
Trinitrophenylation (Trinitrobenzene sulfonic acid)	Lys	1	Slight decrease	Decrease			Tonomura et al.[59]
	Lys	5	Constant	Constant	Constant	Slight decrease	Mühlrad[56]
Succinylation (Succinyl anhydride)	Lys SH	7 1	Decrease	Decrease	Constant	Slight decrease	Mühlrad et al.[55]

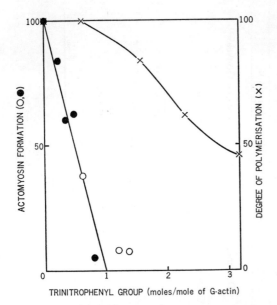

Fig. 4. The effect of trinitrophenylation of G-actin upon the G⇌F transformation and binding with myosin. Trinitrophenylation was carried out by reacting 30 moles of TBS with 1 mole of G-actin at pH 8.3–8.4 and 2–3°C for a suitable time. ×, the extent of polymerisation in 0.1M KCl, 5mM $MgCl_2$, pH 7.0 at room temperature. ○, ●, binding of 0.04mg/ml myosin with 0.01mg/ml F-actin measured by the increase in light-scattering intensity in 0.6M KCl and pH 7.0 at 19°C. ○ and ● indicate results obtained, using different preparations of myosin and actin.

presence of Mg^{2+} decreased only slightly, and the extent of superprecipitation of actomyosin prepared from this actin and myosin was completely unchanged. Thus Martonosi's results on the role of tyrosine in the activity of actin do not agree with those of Mühlrad et al.

The lysine residue in F-actin can be chemically modified with TBS[56,59]. Only 1 mole of TBS binds rapidly with 1 mole of G-actin, and fragmentation of the TNP-actin followed by separation of a TNP-peptide showed that the TBS binds the ε-amino group of a lysine residue in actin. This specific reaction with TBS does not greatly affect the polymerisation of G-actin (Fig. 4), but the binding of TNP-F-actin to myosin is substantially hindered. Yasui et al.[60] also reported that the protective effect of actin against the denaturation of myosin by heat disappears upon its trinitrophenylation. These results suggest that the ε-amino group of 1 mole of lysine per mole of actin participates in the complexing with myosin.

Recently Mühlrad[56] repeated the trinitrophenylation of actin and also studied the succinylation of the lysine residue[55], but his results do not agree with ours. The formation of 5 moles of TNP-lysine per mole of actin decreased the ability to polymerise by 50%, but the ability to bind with myosin, as measured by the sensitivity of actomyosin to ATP, hardly changed. Table II shows that the results of chemical modification of actin by different groups of workers do not all necessarily agree, and this may be due to differences in specificity depending on the reagents and reaction conditions used, and to the effect of different measurement conditions upon the activities of the various functions. At present, we can make no decisive conclusion about the activity of a particular amino acid residue in any function of actin other than binding to the nucleotide, and we do not know whether chemical modification can distinguish between the roles of the different amino acid residues.

4. F-Actin

Hanson and Lowy's[61] electron microscope study has shown how F-actin is formed by polymerisation of the actin monomer (see the electron micrograph in the frontispiece). Figure 5 shows a structural model of F-actin based on the image obtained using a negative staining method. It can be seen clearly that F-actin consists of two mutually intertwining strands crossing at intervals of about 350Å, in which length there are about 13 spherical monomers of diameter about 55Å. The pitch of the actin filament is $350 \times 2 = 700$Å. A more recent study by Hanson[62] showed that the value of the axial periodicity depends to some extent on

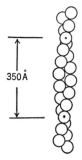

Fig. 5. Electron microscopic model of F-actin (from Hanson and Lowy, Ref. *61*).

the method used for making the negatively stained preparation, and Moore et al.[63] recently improved the negative staining technique and showed that the pitch of paracrystalline F-actin is 360Å. The approximate agreement of the above value for the diameter of the monomer with the diameter estimated from the molecular weight and density supports the conclusion that the monomer of F-actin is G-actin itself, assuming the molecule to be spherical[61]. Moore et al.[63] and Benett et al.[64] have recently shown the probably outline of the G-actin monomer in F-actin by the three-dimensional reconstruction technique of DeRosier and Klug[65]. The electron microscopic image of F-actin is indistinguishable from that of the thin filaments isolated from various types of muscle, and the structure of the thin filaments determined from the X-ray diffraction pattern of muscle[66,67] (chapter 2) is in good agreement with the Hanson-Lowy model of F-actin.

Spencer[68] showed that concentrated solutions of oriented F-actin give a strong equatorial X-ray diffraction maximum, the position of which varies with concentration in the manner expected for an array of regularly packed, mutually repulsive cylinders. Since the spacing was substantially larger than the diameter of the filaments, he postulated the existence of long-range forces between filaments. Elliott[69,71,73], Rome[70,72,73] and Spencer[68,73] all concluded that long-range repulsive forces are responsible for keeping the filaments of certain muscles in a highly regular array, even though they are not in contact. As will be mentioned in chapter 13, section 1, they proposed a molecular mechanism in which repulsive forces play a key role in the contractile process[73]. It must be added that Kawamura and Maruyama[74] have recently reported polymorphic assemblies of F-actin. They also observed that in the presence of ATP and at pH 5.0 a network structure with two fold rotational symmetry is formed.

F-Actin prepared by polymerisation of G-actin is highly polydisperse. Kawamura and Maruyama[75] found the distribution of length in F-actin to be very heterogeneous; thus the number-average length was about 1.5μ and the weight-average length was 3.1μ, for F-actin polymerised at a concentration of 0.5mg/ml. The exponential distribution of F-actin particle length is in accordance with the theories of linear and helical aggregates[76,77]. Isolation of monodisperse F-actin directly from myofibrils has been attempted[78]. Furthermore, Takahashi and Yagi[79] accelerated the polymerisation of G-actin with H-meromyosin and thereby obtained F-actin of relatively uniform length. On the other hand, Kawamura and Maruyama[75] obtained an F-actin preparation of fairly uniform length by polymerisation in 0.1M KCl, in the presence of myosin aggregates and ATP. A new protein which reduces the length of F-actin, β-actinin, was isolated by Maruyama[80–82], but its chemical properties and physiological action have not been elucidated.

5. The G⇌F Transformation of Actin

Actin polymerises on the addition of neutral salts, but the optimum concentration for promoting polymerisation is about 0.06M KCl, while at concentrations above 1M, depolymerisation occurs. The monovalent salts have no specific activity with respect to the polymerisation-depolymerisation equilibrium, but the optimum concentration of divalent cations is about 1mM and they are more efficient than would be expected from the ionic strength alone. In media containing an optimum concentration of KCl, polymerisation is greatest between pH 7.0 to 9.0, while depolymerisation occurs above pH 9.5. Below pH 6.0, a gel-like aggregate is formed, even in the absence of a salt. These results indicate that above pH 7 G-actin is dispersed by repulsions between its net charges, unless salts are added in order to reduce these repulsions and allow intermolecular attractions to dominate, thereby causing the monomers to polymerise to F-actin. Furthermore, the polymerisation of actin is greatly accelerated by adding myosin[83] or H-meromyosin[84]. Cooke and Morales[84a] have recently analysed the mechanism of the acceleration of polymerisation of actin by H-meromyosin, using actin attached by a spin label or a fluorescence dye, and indicated that 4 actin monomers are incorporated into a polymer per H-meromyosin molecule added.

Under conditions in which depolymerisation is not induced by irreversible denaturation of the actin, the rate of polymerisation increases with increasing temperature. Measurements of the effect of pressure on the G⇌F transformation show that there is about an 80ml increase in volume when 1 mole of G-actin polymerises[85], and this was confirmed more recently by dilatometry[86]. These influences of temperature and pressure indicate an entropy effect in the interaction between actin monomers.

Changes in the EPR and UV absorption spectra accompany polymerisation. EPR spectra of actin labelled by a paramagnetic derivative of maleimide reveal a strong immobilisation of the label when the actin polymerises [87]. Higashi and Oosawa[88] reported a change in the UV absorption of tryptophane and tyrosine residues upon polymerisation of actin. From an analysis of the spectral change they concluded that aromatic residues previously exposed on the exterior of the actin monomer are folded inside the molecule during polymerisation. Murphy[29a] reported that the G⇌F transformation of actin accompanies increases in the interaction between aromatic residues and between purine and protein. Environmental changes of aromatic side chains due to polymerisation to F-actin have also been suggested by Lehrer and Kerwar[88a] from a small decrease in the tryptophane and an increase in the tyrosine contribution in the intrinsic fluorescence of actin upon polymerisation.

With respect to the mechanism of the G⇌F transformation, Gergely et al.[89] found that the ratio of the concentrations of G-actin and F-actin is constant and independent of the total concentration of actin, when the external conditions are constant. In contrast, Oosawa et al.[90,91] showed that the formation of F-actin from G-actin is a condensation reaction. Thus they reported that a 'critical' actin concentration, determined by conditions in the medium, is required before polymerisation occurs. Above the critical concentration, excess G-actin is converted into the F-form.

If both G- and F-actin are present, they should be separable by ultracentrifugation, since their sedimentation coefficients are very different. When a low concentration of KCl was added to various concentrations of actin, and when equilibrium had been attained, the ratio of the two components was constant when they were separated by ultracentrifugation[8]. Under the experimental conditions, 40% of the total protein was contained in the supernatant phase. The amount of the light component was measured from the Schlieren pattern (which avoids disturbing the equilibrium), and these results agree with those of Gergely et al.[89] However, we cannot agree with their identification of the supernatant protein as G-actin, since although the actin was initially bound to ATP under the above conditions, this was completely converted to ADP in the supernatant. Moreover, the sedimentation coefficient of the supernatant component was much greater than that of the original G-actin-ATP. This equilibrium can be shown to be G-actin→light component of polymerised actin⇌heavy component of polymerised actin.

The dependence of viscosity on actin concentration in equilibrium systems of actin to which a low concentration KCl had been added has also been studied. There is a drastic change in viscosity in a narrow concentration range of G-actin[8] (Fig. 6), which is reminiscent of the 'critical' concentration for the G⇌F transformation described by Oosawa et al.[90,91] To help elucidate this phenomenon, viscosity and light-scattering measurements were made. The viscosities were measured in the capillary of an

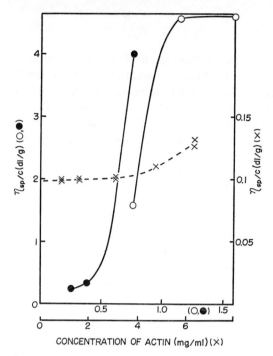

Fig. 6. Dependence of reduced viscosity upon actin concentration. 60 μM free ATP, 15mM borate (pH 7.5), 19.8°C; ×, G-actin; ○, ● indicate polymerisation at 20°C for 24–25 hr using 20mM KCl. Viscosities are those measured in the capillary of the viscometer. The water flow-time of the viscometer was ×, ●, 1,000 sec; ○, 100 sec.

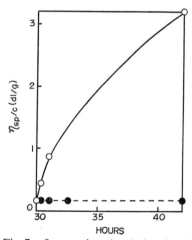

Fig. 7. Increase in reduced viscosity of polymerised actin after passing through capillary viscometer. ○, increase in η_{sp}/c (dl/g) upon standing after passing through capillary. ●, η_{sp}/c immediately after passing through capillary. Actin of 0.138mg/ml was polymerised in 20mM KCl, 2mM dehydroacetate and 70μM ATP at pH 7.1 and 20°C for 30 hr. Water flow-time of viscometer, 1,000 sec.

Ostwald viscometer, but even after reaching equilibrium using lower actin concentrations than the so-called 'critical' concentration, the viscosity gradually increased from 0.4 to 3, or greater, upon standing after passing through the capillary (Fig. 7). Hence, even at low concentrations of actin, there are aggregations of very high viscosity which are destroyed by shear stress, and which regenerate slowly with time. Light-scattering measurements of the molecular weight and radius of gyration under these conditions show that the molecular weight below the so-called 'critical' concentration decreases twofold, while the radius of gyration increases. These results lead us to suggest that, when the actin concentration is low, a polymer with a molecular weight about 1/2 that of the double-stranded polymer is formed, and it can be subsequently broken down by shear stress.

The rate of polymerisation of actin in the presence of KCl or $MgCl_2$ was followed by light-scattering and viscosity measurements and by the generation of P_i from G-actin-ATP accompanying the polymerisation. The rate law is $\dfrac{-d[\text{G-A}]}{dt} = k[\text{G-A}][\Sigma \text{A}]$ (8) where [G-A] and [ΣA] are the concentrations of G-actin and total actin (measured as monomers) respectively.

Figure 8 shows the liberation of protons accompanying the polymerisation of actin, and also the liberation of P_i, which was measured by quenching the reaction with TCA, together with measuring the change of viscosity with time. The generation of P_i and the increase in viscosity followed a first order relationship, but the liberation of protons showed an initial lag[92]. Thus 5 min after the reaction started, the increase in viscosity and

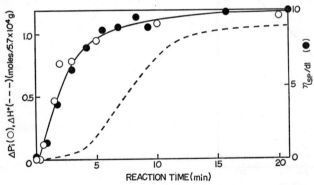

Fig. 8. Liberation of protons and P_i accompanying polymerisation of actin. Polymerisation was carried out at pH 8.16, and 25°C by adding 2mM $MgCl_2$ to 1mg/ml actin.
---, H^+-liberation; ○, P_i-liberation; ●, reduced viscosity.

formation of P_i were almost completed, but only a small amount of protons had been generated. The pH dependence of the relationship between the amounts of P_i and protons produced showed that the proton liberation is itself derived from the liberation of P_i. The reason for this apparent contradiction is not yet completely clear. One possibility is, as described with respect to myosin-ATPase in chapter 3, that the initial phosphorylation occurs at anionic sites of actin, producing F-actin$\genfrac{}{}{0pt}{}{\cdot ADP}{P}$, which is unstable to acid and readily forms P_i with TCA. F-Actin-ADP can then be produced by the liberation of P_i and H^+ from F-actin$\genfrac{}{}{0pt}{}{\cdot ADP}{P}$.

Lastly, we shall briefly mention the hydrolysis of ATP on ultrasonification catalysed by actin. This phenomenon was discovered by Asakura et al.[93,94] who assumed the presence of a linear polymer with bound ATP, and interpreted the decomposition of ATP as involving: helical polymer-ADP + P_i $\overset{ATP}{\rightleftharpoons}$ linear polymer-ATP + H_2O. Asakura and Oosawa based their 'shrinkage theory' on this assumed transformation of actin[95,96]. However, the reaction was later shown by Nakaoka and Kasai[97] to be due to the fragmentation of F-actin by sonication, followed by the replacement of ADP at the end of the fragment by ATP, followed by the reaction: $ATP + H_2O \rightarrow ADP + P_i$ accompanying the recombination of the fragments. This cycle can then be repeated.

REFERENCES

1 F. B. Straub, *Stud. Inst. Med. Chem. Univ. Szeged*, **2**, 3 (1942).
2 F. B. Straub, *Stud. Inst. Med. Chem. Univ. Szeged*, **3**, 23 (1942).
3 J. Hanson and H. E. Huxley, *Symp. Soc. Exp. Biol.*, **9**, 228 (1955).
4 F. B. Straub and G. Feuer, *Biochim. Biophys. Acta*, **4**, 455 (1950).
5 W. F. H. M. Mommaerts, *J. Biol. Chem.*, **198**, 445 (1952).
6 W. Drabikowski and J. Gergely, *J. Biol. Chem.*, **237**, 3412 (1962).
7 K. Laki, K. Maruyama and D. R. Kominz, *Arch. Biochem. Biophys.*, **98**, 323 (1962).
8 J. Yoshimura, H. Matsumiya and Y. Tonomura, *Annu. Rep. Sci. Works, Fac. Sci., Osaka Univ.*, **11**, 51 (1963).
9 M. K. Rees and M. Young, *J. Biol. Chem.*, **242**, 4449 (1967).
10 T. C. Tsao, *Biochim. Biophys. Acta*, **11**, 227 (1953).
11 C. M. Kay, *Biochim. Biophys. Acta*, **43**, 259 (1960).

12 M. S. Lewis, K. Maruyama, W. R. Carroll, D. R. Kominz and K. Laki, *Biochemistry*, **2**, 34 (1963).
13 P. Johnson, C. I. Harris and S. V. Perry, *Biochem. J.*, **105**, 361 (1967).
14 R. S. Adelstein, J. S. Godfrey and W. W. Kielley, *Biochem. Biophys. Res. Commun.*, **12**, 34 (1963).
15 K. K. Tsuboi, *Biochim. Biophys. Acta*, **160**, 420 (1968).
16 I. Sakakibara and K. Yagi, *Biochim. Biophys. Acta*, **207**, 178 (1970).
17 K. Laki and J. Standaert, *Arch. Biochem. Biophys.*, **86**, 16 (1960).
18 M. E. Carsten, *Biochemistry*, **2**, 32 (1963).
19 M. Elzinga, *Biochemistry*, **9**, 1365 (1970).
20 A. M. Asatoor and M. D. Armstrong, *Biochem. Biophys. Res. Commun.*, **26**, 168 (1967).
21 Y. Tonomura and J. Yoshimura, *J. Biochem.*, **51**, 259 (1962).
22 C. J. Justy and H. Fasold, *Biochemistry*, **8**, 2933 (1969).
23 R. H. Locker, *Biochim. Biophys. Acta*, **14**, 533 (1954).
24 R. E. Alving and K. Laki, *Biochemistry*, **5**, 2597 (1967).
25 E. Gaetjens and M. Bárány, *Biochim. Biophys. Acta*, **117**, 176 (1966).
26 R. S. Adelstein and W. M. Kuehl, *Biochemistry*, **9**, 1355 (1970).
27 M. Elzinga, *Biochemistry*, **10**, 224 (1971).
28 J. H. Collins, E. Morkin and M. Elzinga, *Federation Proc.*, **30**, 1148Abs (1971).
29 B. Nagy and W. P. Jencks, *Biochemistry*, **1**, 987 (1962).
29a A. J. Murphy, *Biochemistry*, **10**, 3723 (1971).
29b B. Nagy and H. Strzelecka-Golaszewska, *Arch. Biochem. Biophys.*, **150**, 428 (1972).
30 A. Martonosi, M. A. Gouvea and J. Gergely, *J. Biol. Chem.*, **235**, 1700 (1960).
31 A. Martonosi and M. A. Gouvea, *J. Biol. Chem.*, **236**, 1345 (1961).
32 A. Martonosi, *Biochim. Biophys. Acta*, **57**, 163 (1962).
33 M. R. Iyengar and H. H. Weber, *Biochim. Biophys. Acta*, **86**, 543 (1964).
34 J. J. West, *Biochemistry*, **9**, 3847 (1970).
34a J. J. West, *Biochemistry*, **10**, 3547 (1971).
35 W. F. H. M. Mommaerts, *J. Biol. Chem.*, **198**, 469 (1952).
36 M. Kasai, E. Nakano and F. Oosawa, *Biochim. Biophys. Acta*, **94**, 494 (1965).
37 T. Hayashi and R. Rosenbluth, *Biol. Bull.*, **119**, 294 (1960).
38 T. Hayashi, *J. Gen. Physiol.*, **50**, *Suppl.*, 119 (1967)
39 M. Bárány, A. F. Tucci and T. E. Conover, *J. Mol. Biol.*, **19**, 483 (1966).
40 G. Kuschinsky and F. Turba, *Biochim. Biophys. Acta*, **6**, 426 (1951).
41 K. Maruyama and J. Gergely, *Biochem. Biophys. Res. Commun.*, **6**, 245 (1961).
42 W. Drabikowski and J. Gergely, *J. Biol. Chem.*, **238**, 640 (1963).
43 A. Martonosi, *Federation Proc.*, **20**, 298 (1961).

44 M. Bárány, B. Nagy, F. Finkelman and A. Chrambach, *J. Biol. Chem.*, **236**, 2917 (1961).
45 A. Chrambach, M. Bárány and F. Finkelman, *Arch. Biochem. Biophys.*, **93**, 456 (1961).
46 M. Bárány, F. Finkelman and T. Therattil-Antony, *Arch. Biochem. Biophys.*, **98**, 28 (1962).
47 A. M. Katz and W. F. H. M. Mommaerts, *Biochim. Biophys. Acta*, **65**, 82 (1962).
48 R. C. Strohman and A. J. Samorodin, *J. Biol. Chem.*, **237**, 363 (1962).
49 M. Grubhofer and H. H. Weber, *Z. Naturforsh.*, **16b**, 435 (1961).
50 Y. Tonomura and J. Yoshimura, *J. Biochem.*, **50**, 79 (1961).
51 P. D. Boyer, *J. Amer. Chem. Soc.*, **76**, 4331 (1954).
52 A. Martonosi, *Arch. Biochem. Biophys.*, **123**, 29 (1968).
53 A. Martonosi and M. A. Gouvea, *J. Biol. Chem.*, **236**, 1338 (1961).
54 A. Mühlrad, G. Hegyi and H. Horányi, *Biochim. Biophys. Acta*, **181**, 184 (1969).
55 A. Mühlrad, A. Carsi and A. L. Granata, *Biochim. Biophys. Acta*, **162**, 435 (1968).
56 A. Mühlrad, *Biochim. Biophys. Acta*, **162**, 444 (1968).
57 S. Tokura and Y. Tonomura, *J. Biochem.*, **53**, 422 (1963).
58 H. Portzehl, G. Schramm and H. H. Weber, *Z. Naturforsch.*, **5b**, 61 (1950).
59 Y. Tonomura, S. Tokura and K. Sekiya, *J. Biol. Chem.*, **237**, 1074 (1962).
60 T. Yasui, H. Kawakami and F. Morita, *Agr. Biol. Chem.*, **32**, 225 (1968).
61 J. Hanson and J. Lowy, *J. Mol. Biol.*, **6**, 46 (1963).
62 J. Hanson, *Nature*, **213**, 353 (1967).
63 P. B. Moore, H. E. Huxley and D. J. DeRosier, *J. Mol. Biol.*, **50**, 279 (1970).
64 P. M. Bennett, E. J. O'Brien and J. Hanson, *J. Gen. Physiol.*, **57**, *Suppl.*, 241 (1971).
65 D. J. DeRosier and A. Klug, *Nature*, **217**, 130 (1968).
66 H. E. Huxley and W. Brown, *J. Mol. Biol.*, **30**, 383 (1967).
67 H. E. Huxley, *Science*, **164**, 1356 (1969).
68 M. Spencer, *Nature*, **223**, 1361 (1969).
69 G. F. Elliott, *J. Gen. Physiol.*, **50**, *Suppl.*, 171 (1967).
70 E. Rome, *J. Mol. Biol.*, **27**, 591 (1967).
71 G. F. Elliott, *J. Theor. Biol.*, **21**, 71 (1968).
72 E. Rome, *J. Mol. Biol.*, **37**, 331 (1968).
73 G. F. Elliott, E. Rome and M. Spencer, *Nature*, **226**, 417 (1970).
74 M. Kawamura and K. Maruyama, *J. Biochem.*, **68**, 885 (1970).
75 M. Kawamura and K. Maruyama, *J. Biochem.*, **67**, 437 (1970).
76 P. Flory, "Principles of Polymer Chemistry," Cornell University Press, Ithaca, New York (1957).
77 F. Oosawa and M. Kasai, *J. Mol. Biol.*, **4**, 10 (1962).

78 H. Hama, K. Maruyama and H. Noda, *Biochim. Biophys. Acta*, **102**, 249 (1965).
79 K. Takahashi and K. Yagi, *J. Biochem.*, **64**, 271 (1968).
80 K. Maruyama, *Biochim. Biophys. Acta*, **102**, 542 (1965).
81 K. Maruyama, *Biochim. Biophys. Acta*, **126**, 389 (1966).
82 K. Maruyama, *J. Biochem.*, **69**, 369 (1971).
83 A. Szent-Györgyi, "Chemistry of Muscular Contraction," 2nd ed., Academic Press, New York, p. 71 (1951).
84 K. Yagi, R. Mase, I. Sakakibara and H. Asai, *J. Biol. Chem.*, **240**, 2448 (1965).
84a R. Cooke and M. F. Morales, *J. Mol. Biol.*, **60**, 249 (1971).
85 T. Ikkai and T. Ooi, *Biochemistry*, **5**, 1151 (1966).
86 T. Ikkai, T. Ooi and H. Noguchi, *Science*, **152**, 1756 (1966).
87 D. B. Stone, S. C. Prevost and J. Botts, *Biochemistry*, **9**, 3937 (1970).
88 S. Higashi and F. Oosawa, *J. Mol. Biol.*, **12**, 843 (1965).
88a S. S. Lehrer and G. Kerwar, *Biochemistry*, **11**, 1211 (1972).
89 J. Gergely, M. A. Gouvea and A. Martonosi, *J. Biol. Chem.*, **235**, 1704 (1960).
90 F. Oosawa, S. Asakura, K. Hotta, N. Imai and T. Ooi, *J. Polymer Sci.*, **37**, 323 (1959).
91 F. Oosawa, *J. Polymer Sci.*, **26**, 29 (1957).
92 T. Tokiwa, T. Shimada and Y. Tonomura, *J. Biochem.*, **58**, 577 (1965).
93 S. Asakura, *Biochim. Biophys. Acta*, **52**, 65 (1962).
94 S. Asakura, M. Taniguchi and F. Oosawa, *Biochim. Biophys. Acta*, **74**, 140 (1963).
95 S. Asakura, M. Taniguchi and F. Oosawa, *J. Mol. Biol.*, **7**, 55 (1963).
96 S. Asakura and F. Oosawa, *Protein, Nucleic Acid, Enzyme*, **7**, 63 (1962) (in Japanese).
97 Y. Nakaoka and M. Kasai, *J. Mol. Biol.*, **44**, 319 (1969).

6
THE FORMATION OF ACTOMYOSIN AND ITS DISSOCIATION BY ATP*

Banga and Szent-Györgyi[1] extracted myosin from muscle with a weakly alkaline, concentrated salt solution (Weber-Edsall solution) and found that varying the extraction time gave proteins of differing viscosities and showing different degrees of reduction in viscosity upon the addition of ATP. They called the product from a short extraction myosin A, and that from a long extraction, myosin B. After Straub[2] had discovered actin, it became clear that the differences observed were due to variations in the actin content of the myosin. Purified specimens of myosin and actin, prepared as described in chapters 2 and 5, can be combined to produce a reconstituted actomyosin with properties similar to myosin B, which is a natural actomyosin. The enzymatic activity and physical properties of actomyosin depend markedly on the ionic strength; this discussion will concentrate on the properties at high ionic strength.

* Contributors: Kikuko Takeuchi and Hirofumi Onishi

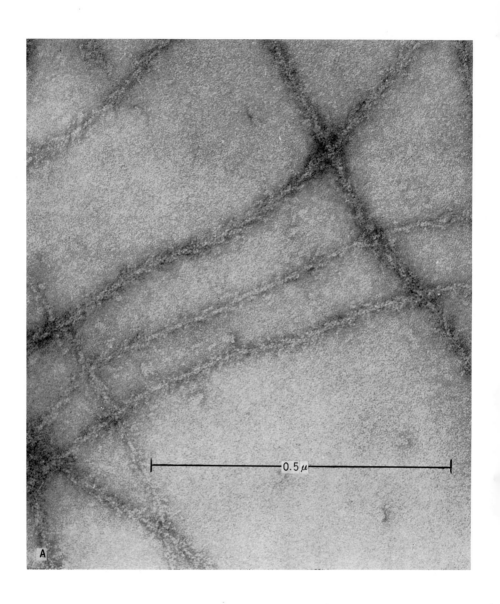

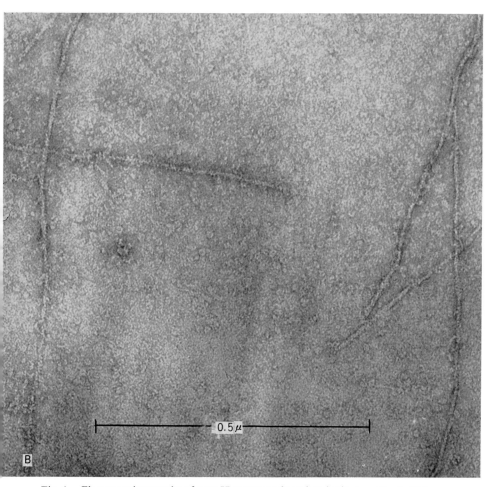

Fig. 1. Electron micrographs of acto-H-meromyosin at low ionic strength. A: Before addition of ATP. B: After addition of ATP. 0.03mg/ml F-actin, 0.12mg/ml H-MM, 1mM $MgCl_2$, 10mM $Tris-HCl_2$, pH 7.6.

F-Actin binds with myosin in 0.6M KCl in proportions to be described later, but because the F-actin is polydisperse the reconstituted actomyosin is also polydisperse (cf. chapter 5, section 4). Solutions of the complex are of high viscosity and show a substantial flow-birefringence and high light-scattering intensity, both of which are considerably decreased upon the addition of ATP[3-13]. This is due to the dissociation of actomyosin into myosin and F-actin by ATP. According to Hanson and H. E. Huxley's sliding model, the three basic reactions of muscle contraction are (1) the specific binding of F-actin with myosin, (2) the translocation of F-actin, coupled with ATP-splitting catalysed by myosin and (3) the release or dissociation of F-actin from myosin (chapter 13). Hence the formation and dissociation of actomyosin are not only of general interest as reactions between proteins, but also have the added significance of being fundamental reactions in muscle contraction. It is therefore important to elucidate their mechanism.

1. Electron Micrographs of Actomyosin

H. E. Huxley's electron microscope studies[14] showed that two kinds of filament constitute the contractile element of skeletal muscle. These are joined by cross-bridges, which appear as projections from the myosin filament and are believed to be the H-MM portions of the myosin molecules (chapter 1). Electron micrographs of actomyosin at high ionic strength resemble those of acto-H-meromyosin at low ionic strength. Figure 1A is an electron micrograph of acto-H-meromyosin at low ionic strength, and shows the 'arrow-head' structure. When ATP is added, a micrograph of the almost completely dissociated complex is obtained, in which the structure of F-actin appears together with some H-MM (Fig. 1B). The particle lengths of myosin B were fairly homogeneous: the number-average length was about 0.5μ and the weight-average length was about 0.6μ[15]. In contrast, the particle lengths of reconstituted actomyosin were heterogeneous, and were much shorter than those of the F-actin used to form actomyosin[15]. This might be due to the loosening of the structure of F-actin when it binds with myosin (see p. 173) and the breakdown of the F-actin structure due to the procedures for negative staining.

2. The Binding Ratio of Actin with Myosin

Binding between different proteins is very important in biological pro-

TABLE I. Binding Ratio of Actin with Myosin or H-Meromyosin

Authors	Method	Measurement conditions	Binding ratio(g/g) myosin or H-MM : actin	Dissociation constant (M)	References
S. S. Spicer and J. Gergely	Ultracentrifugation	0.13M KCl,pH7.5	4 : 1		16
A. G. Szent-Györgyi	Viscosity	0.6M KCl,6.7mM phosphate, pH7.0, 20°C	4.3 : 1		19
J. Gergely and H. Kohler	Light-scattering	0.6M KCl,1mM MgCl$_2$, pH7.0	4 : 1	$K \geqq 10^{-10}(23°C)$	17
Y. Tonomura et al.	Light-scattering	0.6M KCl,pH7.5	3.7 : 1	$K=4.6\times10^{-7}(9°C)$	18
L. B. Nanninga	Ultracentrifugation	0.3M KCl,10mM Tris buffer, pH7.0	3.7 : 1	$K=4.3\times10^{-7}(26°C, 0°C)$	20
K. Sekiya et al.	Light-scattering	Acto-H-MM,4mM KCl, 1mM MgCl$_2$,pH7.5	2.8 : 1		21
M. Young	Analytical ultracentrifugation	Acto-H-MM,acto-S$_1$,0.1 M KCl,1mM MgCl$_2$,pH7.0	1 : 1 (molar ratio)	$K=1.4$–3.4×10^{-6}(H-MM); 1×10^{-6}(S-1)	22
K. Takeuchi and Y. Tonomura	Light-scattering	Acto-H-MM,5mM–0.6M KCl, 10μM–4mM MgCl$_2$,pH7.5, room temp.	4.07 : 1		23
K. Takeuchi and Y. Tonomura	Ultracentrifugation	Acto-H-MM,0.1M KCl,1mM MgCl$_2$,pH7.0–8.8,3–22°C	0.46–0.52 : 1 (molar ratio)	$K=2.6\times10^{-7}(3°$–$4°C)$	23
K. Takeuchi and Y. Tonomura	Light-scattering	Acto-S-1,0.1M KCl,50μM MgCl$_2$,pH7.6,23°C	1 : 1 (molar ratio)		23
K. Takeuchi and Y. Tonomura	Ultracentrifugation	Acto-S-1,3.8mM–0.1M KCl,0.2–1mM MgCl$_2$,pH 7.6,2–16.5°C	0.9–1.08 : 1 (molar ratio)	$K=0.38$–1.22×10^{-6}	23

cesses, including antigen-antibody interactions. The binding between F-actin and myosin has been studied by many researchers[5,6,16-23]. If 1 mole of actin monomer combines with 1 mole of myosin to form actomyosin, the weight ratio of the two proteins in the complex at saturation should be about 1:10, taking the molecular weights as 4.6×10^4 and 4.8×10^5, respectively (chapters 2 and 5). Most of the earlier experimental results indicate a weight ratio of 3.7–4.0 myosin to actin, suggesting that 1 mole of myosin complexes 2 moles of actin monomer (Table I). More recently, Tokiwa[24] found that F-actin decreases the motility of spin labels (N-2,2,6,6-tetramethyl piperidine nitrooxide iodoacetamide) bound to myosin, the maximum change being obtained when the solution contained 2 moles of actin monomer per mole of myosin.

However, H. E. Huxley's electron microscopy[25] and Young's ultracentrifugal analysis[22] have indicated that the complex of actin with H-meromyosin contains 1 mole each of H-MM and actin monomer. Young's results suggest that the earlier discrepancies were due to the lower purity of the F-actin samples used. The same ratio of binding of these two proteins was also obtained by Tawada[26]. Very recently, Rizzino et al.[27] analysed the effect of actin on the K^+-activated ATPase of H-meromyosin by applying the theory of multiple equilibria, and concluded that 1 mole of H-meromyosin stoichiometrically binds each mole of actin monomer. Since there are two opposing views on the value of the binding ratio of the components, we[23] have measured the complexing of F-actin with H-MM under conditions when they bind strongly to each other, in order to estimate this ratio. Our first measurements were by the light-scattering method. We studied the effect of the extraction temperature of F-actin, and compared Young's purification method to others, but the weight ratio of binding with H-MM hardly changed with different preparations of F-actin, and was in the range 1: 4.07 ± 0.085. Assuming that the molecular weights of actin monomer and H-meromyosin are 4.6×10^4 and 3.4×10^5 respectively, this indicates that an actin dimer complexes an H-MM molecule (theoretical ratio 1:3.7).

Secondly, F-actin and H-MM were mixed in a weight ratio of 1:1.5, and an electron micrograph was taken of the acto-H-meromyosin formed. The distribution of the number of H-MM molecules bound per 1μ length of F-actin was then calculated, and the saturation value, which could be seen on the electron micrograph, was about 55. The binding distribution was often approximately random if a binding ratio of 1: 3.6 was assumed (Fig. 2).

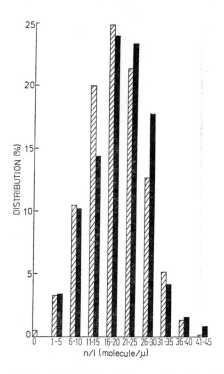

Fig. 2. Distribution of the number of molecules of H-MM bound per 1μ length of F-actin. The number of molecules of H-MM bound to F-actin at 0.03mg/ml F-actin, 0.045mg/ml H-MM, 1mM $MgCl_2$, pH 7.5 was estimated from an electron micrograph. The number of molecules of H-MM visibly bound per 1μ length of F-actin reached a saturation value of 55 if enough H-MM was present. ▬: observed value ($\Sigma 1=209\mu$). ▨: calculated value assuming that H-MM binds randomly with the binding sites of actin, and that the binding ratio is 1 : 3.6 of actin: H-MM (as determined by light-scattering).

Thirdly, a fixed concentration of H-MM was added to various concentrations of F-actin, the mixture was ultracentrifuged and the concentration of protein remaining in the supernatant phase was measured. Under these conditions F-actin and acto-H-meromyosin precipitate, but very little H-MM does so. Therefore, if a correction from similar experiments using only H-MM is made, one can easily determine the amount of bound H-MM. Figure 3 shows the Scatchard plot. If we assume that F-actin complexes each H-MM independently with a fixed dissociation constant, K, the number of bound sites per actin monomer, n, can be obtained from the

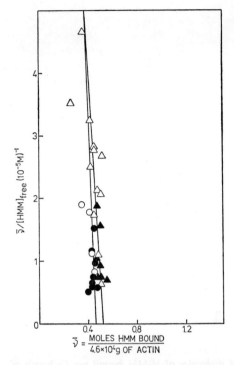

Fig. 3. Scatchard plot for the binding of F-actin with H-MM. 0.1M KCl, 1mM MgCl$_2$, 10mM Tris-HCl or 20mM phosphate. The amount of H-MM bound to F-actin was determined by ultracentrifugation over F-actin concentrations ranging from 0.05–0.9 mg/ml. $\bar{\nu}$ is the number of H-MM (molecular weight 3.4×10^5) bound to an actin monomer (molecular weight 4.6×10^4). ○, 2.9mg/ml H-MM, pH 7.0, 7°C; ●, 3mg/ml H-MM, pH 7.6, 22°C; △, 0.34–6.8mg/ml H-MM, pH 8.8, 4°C; ▲, 3 mg/ml H-MM, pH 7.6, 3.5°C.

Scatchard equation: $\bar{\nu}/[\text{H-MM}] = K(n-\bar{\nu})$, where $\bar{\nu}$ denotes the extent of binding of H-MM to the actin monomer and [H-MM] is the free H-MM concentration. This method shows that the H-MM molecule complexes with the actin dimer. Quite recently, Eisenberg et al.[27a] have also reported the binding of 1 mole of H-MM to 2 moles of actin monomer, based on their analytical ultracentrifugal method of investigation.

We have also investigated the complexing of S-1 (the active subfragment of myosin) with F-actin by ultracentrifugation and light-scattering. The results clearly indicate that the S-1 molecule binds with an actin monomer

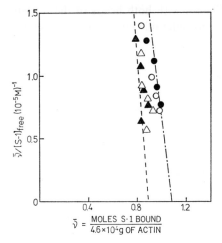

Fig. 4. Complexing of F-actin with S-1. The amount of bound S-1 was determined by ultracentrifugation after addition of various concentrations of F-actin to 2.3mg/ml of S-1. $\bar{v}$ is the number of moles of S-1 (molecular weight 1.2×10^5) bound to an actin monomer (molecular weight 4.6×10^4). 10mM Tris-HCl, pH 7.6. ○, 1mM MgCl$_2$, 0.1M KCl 16.5°C; ●, 0.2mM MgCl$_2$, 0.1M KCl, 2°C; △, 0.2mM MgCl$_2$, 10mM KCl, 16.5°C; ▲, 1mM MgCl$_2$, 3.8mM KCl, 16.5°C.

(Fig. 4, Table I). The electron micrographs of the negatively stained complex of the thin filament with S-1 have recently been analysed by Moore et al.[28] using the three-dimensional reconstruction technique of De Rosier and Klug[29]. They showed that one S-1 molecule attaches to each of the G-actin units in the double helix of F-actin. The S-1 molecules, which appear to be somewhat elongate and curved, are attached to the actin filament in a configuration in which they are both tilted and skew with respect to the filament axis.

Hence we have concluded that each of the two heads of the myosin molecule can combine with F-actin, so that one molecule of myosin complexes an actin dimer. The active site for myosin-ATPase is thought to be present in only one of the two heads of the myosin molecule (chapter 4), but we have not yet succeeded in discriminating between the two S-1's in the binding with F-actin. However, we can tentatively assume that the S-1 containing the ATPase site (S-1B) attaches more strongly to F-actin than the S-1 lacking the site (S-1A). It is important to note that actomyosin 'dissociates' upon reaction with 1 mole of ATP per mole of myosin (section 6) despite the complexing of F-actin with both heads of the myosin molecule. This suggests that there is a considerable interaction between the two heads when bound to F-actin.

3. The Nature of the Binding Forces

F-Actin is present in solution as a macromolecule, containing two long chains which provide many binding sites for myosin. Furthermore, because it is polydisperse, it is difficult to determine by light-scattering the number of myosin molecules bound to F-actin. If we assume that there are n binding sites for myosin per mole of F-actin of which j-1 sites are occupied by myosin, and that another myosin molecule enters to occupy the jth site, then $M + A_n M_{j-1} = A_n K_j$. The change in intensity of 90° angle light-scattering accompanying this reaction will be assumed to be a function of j. If $\varDelta$ is the increment of scattering intensity at 90° when a given amount of myosin is added to a fixed amount of F-actin, and $\varDelta_m$ is the increment when a swamping amount of myosin is added, and if binding 1 molecule of myosin causes a constant increment in light-scattering intensity, then the ratio $\varDelta/\varDelta_m$ should be proportional to the degree of binding of myosin, ν. However, this particular increment varies with the step j in the present case, so that in order to estimate ν it is necessary to multiply $\varDelta/\varDelta_m$ by a factor f, assuming that $\varDelta/\varDelta_m$ is dependent only on j. Since ν can be determined under conditions where the myosin added to F-actin combines completely with the actin, the factor f can be estimated from the relationship between $\varDelta/\varDelta_m$ and ν. Figure 5 presents measurements of $\varDelta$ upon addition of various amounts of myosin to a fixed amount of F-actin under such conditions, showing that f at each value of ν can be derived by obtaining the ratio between the value on the line joining the origin and the point when F-actin is saturated with myosin, and the experimental value $\varDelta$.

Equilibrium and kinetic analysis of the complexing between myosin and F-actin has been attempted by Nanninga[20], Laki[30] and ourselves[18]. The last two groups showed that the association is both temperature dependent and reversible. The equation $[\varSigma M]/\nu = K/(1-\nu) + [\varSigma A]$ should be valid for a reversible binding reaction, where $[\varSigma M]$ is the total molar concentration of myosin, and ν is the degree of binding as determined from the light-scattering measurements described above. Therefore, by plotting $[\varSigma M]/\nu$ versus $1/(1-\nu)$ (Fig. 6) one can obtain the total molar concentration $[\varSigma A]$ of myosin-binding sites on actin and the dissociation constant, K, for the complexing. K increases with increasing ionic strength, but decreases on

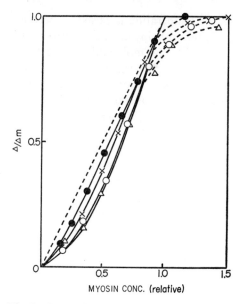

Fig. 5. Determination of the binding ratio of actin to myosin and the factor f for deriving [AM]/[ΣM] from Δ/Δ_m. f is obtained as the ratio of the value of Δ/Δ_m on the straight line joining the origin to the binding saturation point (------) to that on the experimental curve. 0.03mg/ml F-actin, 0.6M KCl, pH 7.5, 25°C. ○, △, ×, ● indicate different samples of myosin and F-actin.

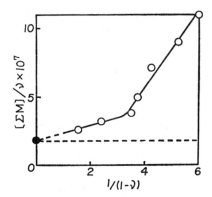

Fig. 6. Plot of [ΣM]/ν versus $1/(1-\nu)$. $\nu = f(\Delta/\Delta_m)$ is the degree of occupation of myosin binding sites by actin. (Δ/Δ_m) was measured in 0.6M KCl at pH 6.5 and 16.5°C. 0.029mg/ml F-actin, 0.05–0.50mg/ml myosin. ○, experimental values; ●, extrapolated one. The upward bend at high values of $1/(1-\nu)$ is probably due to interaction between myosin molecules bound on F-actin.

the addition of ethanol, and when the temperature is raised, K becomes very small. By plotting pK versus $1/T$, the enthalpy (ΔH) and entropy (ΔS) of this equilibrium were found to be $+57.5$ kcal/mole and $+237$ cal/deg/mole, respectively, despite the fact that a complex is formed.

Measurement of the rate of increase in light-scattering intensity after mixing myosin with actin in various concentrations showed that the formation of actomyosin follows the rate law: $v = k[A][M]$. The reaction is not appreciably affected by EDTA or EGTA[31]. The rate constant, k, was 1.9×10^5 M^{-1}sec^{-1} in 0.6M KCl at 9°C, and the temperature dependence of k shows the activation enthalpy $\Delta H^{\ddagger} = \Delta E^{\ddagger} - RT$ to be $+11.5$ kcal/mole. According to the absolute reaction rate theory[32], $k = RT/\kappa Nh \exp(T\Delta S^{\ddagger} - \Delta H^{\ddagger})/RT$, where N is Avogadro's number, and h is Planck's constant. If the usual assumption is made that the transmission coefficient κ equals 1, then $\Delta S^{\ddagger}$ is found to be $+6.2$ cal/deg/mole. These $\Delta H^{\ddagger}$ and $\Delta S^{\ddagger}$ values are much smaller than ΔH and ΔS for the overall reaction.

According to Kauzmann[33], there are three factors which contribute to intermolecular and intramolecular binding of proteins—electrostatic forces, hydrogen bonding, and hydrophobic bonding. Of these, electrostatic and hydrophobic bondings are accompanied by a positive ΔS. The association of F-actin with myosin is weakened by increasing ionic strength and enhanced by decreasing the dielectric constant of the solvent, thereby indicating that electrostatic factors are involved. However, the positive ΔS accompanying electrostatic bonding is not usually as large as that found for the complexing of actin with myosin, so that other factors must also be effective. Two mechanisms have been advanced to explain the positive ΔS when proteins complex low molecular weight materials and when two proteins associate. One states that water molecules bound to the protein are desolvated as complexing takes place, and the other that a portion of the secondary structure of the protein changes from a helix to a random coil. Optical rotatory dispersion measurements of the myosin-actin system[18] show that a fusion of about 2–5% of the helical structure accompanies the binding reaction (chapter 2, section 8). H-MM is an active fragment obtained by removing the L-MM—which has a high helical content—from myosin, so that when H-MM binds to actin there is a larger decrease in helical content than when myosin binds to actin[34]. However the optical rotatory dispersion results could be ambiguous, since the effect of light-scattering on the measurements is not negligible. On the other hand, Tokiwa[24] has found that F-actin decreases the mobility of spin labels bound to myosin, as mentioned in section 2. Fujime and Ishiwata[35] have recently concluded from dynamic studies by quasi-elastic scattering of laser light that the flexibility of F-actin increases upon binding of H-MM at a molar ratio of actin monomer:H-MM of 6:1, and then decreases to the intrinsic value for F-actin on further binding of H-MM. These results indicate that one cause of the positive ΔS measured by thermodynamic analysis of the complexing is a conformational change in myosin and/or F-actin which accompanies binding.

The electrostatic component of the association between F-actin and myosin may involve a myosin-SH group (chapter 4, section 4)[36-38]. Furthermore, the pH-dependence of the rate constant for the reaction between actin and myosin indicates the presence of an amino group which changes from being positively charged to neutral at pH 8.0–9.0 or greater[18]. As already described in chapter 5, the specific reaction of TBS with 1 mole of lysine per mole of G-actin almost completely inhibits the binding with myosin, even though the G$\rightleftharpoons$F transformation is virtually unaffected[18]. Hence one can suggest that one mole of amino groups per mole of actin is needed for the binding, and that electrostatic bonds such as M-S$^-$.... H$_3^+$N-A play an important role.

4. Dissociation of Actomyosin by ATP

We shall now describe the effect of ATP on actomyosin in solutions containing high concentrations of KCl. Lipmann[38a] showed that ATP is a key substance in biological energy transformation, and many researchers were attracted by the discovery of Szent-Györgyi[6] and Dainty et al.[4] that ATP decreases the viscosity and flow-birefringence of an actomyosin solution. The molecular mechanism of this reaction was studied with the expectation that it would be directly related to the molecular mechanism of muscle contraction[3,10-13,17,39-43]. We shall first describe the changes which occur when ATP or PP$_i$ is added to myosin B. Jordan and Oster[44] first showed that the light-scattering intensity of myosin B decreases on addition of ATP. This can be interpreted as a decrease in the weight-average molecular weight, $\bar{M}_w$, due to the dissociation of myosin B into its components myosin and F-actin, or as a decrease in the factor P(θ), which depends on the molecular shape, in the equation for the light-scattering intensity: $R_\theta = Kc\bar{M}_w P(\theta)$ (chapter 2).

We showed in chapter 3 that PP$_i$ is a competitive inhibitor, but PP$_i$ induces a change in myosin B in 0.6M KCl similar to that induced by ATP. Therefore, in our study[45] of the mechanism of the change in myosin B,

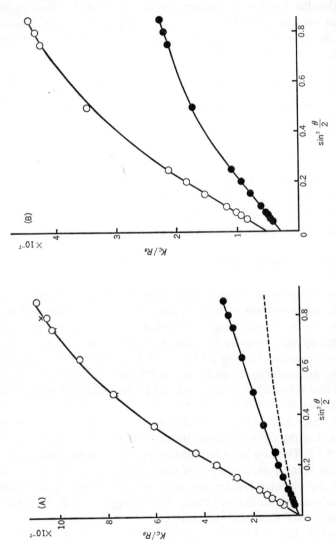

Fig. 7. A: Zimm plot of the light-scattering intensity of myosin B. The dotted line represents the Zimm plot for a monodisperse rod-shaped molecule. 0.6M KCl, 1mM MgCl$_2$, pH 7.2, 23°C. ●, control; ○, 0.5mM PP$_i$; ×, 0.5mM Salyrgan, an SH poison. B: Zimm plot of myosin B solution centrifuged for 3 hr at 3.5×10^4 rpm at a concentration of 0.05mg/ml. 0.6M KCl, 1mM MgCl$_2$, pH 7.4, 20°C. ●, control; ○, 1mM PP$_i$.

we have used mainly PP_i, on account of its simpler structure. Figure 7A shows a Zimm plot of the light-scattering intensity of the myosin B-PP_i system. It indicates that the weight-average molecular weight of myosin B does not change, but that only $P(\theta)$ decreases, from which it was tentatively concluded that the molecule increases in size[45]. Blum and Morales[10] first arrived at this result using the myosin B-ATP system, and this was later confirmed by them[13] and by ourselves[45].

It cannot be concluded from light-scattering measurements alone that the length of all myosin B particles is extended upon addition of ATP. The light-scattering measurement gives $\bar{M}_w$, and the above results indicate that those particles of high molecular weight may become extended with the binding of ATP or PP_i. Myosin B, when ultracentrifuged after standing at 0°C for several days or at very low concentrations, can be freed from large particles. Such samples of myosin B gave the results of Fig. 7B, which show clearly that $\bar{M}_w$ decreases when ATP or PP_i is added[12,45]. Hence there must be two kinds of myosin B particle, which behave differently when they bind with PP_i or ATP. One, of very high molecular weight, expands and the other, of lower molecular weight, dissociates into myosin and F-actin.

Gergely and Kohler[17] clearly showed by light-scattering measurements at low concentrations that reconstituted actomyosin dissociates with ATP into myosin and F-actin. In the case of myosin B and reconstituted actomyosin the ATPase in 0.6M KCl can hardly be distinguished from the ATPase of the component myosin. The two apparently contradictory changes with PP_i or ATP—dissociation and extension of the molecule—probably both result from scission of the bond between F-actin and myosin by PP_i or ATP, so that molecular extension takes place when the bonds are relatively strong and only partially broken by ATP or PP_i, while molecular dissociation takes place if they are weak and break completely. In order to validate such a concept, it is necessary to study the manner of these changes at the molecular level by some other methods. We have investigated the shape of myosin B by orienting myosin B particles in concentrated solution by applying a flow shearing stress, abruptly terminating the flow, and measuring the relaxation time of its birefringence[7,46]. If the only effect which complexing with ATP has on the myosin B particles is to increase their length, then their rotation should slow down if they are long rigid rods, but instead, the addition of ATP induces a very fast rotation (Fig. 8). Hence we conclude that myosin B particles loosen and become ex-

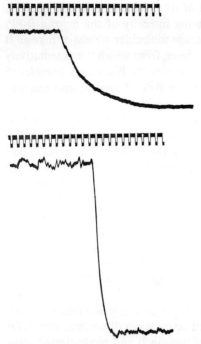

Fig. 8. Relaxation of flow-birefringence of myosin B at high velocity gradients. Upper trace: 0.55M KCl, pH 6.3, 20°C, 5.94mg/ml myosin B. Lower trace: 1.1mM ATP, 1.5mM Mg^{2+}, 0.6M KCl, pH 6.7, 20°C, 1.5mg/ml myosin B. Time scale: 400 cps.

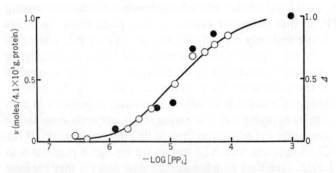

Fig. 9. Binding of PP_i to myosin B. 0.6M KCl, 0.3mM $MgCl_2$, pH 7.5, 5°C. ○, degree of binding; ●, light-scattering intensity. The observed value of ν was approximately given by $\nu = \dfrac{1}{1+10^{-4.7}/[PP_i]}$.

tended on binding with ATP and undergo degradation upon application of a shearing stress.

It is desirable to find a physical or chemical property which is proportional to the amount of bound PP_i and can be followed easily in order to analyse the mechanism of reaction of myosin B with PP_i. The degree of binding of PP_i should then be easy to determine, despite the complexity of the reaction. Equilibrium dialysis can be used to measure the binding of $^{32}PP_i$ with myosin B, and this method was used to determine the amount of bound PP_i, and to compare it with the value indicated by change in the light-scattering intensity under the same conditions. Both methods give similar estimates (Fig. 9) and the maximum change in light-scattering intensity occurs when 1 mole of PP_i binds $5-6 \times 10^5$ g of myosin B[47]. Thus the maximum change in light-scattering intensity occurs when 1 mole of PP_i binds with 1 mole of the myosin constituent in myosin B, despite the fact that 2 moles of PP_i can bind with 1 mole of myosin (chapter 4). More recently, Nauss et al.[48] obtained similar results for the reaction of reconstituted actomyosin with PP_i. Neither these results nor the above physicochemical measurements can be explained by the conventional model which assumes that actomyosin is completely dissociated into F-actin and myosin in the presence of ATP or PP_i at high ionic strength. Thus the nature of the changes in actomyosin induced by ATP or PP_i is rather complicated, as will be discussed again in section 7, but henceforth we shall tentatively use the customary term 'dissociation.'

5. Kinetic Studies of the Reaction of Actomyosin with PP_i

As described in the previous section, the intensity of light-scattering by actomyosin varies proportionately to the amount of bound PP_i, so that this technique is easy to use to follow the reaction[49,50]. A divalent metal ion is essential to the reaction with PP_i and the necessary amount of Ca^{2+} was added initially in each of the following experiments. In the reaction of PP_i with myosin B, the rate law was found to be $-\ln(1 - \Delta_t/\Delta_m) = \bar{k}t$, where $\bar{k}$ is a constant and Δ_m is the maximum decrease in light-scattering intensity at high concentrations of PP_i when the myosin B-PP_i complex has undergone its maximum change in shape. $1/\bar{k}$ also varied linearly with $1/[PP_i]$. The simplest mechanism satisfying these relationships is $MB + PP_i \underset{k_{-1}}{\overset{k_{+1}}{\rightleftharpoons}} MB.PP_i \underset{k_{-2}}{\overset{k_{+2}}{\rightleftharpoons}} MB^*.PP_i$, where MB represents the unit of myosin

TABLE II. Dissociation and Rate Constants for the Reaction of Myosin B with PP_i

Organic solvent added	D	$10^5 K$ (M)	$10^4 K_1$ (M)	$10^2 K_2$	$10^2 k_{+2}$ (sec^{-1})	$10^3 k_{-2}$ (sec^{-1})	Δ_c/Δ_m (%)	$\dfrac{1}{(1+K_2)}$ (%)
None	86.30	36	6.0	150	0.84	13	44	40
Acetone	85.20	24	7.2	51	1.9	9.7	70	66
Dioxane	84.30	15	8.9	21	2.9	6.1	86	83
Dioxane	83.50	8.7	10	9.5	5.0	4.8	96	92
Dioxane	82.56	4.0	12	3.6	9.0	3.2	100	97

0.6M KCl, 0.5mM Ca^{2+}, pH 7.9, 5°C

TABLE III. Thermodynamic Functions for the Elementary Steps in the Reaction of Myosin B with PP_i

Prep. No.	Ca^{2+} (mM)	Step 1				Step 2		
		ΔG (kcal/mole)	ΔG_{elec} (kcal/mole)	ΔH (kcal/mole)	ΔS (cal/deg/mole)	ΔG (kcal/mole)	ΔG_{elec} (kcal/mole)	ΔH (kcal/mole)	ΔS (cal/deg/mole)
MO 2	0.5	−3.7	11.6	−9.7	−20	−1.7	−44	−12.1	−37
MO 3	0.1	−4.7	9.6	−7.8	−11.3	−1.15	−21	− 8.6	−26.8
MO 4	0.3–0.7	−4.3				−1.25			
MO 5	0.5	−4.1	12.4	−10.0	−21	−1.5	−39	−10	−31

The values of ΔG and ΔS were measured at 0.6M KCl, pH 7.88, 5°C, $D=82.56$

B which binds with 1 mole of PP_i, and $MB.PP_i$ and $MB^*.PP_i$ represent an undeformed and deformed complex, respectively. Δ_t/Δ_m then equal $[MB^*.PP_i]/[\Sigma MB]$, since the decrease in light-scattering intensity is proportional to the amount of PP_i bound to the myosin B. According to this mechanism, $-\ln(1-\Delta_t/\Delta_m) = \bar{k}\ t = k_{+2}/\left(1+\dfrac{K_b}{[PP_i]}\right)\cdot t$, where $K_b = \dfrac{k_{-1}+k_{+2}}{k_{+1}}$. Hence k_2 and K_b can be estimated from the relationship between $1/\bar{k}$ and $1/[PP_i]$. The k_{+2} value thus derived is greater at high Ca^{2+} concentrations, but K_b shows almost no change so that $K_b = k_{-1}/k_{+1} = K_1$. At equilibrium, $\Delta/\Delta_c = \left(\dfrac{1}{(1+K_2)}\right)/\left(1+\dfrac{K_1 K_2}{(1+K_2)[PP_i]}\right)$ where $K_2 = k_{-2}/k_{+2}$, Δ_c is the value of the decrease in light-scattering intensity, Δ, at high concentrations of PP_i under certain specific conditions, and is usually equal to $\Delta_m/(1+K_2)$ because of the second equilibrium. The combination of these two calculations permits estimation of k_{+2}, K_1 and K_2. Table II shows some typical values, and also shows that the Δ_c value can be reasonably explained relative to the equilibrium constant K_2 in step 2, and this supports the above mechanism. Further kinetic measurements of the recovery of light-scattering intensity when the PP_i complexed to myosin B is rapidly decomposed by a large amount of pyrophosphatase also agree well with the above mechanism.

Karush[51], Linderstrøm-Lang and Schellman[52] and Lumry and Eyring[53] have pointed out that a configurational change when a protein binds with a low molecular weight molecule will be accompanied by thermodynamic changes. The configurational change of myosin B upon addition of PP_i can be followed, so that the accompanying thermodynamic change due to configurational change can be distinguished from that due solely to the binding. Thus, if the above mechanism be assumed, one can estimate ΔH and ΔS for each of the reaction steps from the change in the dissociation constants K_1 and K_2 with temperature. Also, the electrostatic contribution to the free energy change for each reaction step, ΔG_{elec}, can be determined from $\dfrac{\ln K}{1/D_{eff}} = \dfrac{\Delta G_{elec}}{RT}$, where D_{eff} is the effective dielectric constant in the microscopic region directly involved in the reaction, and not the bulk D. However, the binding of PP_i results in the extension of the particle, so that in theory the bulk D is proportional to D_{eff}. Consequently, we have used bulk D to estimate ΔG_{elec}. The thermodynamic constants derived for the two steps of the reaction are shown in Table III. The values of ΔG_{elec} (-35 kcal/mole) and ΔS (-31 cal/mole/deg) for the deformation of myosin B after it has complexed PP_i are very large. This indicates that the basic shape of the complex of PP_i with myosin B is determined by the entropy and electrostatic repulsion of both species, thereby supporting the concept of Morales and Botts[54].

As described in the following section, change in the configuration of actomyosin is caused by the reaction of ATP with site 2 of myosin. Therefore it is reasonable to conclude that PP_i also changes the configuration of actomyosin by binding at site 2. Usually if an ATP analogue prefers site 1 to site 2 in myosin, it becomes a competitive inhibitor of the ATPase, making the change in shape of actomyosin more difficult (see chapter 3, section 4). An example is ADP. Conversely, if binding is more difficult at site 1 than at site 2, the analogue activates the ATPase at low concentrations, becomes a competitive inhibitor at high concentrations and then configurational changes of the actomyosin occur easily. PP_i is an example of this second type, since it is known to activate the ATPase under certain conditions[55]. The above analysis of the reaction of actomyosin with PP_i indicated the presence of equilibrium between undeformed and deformed actomyosin-PP_i complexes. Hence the binding site for actin on myosin is different from that for PP_i, so that if PP_i binds with site 2, actin binds elsewhere.

6. Kinetic Studies of the Dissociation of Actomyosin with ATP

In 1952 we[9,56] first tried to perform kinetic measurements of the change in size and shape of actomyosin with ATP by following the change in light-scattering intensity. Mommaerts also worked on this later[11,57,58]. The rate of change in light-scattering intensity when ATP is added to myosin B solution is shown in Fig. 10. There are three phases—a rapid decrease in intensity immediately after the addition of ATP, a period in which the intensity stays depressed and then a slow return to the original intensity.

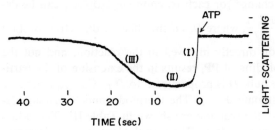

Fig. 10. Rate of change in light-scattering intensity after addition of ATP to myosin B solution.

The initial rapid decrease is complete within a few seconds, and is caused by a change in the configuration of myosin B with ATP. The second phase is observed only when more than a specific amount of ATP is added, and this phase can be extended by increasing the amount of ATP. The final recovery of light-scattering intensity is independent of the initial concentration of ATP, and follows approximately first order kinetics[9]. These results can be interpreted simply by $MB+S \rightleftharpoons MB^*.S \rightarrow MB+P_i+ADP$[59], where $MB^*.S$ indicates the Michaelis complex of myosin B, of which configuration is changed by binding with S. In the presence of Mg^{2+}, the rapid decrease in light-scattering intensity with ATP can be reduced by addition of a large amount of EDTA, since the Michaelis complex, $MB^*.S$, is formed by the equilibrium $MB+S \rightleftharpoons MB^*.S$, which depends on Mg^{2+}. However, the rate of the recovery of light-scattering intensity in the presence of Mg^{2+} is unchanged by adding a large amount of EDTA[58]. The recovery rate is much lower than the rate of binding of actin to myosin under identical conditions (cf. section 3). This effect of EDTA seemed difficult to be explained by the simple mechanism above.

Consequently, we have attempted to elucidate the mechanism of dissociation of actomyosin with ATP by performing kinetic measurements—using the light-scattering method—on the reaction of reconstituted actomyosin with ATP in a rapid mixing chamber[60,61]. Our aim was to investigate the relationship between the dissociation of actomyosin with ATP and the reaction of myosin-ATPase with ATP, which is discussed in chapter 3.

Figure 11 shows how light-scattering intensity decreases when actomyosin is mixed with ATP, and how the half life, $\tau_{1/2}$, depends on the ATP concentration. At high ionic strengths and high concentrations of Mg^{2+}, the decrease in light-scattering intensity increases with ATP concentration up to 1 mole per 4×10^5 g of myosin. Further increase in ATP concentration has no effect. Moreover, its rate is almost equal to the rate of the stoichiometric initial burst of protons when ATP is added to myosin (cf. chapter 3). However, these measurements were carried out at room temperature, so that $E_2{}^1_{ADP \atop P}$ is formed immediately from $E_{2S}{}^1$, (cf. p. 87), and hence we could not decide which species, by its formation, caused the decrease in light-scattering intensity.

At constant F-actin concentration, changes in myosin concentration did not affect the rate of decrease in light-scattering intensity, but the amount of ATP necessary to induce the maximum decrease increased proportionately to the amount of myosin[60] (compare Fig. 11A with Fig. 11B). Hence

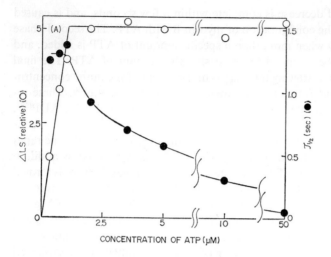

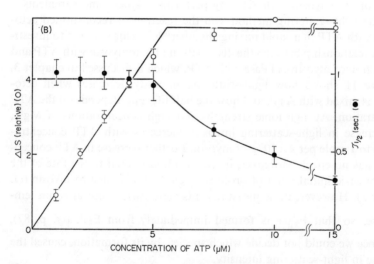

Fig. 11. Dependence of decrease in light-scattering intensity (ΔLS) and of its half-life ($\tau_{1/2}$) on ATP concentration in the presence of a high concentration of Mg^{2+}. A: 0.4 mg/ml myosin, 0.2mg/ml F-actin. B: 2mg/ml myosin, 0.2mg/ml F-actin. 2M KCl, 2mM $MgCl_2$, pH 8.2, 22°C. ○, decrease in light-scattering intensity; ●, half-life for the decrease in light-scattering intensity.

it was concluded that ATP reacts with myosin, that the rate-determining step in the dissociation of actomyosin by ATP involves the formation of E_{2S}^1 or $E_2^1{}_P^{ADP}$ and that the rate of formation of E_{2S}^1 or $E_2^1{}_P^{ADP}$ in the actomyosin-ATP system is hardly affected by F-actin.

At high ionic strength and high Mg^{2+} concentration, the myosin contained in actomyosin is rapidly transformed to $E_2^1{}_{\cdots P}^{\cdots ADP}$ after its initial conversion to $E_2^1{}_{\diagdown P}^{\cdots ADP}$ (chapter 3). Therefore, experiments such as those above performed at room temperature, high ionic strength and high concentrations of Mg^{2+}, cannot distinguish which of the species E_{2S}^1, $E_2^1{}_{\diagdown P}^{\cdots ADP}$ or $E_2^1{}_{\cdots P}^{\cdots ADP}$ myosin changes into upon the dissociation of actomyosin. However, at low Mg^{2+} concentrations, the 'extra-burst' of liberation of P_i and protons is observed in the reaction of myosin with ATP (chapter 3, section 8). Under these conditions, the amount of ATP necessary to attain the maximum decrease in light-scattering intensity was more than the 1 mole per 4×10^5 g of myosin which corresponds to the extra-burst

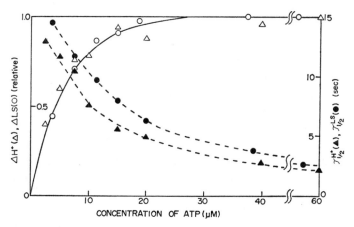

Fig. 12. Dependence of decrease in light-scattering intensity and of its half-life upon ATP concentration for reconstituted actomyosin at low Mg^{2+} concentration. 0.4mg/ml myosin, 0.2mg/ml F-actin, 0.6M KCl, 5µM $MgCl_2$, pH 8.2, 22°C. ○, decrease in light-scattering intensity (ΔLS); ●, half-life for the decrease in light-scattering ($\tau_{1/2}{}^{LS}$); △, the amount of the 'extra-burst' of protons liberated in the myosin-ATP system under the same condition (ΔH^+); ▲, its half-life ($\tau_{1/2}{}^{H^+}$).

(Fig. 12). The rate of decrease in light-scattering intensity also agrees with the rate of liberation of P_i and H^+ [61]. According to the mechanism shown in chapter 3, section 8, these results indicate that the reaction path responsible for the extra-burst—the direct decomposition of $E_2^1 \cdots_P^{ADP}$ myosin, ADP and P_i— is independent of the decrease in light-scattering intensity, and that the light-scattering intensity does not decrease significantly until most of the myosin has been converted to $E_2^1 \cdots_P^{ADP}$ by the added ATP. $\tau^1/_2$ for the extra-burst of protons in Fig. 12 is slightly shorter than $\tau^1/_2$ for the reduction in light-scattering intensity. This result is consistent with the mechanism that the dissociation of actomyosin on adding ATP at a low concentration occurs after the formation of $E_2^1 \cdots_P^{ADP}$.

The important problem is the relationship between the mechanism of dissociation of actomyosin by ATP and that by PP_i, which has been treated in detail in the preceding section. The 'clearing response,' which is thought to be derived from dissociation of actomyosin by ATP at low ionic strengths (chapter 7), occurs only at high concentrations of ATP, and at low temperatures the concentration of ATP required for the 'clearing response' becomes low. As discussed in chapter 3, section 6, at low concentrations of ATP and room temperatures, the rate of conversion of E_2^{1S} to $E_2^1 \cdots_P^{ADP}$ is much higher than that of conversion of E_2^{1S} to E_{2S}^1, and the dissociation of actomyosin is induced by the formation of $E_2^1 \cdots_P^{ADP}$. However, at high concentrations of ATP the conversion of E_2^{1S} to E_{2S}^1 is accelerated, and especially at low temperatures the concentration of E_{2S}^1 increases. Thus, there are possibilities that at high concentrations of ATP actomyosin dissociates on the formation of E_{2S}^1, as in the case of actomyosin-PP_i system, and that this type of dissociation corresponds to the 'clearing response' of actomyosin at low ionic strengths, as will be mentioned in chapter 7. Furthermore, to explain the substrate inhibition of actomyosin-ATPase by this mechanism, we must assume that at high concentrations of ATP the conversion of E_{2S}^1 to $E_2^1{}_P^{ADP}$ is inhibited by the binding of ATP to a regulatory site of myosin ($E_{2S}^1 + S \rightleftharpoons {}^S E_{2S}^1$), and this assumption is consistent with the observation made by Bárány [61a] that myosin in living muscle at relaxed state contains bound ATP but not ADP.

The rate of recovery of light-scattering intensity was measured under conditions of high ionic strength, high Mg^{2+} concentration and room temperature, when only $E_2^1{:::}_{\cdot\cdot P}^{ADP}$ is formed upon the addition of 1 mole of ATP per mole of myosin contained in actomyosin. The rate constant was about 1 min^{-1}, about five times that for the complete decomposition of $E_2^1{:::}_{\cdot\cdot P}^{ADP}$ into myosin which has returned to the original conformation, ADP and P_i[60]. However, it agrees quite well with the rate constant for the liberation of ADP and P_i from $E_2^1{:::}_{\cdot\cdot P}^{ADP}$ to form °E_2^1[62]. The rate of binding of myosin with F-actin was considerably faster than the above reactions under these conditions, and the rate constant for the recovery of light-scattering intensity was independent of the F-actin concentration[54]. These results suggest that the liberation of ADP and P_i from $E_2^1{:::}_{\cdot\cdot P}^{ADP}$ is rate-determining for the recombination of myosin with F-actin which occurs in the recovery phase.

During the second phase, actomyosin shows the activity of myosin-type ATPase at steady-state. Consequently, this stage is thought to involve the step:

$$E_2^1{:::}_{\cdot\cdot P}^{ADP} + S + H_2O \rightleftharpoons E_2^{1S}{:::}_{\cdot\cdot P}^{ADP} + H_2O \rightarrow E_2^1{:::}_{\cdot\cdot P}^{ADP} + ADP + P_i.$$

Based on these results, the following mechanism involving the dissociation of actomyosin to F-actin and myosin can be proposed for the reaction of ATP with reconstituted actomyosin:

$$\begin{array}{c} F\text{-}A\text{-}E_2^1 + S \rightleftharpoons F\text{-}A + E_2^{1S} \xrightarrow{\text{(simple hydrolysis)}} F\text{-}A\text{-}E_2^1 + ADP + P_i \\ \updownarrow \\ F\text{-}A\text{-}E_{2S}^1 \\ \updownarrow \\ F\text{-}A\text{-}E_2^1{:::}_{\smallsetminus P}^{\cdot ADP} \longrightarrow F\text{-}A + E_2^1{:::}_{\cdot\cdot P}^{\cdot ADP} \longrightarrow F\text{-}A\text{-}°E_2^1 + ADP + P_i. \end{array}$$

(dissociation of actomyosin)

Reconstituted actomyosin dissociates into F-actin and myosin when myosin is converted to $E_2^1{:::}_{\cdot\cdot P}^{ADP}$, and recombination occurs when the ADP and P_i are liberated from $E_2^1{\cdots}_{\cdot\cdot P}^{ADP}$. At low ionic strength, when F-actin binds strongly to myosin, or in low concentrations of Mg^{2+} even at high ionic strengths, the initial rapid reaction involves mainly the direct decomposition of $E_2^1{\cdots}_{\cdot\cdot P}^{ADP}$ into myosin, ADP and P_i. The detailed mechanism of the decomposition of ATP by myosin in the presence of F-actin will be discussed in the next chapter.

7. *A Molecular Model for the Dissociation of Actomyosin, Acto-H-meromyosin and Acto-S-1 Complexes by ATP*

Before discussion of the model for the dissociation of actomyosin by ATP, we will briefly mention our results[23] on the dissociation of acto-H-meromyosin and acto-S-1 complexes by ATP. As described in section 2, one H-meromyosin and S-1 molecule combines with actin dimer and monomer respectively. As shown in Fig. 13A, the amount of ATP necessary to cause maximum dissociation of acto-H-meromyosin was 2.08 moles per mole of H-meromyosin. On the other hand, the acto-S-1 complex dissociated upon addition of 1 mole of ATP per mole of S-1 (Fig. 13B). Furthermore, the amount of the stoichiometric burst of P_i-liberation by H-meromyosin and S-1 was 1 and 0.5 mole per mole of protein respectively, as described in chapter 3, section 6 and chapter 4, section 2. These results are summarised in Table IV.

S-1 is composed of two species, S-1A and S-1B, both of which complex with actin, but only the second contains an active site for the ATPase (chapter 2). Therefore in acto-H-meromyosin and the acto-S-1 complex, the bonds to actin are cleaved by the formation of $E_2^1{\cdots}_{\cdot\cdot P}^{ADP}$ by the site in S-1B, and by simple binding of ATP to the site in S-1A. This results in the complete dissociation of H-meromyosin or S-1 from F-actin (Fig. 14).

We suggest the scheme shown in Fig. 14 for the reaction of ATP with actomyosin[23,63], although it is difficult to construct a completely satisfactory model. When ATP binds to myosin in actomyosin to form $E_2^1{\cdots}_{\cdot\cdot P}^{ADP}$, the bond between actin and S-1B, which is the stronger of the two bonds between actin and the S-1 of myosin, is cleaved, since S-1B contains the ATPase active site (section 6). The other bond—between actin and S-1A— still remains uncleaved, probably because of interactions between the two

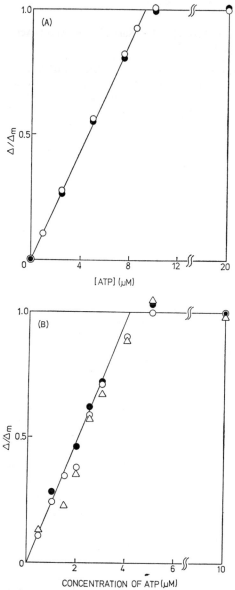

Fig. 13. Decrease in light-scattering intensity of acto-H-MM and the acto-S-1 complex on adding various concentrations of ATP. 1mM $MgCl_2$, 20mM Tris-HCl at pH 7.6, 21–22″C Δ/Δ_m is the relative value of decrease in light-scattering. A: 1.36mg/ml H-MM, 0.368mg/ml F-actin. ○, 0.5M KCl; ●, 0.1M KCl. B: 0.5mg/ml S-1, 0.192mg/ml F-actin. ○, 50mM KCl; ●, 0.1M KCl; △, 0.5M KCl.

TABLE IV. Formation of Complexes of Myosin and Its Subfragments with F-Actin, and Their Dissociation upon Adding ATP

Protein	Binding ratio with F-actin (mole of actin monomer/ mole of protein)	Amount of initial burst (mole/mole of protein)	Amount of ATP necessary for dissociation (mole/mole of protein)
Myosin	2	1	1
H-MM	2	1	2
S-1	1	0.5	1

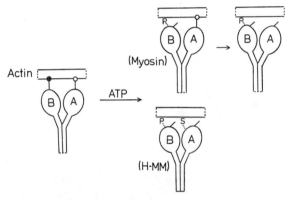

Fig. 14. Model for the reaction between H-MM, F-actin and ATP(S). ···P, protein-phosphate-ADP complex; ···S, simple binding of ATP (see text for explanation).

globular heads S-1A and S-1B in the myosin molecule, so that the other myosin binding site for ATP on S-1A is blocked (sections 4 and 6). Especially in the case of reconstituted actomyosin in which the binding between actin and myosin is relatively weak, dissociation of the actomyosin into F-actin and myosin can now take place (even if ATP does not bind to S-1A) when the equilibrium is shifted by ultracentrifugation[40], or when the concentration of actomyosin is extremely low, as in the light-scattering measurements[17].

It is not yet known how this complicated reaction, which we have studied in 0.6M KCl, is modified under the physiological conditions in which myosin forms filaments. The detailed role of the reaction in the mechanism of muscle contraction remains to be elucidated.

THE FORMATION OF ACTOMYOSIN AND ITS DISSOCIATION

REFERENCES

1 I. Banga and A. Szent-Györgyi, *Stud. Inst. Med. Chem. Univ. Szeged*, **1**, 5 (1941).
2 F. B. Straub, *Stud. Inst. Med. Chem. Univ. Szeged*, **2**, 3 (1942).
3 W. F. H. M. Mommaerts, *J. Gen. Physiol.*, **31**, 361 (1948).
4 M. Dainty, A. Kleinzeller, A. S. C. Lawrence, M. Miall, J. Needham, D. M. Needham and S. C. Shen, *J. Gen. Physiol.*, **27**, 355 (1944).
5 H. Portzehl, G. Schramm and H. H. Weber, *Z. Naturforsch.*, **5b**, 61 (1950).
6 A. Szent-Györgyi, "Chemistry of Muscular Contraction," 1st ed. (1947) & 2nd ed. (1951), Academic Press, New York.
7 Y. Tonomura and H. Matsumiya, *Biochim. Biophys. Acta*, **27** 311 (1958).
8 H. Noda and K. Maruyama, *Biochim. Biophys. Acta*, **30**, 598 (1958).
9 Y. Tonomura, S. Watanabe and K. Yagi, *J. Biochem.*, **40**, 27 (1953).
10 J. J. Blum and M. F. Morales, *Arch. Biochem. Biophys.*, **43**, 208 (1953).
11 W. F. H. M. Mommaerts and J. Hanson, *J. Gen. Physiol.*, **39**, 831 (1956).
12 J. Gergely, *J. Biol. Chem.*, **220**, 917 (1956).
13 M. F. Gellert, P. H. von Hippel, H. K. Schachman and M. F. Morales, *J. Amer. Chem. Soc.*, **81**, 1384 (1959).
14 H. E. Huxley, *J. Biophys. Biochem. Cytol.*, **3**, 631 (1957).
15 M. Kawamura and K. Maruyama, *J. Biochem.*, **66**, 619 (1969).
16 S. S. Spicer and J. Gergely, *J. Biol. Chem.*, **188**, 179 (1951).
17 J. Gergely and H. Kohler, in "Conf. Chem. Muscular Contraction," Igaku Shoin, Tokyo, p. 14 (1957).
18 Y. Tonomura, S. Tokura and K. Sekiya, *J. Biol. Chem.*, **237**, 1074 (1962).
19 A. G. Szent-Györgyi, *J. Biol. Chem.*, **192**, 361 (1951).
20 L. B. Nanninga, *Biochim. Biophys. Acta*, **82**, 507 (1964).
21 K. Sekiya, K. Takeuchi and Y. Tonomura, *J. Biochem.*, **61**, 567 (1967).
22 M. Young, *Proc. Natl. Acad. Sci. U. S.*, **58**, 2393 (1967).
23 K. Takeuchi and Y. Tonomura, *J. Biochem.*, **70**, 1011 (1971).
24 T. Tokiwa, *Biochem. Biophys. Res. Commun.*, **44**, 471 (1971).
25 H. E. Huxley, *J. Mol. Biol.*, **7**, 281 (1963).
26 K. Tawada, *Biochim. Biophys. Acta*, **172**, 311 (1969).
27 A. A. Rizzino, W. W. Barouch, E. Eisenberg and C. Moos, *Biochemistry*, **9**, 2402 (1970).
27a E. Eisenberg, L. Dobkin and W. W. Kielley, *Federation Proc.*, **30**, 1310 Abs (1971).
28 P. B. Moore, H. E. Huxley and D. J. DeRosier, *J. Mol. Biol.*, **50**, 279 (1970).
29 D. J. De Rosier and A. Klug, *Nature*, **217**, 130 (1968).
30 K. Laki, S. S. Spicer and W. R. Carroll, *Nature*, **169**, 328 (1952).
31 Y. Tonomura, S. Watanabe and M. Morales, *Biochemistry*, **8**, 2171 (1969).
32 S. Glasstone, K. J. Laidler and H. Eyring, "The Theory of Rate Processes," McGraw-Hill Book Company, New York (1941).

33 W. Kauzmann, *Advan. Protein Chem.*, **14**, 1 (1959).
34 Y. Tonomura, K. Sekiya and K. Imamura, *Biochim. Biophys. Acta*, **78**, 690 (1963).
35 S. Fujime and S. Ishiwata, *J. Mol. Biol.*, **62**, 251 (1971).
36 G. Kuschinsky and F. Turba, *Biochim. Biophys. Acta*, **6**, 426 (1951).
37 A. Stracher, *J. Biol. Chem.*, **239**, 1118 (1964).
38 M. Bárány and K. Bárány, *Biochim. Biophys. Acta*, **35**, 293 (1959).
38a F. Lipmann, *Advan. Enzymol.*, **1**, 99 (1941).
39 H. H. Weber and H. Portzehl, *Advan. Protein Chem.*, **7**, 161 (1952).
40 A. Weber, *Biochim. Biophys. Acta*, **19**, 345 (1956).
41 Y. Tonomura, *J. Res. Inst. Catalysis, Hokkaido Univ.*, **4**, 87 (1956).
42 Y. Tonomura and F. Morita, *J. Res. Inst. Catalysis, Hokkaido Univ.*, **7**, 126 (1959).
43 Y. Tonomura, *J. Res. Inst. Catalysis, Hokkaido Univ.*, **16**, 323 (1968).
44 W. K. Jordan and G. Oster, *Science*, **108**, 188 (1948).
45 T. Nihei and Y. Tonomura, *J. Biochem.*, **46**, 1355 (1959).
46 H. Matsumiya and M. Tanisaki, *J. Biochem.*, **45**, 333 (1958).
47 Y. Tonomura and F. Morita, *J. Biochem.*, **46**, 1367 (1959).
48 K. M. Nauss, S. Kitagawa and J. Gergely, *J. Biol. Chem.*, **244**, 755 (1969).
49 F. Morita and Y. Tonomura, *J. Amer. Chem. Soc.*, **82**, 5172 (1960).
50 Y. Tonomura, F. Morita and K. Yagi, *J. Phys. Chem.*, **61**, 605 (1957).
51 F. Karush, *J. Amer. Chem. Soc.*, **72**, 2705 (1950).
52 K. U. Linderstrøm-Lang and J. A. Schellman, *in* "The Enzymes,"ed. by P. D. Boyer, H. Lardy and K. Myrbäck, Academic Press, New York, Vol. 1, p. 443 (1959).
53 R. Lumry and H. Eyring, *J. Phys. Chem.*, **58**, 110 (1954).
54 M. F. Morales and J. Botts, *Arch. Biochem. Biophys.*, **37**, 283 (1952).
55 P. M. Gallop, C. Franzblau and E. Meilman, *Biochim. Biophys. Acta*, **24**, 644 (1957).
56 Y. Tonomura and S. Watanabe, *Nature*, **169**, 112 (1952).
57 W. F. H. M. Mommaerts, *J. Gen. Physiol.*, **39**, 821 (1956).
58 W. F. H. M. Mommaerts, *in* "Enzymes: Units of Biological Structure and Function," ed. by O. H. Gaebler, Academic Press, New York, p. 317 (1956).
59 Y. Tonomura, H. Matsumiya and S. Kitagawa, *Biochim. Biophys. Acta*, **24**, 568 (1957).
60 H. Onishi, H. Nakamura and Y. Tonomura, *J. Biochem.*, **64**, 769 (1968).
61 N. Kinoshita, T. Kanazawa, H. Onishi and Y. Tonomura, *J. Biochem.*, **65**, 567 (1969).
61a M. Bárány, personal communication.
62 N. Kinoshita, S. Kubo, H. Onishi and Y. Tonomura, *J. Biochem.*, **65**, 285 (1969).
63 Y. Hayashi and Y. Tonomura, *J. Biochem.*, **68**, 665 (1970).

7
THE SUPERPRECIPITATION OF ACTOMYOSIN WITH ATP, AND THE ACTOMYOSIN-ATPase SYSTEM*

In the previous chapter we described the reaction of actomyosin with ATP under conditions of high ionic strength, when it is easy to make physicochemical measurements. We shall now discuss the reaction occurring at low ionic strengths, when actomyosin is not soluble but forms a suspension which, upon addition of a low concentration of Mg-ATP, suddenly becomes extremely turbid and yields a small precipitate. This phenomenon was called 'superprecipitation' by Szent-Györgyi, to distinguish it from the simple precipitation which occurs upon changing the KCl concentration. Under conditions in which superprecipitation occurs, actomyosin shows a much higher ATPase activity than myosin, and myosin-ATPase activated with actin is called actomyosin-ATPase. As a result of the studies of Szent-Györgyi[1] and Weber and Portzehl[2], there is now considerable evidence that actomyosin-ATPase is directly coupled with muscle contraction.

* Contributors: Hirofumi Onishi and Takamichi Shimada

Hence an understanding of its mechanism would greatly facilitate the elucidation of the molecular mechanism of muscle contraction.

First we shall discuss the mechanism of the actomyosin-ATPase in relation to the phosphorylation of myosin by ATP, and then describe the physiological significance of superprecipitation. Next we shall discuss the effects of factors such as temperature and KCl and Mg-ATP concentrations, which control the colloidal properties and ATPase activity of actomyosin[3,4,4a]. In living organisms muscle contraction and relaxation is regulated by a complex mechanism involving a protein factor (the regulatory protein factor) and Ca^{2+} (chapter 10). However, the ATPase activity of actomyosin and its colloidal properties are affected not only by the regulatory protein and Ca^{2+}, but also by the KCl and Mg-ATP concentrations and temperature. Lastly we will consider the function of the myosin-actin-ATP system in cell motility in general.

1. Actomyosin-ATPase

As described in the previous chapter, the addition of ATP to actomyosin at high ionic strength partially cleaves the bonding between actin and myosin, and F-actin does not influence the myosin-ATPase. In contrast, at low ionic strength F-actin considerably increases the ATPase activity of myosin.

The effect of divalent cations on the ATPase activity of actomyosin was shown previously (chapter 3, Fig. 2). The curve relating ATPase activity to the cationic radius is bell shaped, with a maximum at 0.95Å, not only for myosin-ATPase in 0.6M KCl, but also for actomyosin-ATPase in 0.075M KCl. However, the ATPase activity declines more slowly at lower than at higher ionic strengths[5].* More recently, Schaub and Ermini[6] measured the stimulation of the ATPase activity of actomyosin by various divalent cations under better defined conditions using desensitised actomyosin, *i.e.*, actomyosin free from the regulatory protein (chapter 10, section 4), and found that the ATPase is almost independent of ionic radius in the range 0.65–1.0Å. In any case, myosin-ATPase is markedly activated by Ca^{2+} and largely inactivated by Mg^{2+}, but at low ionic strengths actomyosin-ATPase shows high activity even in the presence of Mg^{2+}[7,8].

The decomposition of ATP by actomyosin-ATPase is directly coupled to muscle contraction[1,2,9]. The contraction of muscle models by ATP re-

* The system used for these measurements was contaminated with a minute amount of Mg^{2+}.

TABLE I. Comparison of the Liberation of the Initial Burst of P_i from the Reaction of Myosin with ATP with the Contraction of a Muscle Model and the Actomyosin-ATPase

	Myosin-ATPase at steady-state	Initial burst of P_i	Contraction of model muscle actomyosin-ATPase
PCMB	Activation at low concentrations Inhibition at high concentrations	Inhibited	Inhibited
EDTA	Activation at high ionic strength Inhibition at low ionic strength	Inhibited	Inhibited
Mg^{2+}	Inactivation	Necessary	Necessary
9-(2'-Hydroxyethyl) -6-aminopurine 2'-TP	Hydrolysis	None	Does not contract muscle model
p-Nitrothiophenylation of myosin	No change	Disappears	Actomyosin-ATPase and superprecipitation suppressed
pH-Activity curve	A maximum and minimum at pH 6.0 and 7.5, respectively	Increases with increasing pH. The activity is 1/2 the maximum value at pH 6.4.	Increases with increasing pH. The activity is 1/2 the maximum value at pH 6.1.
$\Delta H^{\ddagger}$ (kcal/mole)	12.6	23.7	30.1 (AM-ATPase)
$\Delta S^{\ddagger}$ (cal/mole/degree)	−24.8	+21.6	+45.5 (AM-ATPase)

TABLE II. Comparison of Conditions Required for Optimal ^{18}O-Exchange and Optimal Contraction

Condition	Optimal ^{18}O-exchange		Contraction (optimal tension or shortening)	
	Results	Protein	Actomyosin (myosin B)	
1. Varying substrate	ATP>CTP>ITP~UTP>GTP>TP	H-MM, myosin, actomyosin	ATP>CTP>UTP>ITP>GTP>(TP)	
2. pH-Dependence	None from pH 5.5 to 9.3 Same at pH 7.3 and 9.0	H-MM Myosin	~pH 7.6	
3. Divalent metal ions	Mn^{2+}>Co^{2+}>Mg^{2+}>Ni^{2+} Mn^{2+}>Mg^{2+}>Co^{2+}>Ni^{2+}	H-MM Myosin	Mg^{2+}~Mn^{2+}~Co^{2+}	
4. Mg^{2+} concentration	>2.5mM	Myosin	>1mM	
5. KCl concentration	<0.3M	Myosin	0.05 to 0.12M	

(Yount and Koshland[26])

quires the presence of Mg^{2+} and is hindered by EDTA and SH reagents such as PCMB, which substantially inhibit the actomyosin-ATPase activity[10-14]. The degree of contraction of a muscle model is proportional to the amount of ATP decomposed, and the tension developed is proportional to the rate of decomposition[15-17] (chapter 8).

We believe that the phosphorylation of myosin by ATP is basic to muscle contraction[18-22]. There is substantial evidence that the initial burst of P_i observed immediately after the addition of ATP to myosin (chapter 3) is closely related to the contraction of muscle models and the superprecipitation and ATPase activity of actomyosin (Table I). Firstly, the initial burst of P_i is suppressed by EDTA and SH reagents such as PCMB and oxarsan[23,24]. Secondly, the initial burst is eliminated by thorough dialysis of myosin with KCl solution, and largely restored by the addition of Mg^{2+}. There is also some recovery with Mn^{2+}, but none with Ca^{2+}[24]. Likewise, the contraction of muscle models by ATP also requires Mg^{2+} and unmodified SH groups. Thirdly, analogues of ATP, which can dissociate actomyosin in 0.6M KCl and which are decomposed by myosin B at various ionic strengths at rates similar to ATP, still can neither induce muscle contraction (chapter 9) nor liberate an initial burst of P_i with myosin[25]. As described in p. 108, Koshland et al. found that ^{18}O-exchange occurs in the reaction of myosin with ATP, and this may be due to the formation of phosphoryl myosin. The conditions for ^{18}O-exchange agree well with those for muscle contraction[26] (Table II).

As has been repeatedly suggested in chapter 3, the initial burst occurs after the myosin has been phosphorylated by ATP. Hence it is postulated that ATP is hydrolysed by actomyosin-ATPase mainly via phosphoryl myosin[19,20,27]. In order to confirm this hypothesis, the following four experiments were performed.

(1) The rate of actomyosin-ATPase at low ionic strength in the presence of Mg^{2+} was compared with that of the initial burst of P_i-liberation of myosin-ATPase[28]. As shown in Fig. 1, they agreed with each other at least in the presence of low concentrations of ATP. Furthermore, the ATPase activity of actomyosin at high ATP concentrations was much higher than expected from the Michaelis-Menten relation at low ATP concentrations. Kominz[29] has reported that V_{max} and K_m for myofibrillar ATPase at high ATP concentrations are, respectively, 4-5 and 20 times higher than at low ATP concentrations. As already described in chapter 3, section 6, the rate of the initial burst of P_i-liberation also shows a similar dependence on high concentrations of ATP.

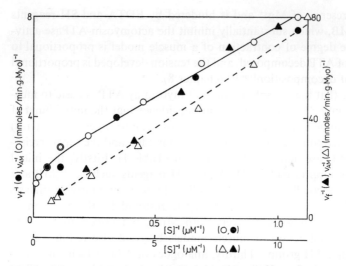

Fig. 1. Lineweaver-Burk plots of the rate of the actomyosin-ATPase reaction, v_{AM}, and the rate of the intial burst of P_i-liberation from myosin-ATPase, v_f. 50mM KCl, 2mM $MgCl_2$, pH 7.0, 25°C. Myosin was used after treatment with PCMB and β-mercaptoethanol to strengthen its bonding to F-actin. ○, △, v_{AM}; ●, ▲, v_f.

(2) The rate of liberation of protons, upon adding ATP in an approximately equimolar amount to the myosin contained in actomyosin, was measured at low ionic strength in the presence of Mg^{2+} and found to be very rapid[30]. Since the protons are mainly derived from the decomposition of a myosin-phosphate-ADP complex and the return of the conformation of myosin to the original state (chapter 3), F-actin must accelerate considerably the decomposition of $E_2^1 \cdots _{ADP}^P$ and the return of the changed conformation of myosin (°$E_2^1 \rightarrow E_2^1$) compared with the myosin-ATP system.

(3) Treatment of myosin with NTP under conditions where the initial stoichiometric burst of P_i-liberation can occur did not affect the steady-state rate of reaction with ATP, but completely inhibited the initial burst (chapter 3). Acto-NTP (1)-myosin reconstituted from this NTP (1)-myosin shows only myosin-ATPase activity at low ionic strength and in the presence of Mg^{2+} (Table III). In the presence of EDTA, which prevents the formation of phosphoryl myosin, the ATPase activity of the control, untreated actomyosin was equal to that of acto-NTP (1)-myosin (Fig. 2A).

TABLE III. Effect of p-Nitrothiophenylation of Myosin

Myosin	p-Nitrothiophenylation		ATPase of myosin			Actomyosin	
	Condition	Bound NTP (mole/mole of myosin)	Initial stoichiometric burst	Initial extra-burst	Steady-state	Super-pptn.	ATPase
Control myosin	In absence of ATP	0					
NTP(1)-myosin	High MgCl$_2$, ATP and KCl (condition for initial stoichiometric burst)	1	+	+	++	+	+
			−		++	−	−
NTP(2)-myosin	Low MgCl$_2$ and ATP, high KCl, room temp. (condition for initial extra-burst)	2	+	−	+	+ (actin-myosin binding weak)	− (actin-myosin binding very strong)

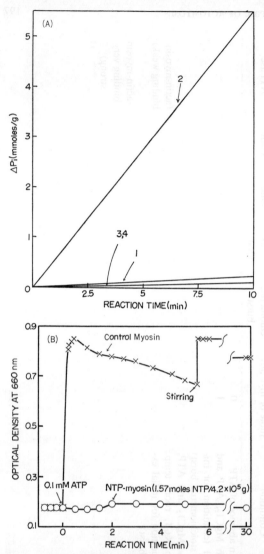

Fig. 2. Effect of *p*-nitrothiophenylation (1) on the ATPase activity and superprecipitation of actomyosin. Actomyosin was obtained by mixing NTP (1)-myosin (1.57 moles of NTP/4×10^5g myosin) or control myosin with F-actin in a weight ratio of 3:1. A: ATPase activity of actomyosin. 0.01M KCl, 0.2mM ATP, pH 7.0, 25°C. In the presence of 2mM $MgCl_2$: 1, acto-NTP (1)-myosin; 2, control actomyosin. In the presence of 1mM EDTA: 3, acto-NTP (1)-myosin; 4, control-actomyosin. B: Superprecipitation of actomyosin with ATP. 0.5mg/ml actomyosin, 0.06M KCl, 4mM $MgCl_2$, 0.1mM ATP, 5mM Tris-maleate at pH 7.0, 20°C. ×, control actomyosin; ○, acto-NTP (1)-myosin. The first and second arrows indicate addition of ATP and stirring of the suspension, respectively.

Superprecipitation, as measured by turbidity, was completely suppressed by *p*-nitrothiophenylation (1) of myosin (Fig. 2B)[31,32].*

(4) The rate of hydrolysis of ATP by actomyosin and the rate of the initial burst of P_i-liberation increased with increasing pH, and were at half their maxima near neutral pH (Table I). Both reactions had positive activation entropies[35]. Chapter 3, section 4 described the complicated pH-activity curve, which has a maximum at pH 6.0 and a minimum at pH

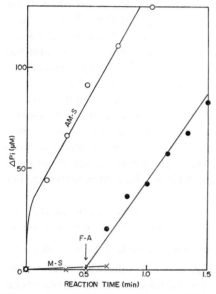

Fig. 3. Rate of liberation of P_i when ATP is added to actomyosin, and that when F-actin is added later to the myosin-ATP system. 0.05M KCl, 2mM $MgCl_2$, 10μM $CaCl_2$, 20mM Tris-maleate at pH 7.0, 23°C. ○, addition of 0.5mM ATP (S) to actomyosin (AM) reconstituted from 0.3mg/ml of moysin and 0.15mg/ml F-actin. ×, addition of 0.5mM ATP to 0.3mg/ml myosin (M). ●, addition of 0.5mM ATP to 0.3mg/ml of myosin, followed after 30 sec by the addition of 0.15mg/ml of F-actin (F-A).

---

* Yazawa and Yagi[33] reported the absence of the intitial burst with Nagarse-treated S-1 (S-1N), although there was a low actomyosin-ATPase activity. However, more recent studies by themselves[34] showed that the amount of the initial burst of S-1N is not zero and about 20% of that of the usual S-1, thus supporting the mechanism that ATP is hydrolysed by actomyosin-ATPase mainly *via* phosphoryl myosin.

7.5, of myosin-ATPase at steady-state. In this case, the activation entropy was negative[35] (Table I). These results indicate that the formation of phosphoryl myosin and a myosin-phosphate-ADP complex is rate-determining in the reaction of actomyosin with ATP.

2. Mechanism of the Reaction of F-Actin with Myosin-ATPase

Immediately after ATP has been added to actomyosin at low ionic strengths, the ATPase activity is several times greater than it is at steady-state (Fig. 3)[32]. The rapid initial burst of P_i liberated by myosin lasts for only a few seconds (chapter 3), but that generated from ATP by actomyosin may continue for several tens of seconds, during which 10–100 moles of P_i per mole of myosin may be produced. This phenomenon was initially discovered in myofibrillar ATPase by Weber and Hasselbach[36], and was later studied by Bendall[37], Bowen et al.[38] and ourselves[20]. However, the molecular mechanism has received little attention. We recently measured the rate of liberation of P_i upon the addition of F-actin after myosin and ATP had first been mixed for a suitable time. Figure 3 clearly shows that as soon as F-actin is added, the ATPase activity approximates that of actomyosin at steady-state. There was neither initial rapid liberation of P_i, as is observed when ATP is added to actomyosin, nor any lag phase[32].

These results suggest that at low ionic strengths F-actin reacts with myosin-ATP in two ways. One is to accelerate the decomposition of $E_2^1 \cdots {}^{ADP}_P$ and the return of the changed conformation, ${}^oE_2^1 \rightarrow E_2^1$ (see the reaction scheme shown in chapter 3, Fig. 7, p. 82); otherwise, after myosin is mixed with ATP, addition of F-actin should cause a lag in the liberation of P_i because the cycle is initiated only after the decomposition of the intermediates of myosin-ATPase and the return of the changed conformation to that to the original free enzymes. The other way is to bind strongly with myosin, which suppresses the conversion of $E_2^1 \cdots {}^{ADP}_P$ to $E_2^1 \cdots {}^{ADP}_P$ (see p. 204, step 5; Fig. 7 in chapter 3); this can be concluded from the dissociation of actomyosin upon the formation of $E_2^1 \cdots {}^{ADP}_P$ (chapter 6). Therefore, in the presence of F-actin at low ionic strengths, ATP is hydrolysed by the direct decomposition of $E_2^1 \cdots {}^{ADP}_P$ into myosin (E_2^1), ADP and P_i (step 8 of p. 204). When actomyosin is mixed with ATP, there is a rapid initial decomposition of ATP mainly *via* this direct hydrolysis of $E_2^1{\cdots}^{ADP}_{P}$ until the actomyosin is partially dissociated and the ATPase gradually attains a steady-state value. However, if the ATP is mixed with myosin first, most of the myosin is converted into $E_2^1{\cdots}^{ADP}_{P}$ and $°E_2^1$, which are restored rapidly to E_2^1 upon addition of F-actin, so that the steady-state ATPase rate appears immediately.

Dissociation of actomyosin into myosin and F-actin occurs only after myosin changes from $E_2^1{\cdots}^{ADP}_{P}$ to $E_2^1{\cdots}^{ADP}_{P}$ (chapter 6, section 6) and this transformation (step 5) is hindered by the binding of F-actin. As described above, one must assume that F-actin accelerates the decomposition of $E_2^1{\cdots}^{ADP}_{P}$ and the return of $°E_2^1$ to E_2^1. Therefore, variations in the strength of binding of F-actin to myosin under different experimental conditions should result in different contributions to the actomyosin-ATPase activity from the two routes of hydrolysis of ATP which proceed *via* phosphoryl myosin (chapter 3, Fig. 7, p. 82).

The *p*-nitrothiophenylation of myosin at room temperature under conditions allowing liberation of the extra-burst of P_i from ATP gives NTP (2)-myosin, which has lost only the extra-burst at low Mg^{2+} concentrations without alteration of the steady-state ATPase activity or of the amount of the initial stoichiometric burst of P_i (Table III)[39] (see chapter 3, section 8). This indicates that *p*-nitrothiophenylation (2) inhibits only the direct hydrolysis of phosphoryl myosin. Acto-NTP (2)-myosin reconstituted from NTP (2)-myosin is only slightly suppressed under conditions of relatively weak binding between myosin and actin, but under conditions of very strong binding, the ATPase activity is substantially suppressed (Table III, Fig. 4)[39]. This result supports the above conclusion that when the binding between actin and myosin is weak, ATP is decomposed largely *via* $E_2^1{\cdots}^{ADP}_{P}$ and $°E_2^1$, while when the strength of binding is increased this route is inhibited and decomposition is effected through the direct hydrolysis of phosphoryl myosin.

The ATPase activity of actomyosin at low ionic strength is maximum at an Mg^{2+} concentration of 10^{-5} M, when the rate of superprecipitation is very high[1,40]. Increasing the Mg^{2+} concentration above this value grad-

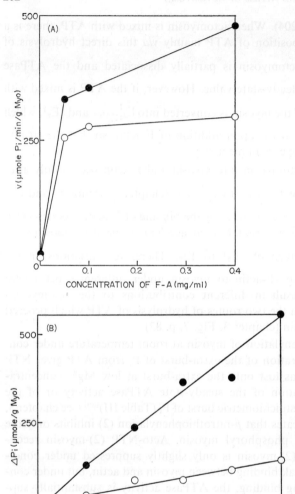

Fig. 4. Effect of *p*-nitrothiophenylation (2) on actomyosin-ATPase. NTP (2)-myosin was prepared by the procedure described in Ref. *32* (about 3.1 moles NTP/4×10^5g myosin). A: Condition under which there is weak bonding between actin and myosin. 0.2mg/ml myosin, 0.4mM ATP, 0.05M KCl, 10μM MgCl$_2$, 10μM CaCl$_2$, pH 7.0, 25°C. ●, control actomyosin; ○, acto-NTP (2)-myosin. B: Condition under which bonding between actin and myosin is strong. 0.07mg/ml myosin, 0.14mg/ml F-actin, 0.1mM ATP, 0.01M KCl, 10μM MgCl$_2$, 2μM CaCl$_2$, pH 7.0, 25°C. ●, control actomyosin; ○, acto-NTP (2)-myosin.

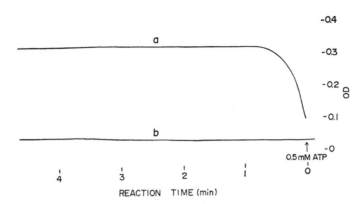

Fig. 5. Effect of *p*-nitrothiophenylation (2) of myosin on the superprecipitation of actomyosin with ATP. The NTP (2)-myosin is that used in Fig. 4. 0.2mg/ml myosin, 0.1mg/ml F-actin, 0.05M KCl, 10μM MgCl$_2$, 10μM CaCl$_2$, 20mM Tris-maleate at pH 7.0, 21°C. The rate of superprecipitation was followed by the increase in OD at 660nm after the addition of 0.5mM ATP (denoted by ↑) at 21°C. a, control actomyosin; b, acto-NTP (2)-myosin.

ually decreases the activity[32]. The dependence of the ATPase activity and superprecipitation of actomyosin on Mg^{2+} concentration is in approximate agreement with that for the extra-burst of P$_i$ from myosin-ATPase[32]. Acto-NTP (2)-myosin reconstituted from NTP (2)-myosin, in which only the extra-burst is inhibited, either shows hardly any superprecipitation or else undergoes it extremely slowly (Fig. 5)[39]. Thus both inhibition of the formation of phosphoryl myosin (by *p*-nitrothiophenylation (1)) (Fig. 2) and of the direct hydrolysis of phosphoryl myosin (by *p*-nitrothiophenylation (2)) suppress the superprecipitation of actomyosin. Although the superprecipitation of actomyosin cannot be considered a direct model for muscle contraction (see sections 4 and 6), these results strongly suggest that during muscle contraction myosin is activated by the rapid decomposition of ATP *via* the direct hydrolysis of $E_2^{1} \cdots {}^{ADP}_{P}$. In support of our conclusions, there is very little extra-burst of P$_i$ liberated by the reaction of myosin with ITP[39] and ITP is also much less efficient than ATP in inducing superprecipitation of actomyosin[41].

Based on the above results we can represent the reaction mechanism of actomyosin with ATP at low ionic strengths by the following scheme:

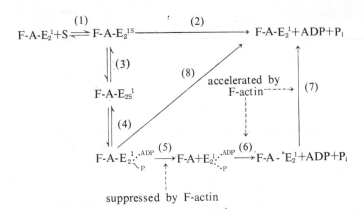

There are two main paths for the decomposition of ATP by actomyosin. One passes through $E_2^1 \cdots ADP \atop P$ (step (8)) and one through $E_2^1 \cdots ADP \atop \cdots P$ and $°E_2^1$ (steps (5), (6) and (7)), the latter route containing processes for both the dissociation and recombination of actomyosin. Step (8) is also thought to be involved in the activation of myosin during muscle contraction, and a molecular model of muscle contraction incorporating the above results will be proposed in chapter 13. In connection with the activation of myosin during muscle contraction, Gillis and Maréchal[41a] made recently a very interesting observation that P_i is incorporated into ATP during contraction of glycerol-treated muscle fibres and the amount of P-incorporation is roughly proportional to the tension-time integral. This suggests that the reaction, $F\text{-}A\text{-}*E_2^1 + ADP + P_i \rightarrow F\text{-}A\text{-}E_2^1 \cdots ADP \atop P$ can occur, though extremely slowly, where $*E_2^1$ indicates the activated myosin.*

A biochemically interesting process in the reaction of F-actin with myosin and ATP would be the transfer of a phosphoryl group from phosphoryl myosin to the ADP bound to F-actin, followed by its conversion to G-actin-ATP (chapter 5), which could then revert rapidly to the original F-actin-ADP:

* Recently we showed that the P-exchange reaction is not catalysed by purified actomyosin[41b].

$$E_2^i \overset{\text{ADP}}{\underset{P}{\cdots}} + \text{F-A-ADP} \xrightarrow{\uparrow} E_2^1 + \text{G-A-ATP} + \text{ADP}.$$
$$+ P_i$$
$$+ H_2O$$

However, removal of the nucleotides from F-actin has no effect on its physiological activity (*vide infra*) (section 5), showing that this sequence does not occur.

Consequently, we shall now consider general allosteric effects on the complexing of F-actin with myosin. If such effects occur, reagents which bind to sites on myosin other than the binding site of F-actin may activate the myosin-ATPase route passing through phosphoryl myosin. There is considerable activation of myosin-ATPase in the presence of Mg^{2+} when myosin is trinitrophenylated. This was shown to derive from the accelerated decomposition of phosphoryl myosin upon trinitrophenylation, since while *p*-nitrothiophenylation of normal myosin reduced the initial burst of P_i but caused no change in the steady-state ATPase rate (chapter 4, section 3), *p*-nitrothiophenylation (1) of trinitrophenyl myosin decreased both the steady-state rate and the initial burst, so that when the initial burst dropped to zero there was no activation of the ATPase by trinitrophenylation[42] (*cf.* p. 125). Thus the effect of F-actin on myosin-ATPase can be simulated by trinitrophenylation. Trinitrophenyl myosin also binds with F-actin[43] and, as described in chapter 4, section 4 and chapter 6, section 5, the ATPase active site of myosin is different from the site of binding with F-actin (*cf.* Ref. *44*). In conclusion, it appears that F-actin is an allosteric effector for the reaction of myosin with ATP.

3. Mechanism of the Reaction of Acto-H-meromyosin with ATP

We shall now discuss the nature of the binding between F-actin and myosin under conditions when F-actin activates myosin-ATPase. Addition of ATP to actomyosin causes superprecipitation, and hence the binding between actin and myosin cannot then be investigated in detail because of the complexity of the system. Since the study by Leadbeater and Perry[45], therefore, many workers have employed the acto-H-meromyosin-ATP system instead, because acto-H-meromyosin is very soluble even at low ionic strengths and there is no precipitation on addition of ATP.

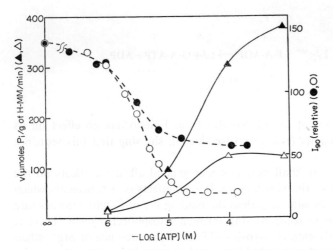

Fig. 6. Dependence of the ATPase activity of acto-H-meromyosin and the decrease in light-scattering intensity induced by ATP upon ATP concentration. 0.15mg/ml H-meromyosin, 0.05mg/ml F-actin, 28μg/ml pyruvate kinase, 0.4mM phosphoenol pyruvate, 1mM MgCl$_2$, 10mM Tris-HCl at pH 7.5, 25°C. Ionic strength due to KCl+Mg-ATP^{2-} was fixed at 5mM. ○ and ●, 90° angle light-scattering intensity, I$_{90}$; △ and ▲, ATPase activity. ○ and △, control H-MM; ● and ▲, H-MM treated with PCMB and β-merocaptoethanol.

Addition of ATP to acto-H-meromyosin rapidly decreases the light-scattering intensity. The intensity remains low so long as ATP remains present, but as the ATP disappears from solution, the light-scattering intensity returns to its original value. This decrease in intensity cannot be attributed simply to complete dissociation into actin and H-meromyosin. Comparison of the ATPase activity and decrease in light-scattering intensity at different ATP concentrations indicates that as the light-scattering intensity decreases, the ATPase activity increases[46] (Fig. 6). If this is so, we must enquire what structural changes occur when ATP is added to acto-H-meromyosin at low ionic strength. Leadbeater and Perry[45] suggested that there are two types of bond between actin and myosin, one detectable by viscosity and turbidity measurements and the other not, since addition of ATP to acto-H-meromyosin is shown by viscosity and turbidity measurements to cause dissociation into actin and H-MM, while at the same time the high ATPase activity of acto-H-meromyosin is observed[45,47]. Detailed kinetic studies of the activation of H-MM-ATPase by actin have been made more recently by Szentkiralyi and Oplatha[48] and Eisenberg and Moos[49,50].

In order to understand this phenomenon, we have studied the angular distribution of light-scattering intensity of mixtures of acto-H-meromyosin and ATP. On mixing acto-H-meromyosin with ATP $\bar{M}_w$ decreases considerably, but our results also suggest that the z-average radius of gyration increases over that when ATP is absent[46]. This indicates the possibility of substantial binding between H-MM and actin, even when the light-scattering intensity has decreased. In addition, we observed that the lower the ionic strength and the greater the F-actin concentration, the greater was the ATP concentration required to attain half the maximum decrease in light-scattering intensity and half the maximum ATPase activity. We also showed that the activation of H-MM-ATPase by F-actin does not depend on the weight ratio of H-MM to actin, but only on the concentration of F-actin[46]. Based on these results and those for the actin-myosin-ATP system at high ionic strengths (chapters 3 and 4), we can propose the following mechanism for the reaction of acto-H-meromyosin (F-A-E) with ATP (S):

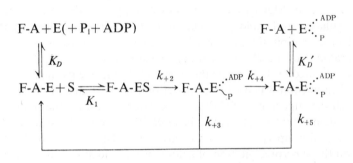

where k_{+2} depends on ATP concentration (cf. chapter 3). K_D is quite small, while K_D' is large, so that F-actin (F-A) and H-meromyosin (E) are almost completely bound and F-A-E∷$^{ADP}_P$ is almost completely dissociated. At steady-state the rate of hydrolysis of ATP, v, and the decrease in light-scattering intensity, Δ, are given by

$$v \approx \frac{(k_{+3}+k_{+4})\varepsilon}{1+\dfrac{k_{+4}}{k_{+5}[A]}+\dfrac{k_{+3}+k_{+4}}{k_{+2}}(1+\dfrac{K_1}{[S]})},$$

and

$$\Delta = \frac{\dfrac{k_{+4}}{k_{+5}[A]}}{1+\dfrac{k_{+4}}{k_{+5}[A]}+\dfrac{k_{+3}+k_{+4}}{k_{+2}}(1+\dfrac{K_1}{[S]})}$$

respectively. Thus the result that a decrease in light-scattering intensity with added ATP approximately proportional to the activation of the ATPase can be easily explained by this mechanism.

The findings of Jones and Perry[51] and Yagi and Yazawa[52] that the ATPase of subfragment-1 in the presence of Mg^{2+} is activated only slightly by F-actin may be due to much weaker binding between S-1 and actin than in the case of myosin or H-MM. Actually, Eisenberg et al.[53] showed that the ATPase of S-1 can be activated similarly to that of H-MM by the addition of a large amount of actin.

4. Superprecipitation and the Actomyosin-ATPase Activity

Szent-Györgyi[1] discovered that under normal physiological conditions (0.1M KCl, 5mM $MgCl_2$) the reaction of actomyosin with ATP results in its coagulation and rapid precipitation (vide supra). This phenomenon is called superprecipitation, and has been generally accepted as an unorganised analogue of muscle contraction. The superprecipitation of myosin B can be inhibited by EDTA or other metal-chelating agents such as EGTA. This phenomenon was used as a model by Ebashi[54] and A. Weber[55] to establish the importance of Ca^{2+} as a regulatory factor for the actomyosin-ATP system (chapter 10). Watanabe et al.[41,56] also established that Mg^{2+} is essential for superprecipitation of myosin B. These effects of Mg^{2+} and Ca^{2+} on the superprecipitation of actomyosin are identical to those on the contraction of glycerol-treated muscle fibres by ATP, strongly indicating that the superprecipitation of actomyosin is a model for muscle contraction.

However, many problems remain as to whether one can consider superprecipitation equivalent to muscle contraction. In 1948, Perry et al.[57] concluded that the superprecipitation of actomyosin is a syneresis phenomenon (such as is often observed in colloids) induced by ATP. The superprecipitation of actomyosin induced by Mg-ATP in the presence of Ca^{2+} is usually not reversed by removal of Ca^{2+} with EGTA, although the increase in turbidity induced by Mg-ATP in the suspension of myofibrils[58] and in the complex of myosin with the myosin aggregating factor[59] (a complex of F-actin with various regulatory proteins[60]) are, respectively, partially and almost completely reversed on adding EGTA. The recent electron microscopy studies of Takahashi and Yasui[61], Ikemoto et al.[62] and Nihei and Yamamoto[63] showed that morphological changes in actin and myosin filaments during superprecipitation involve a different phase from that of the contraction of a myofibril (section 6). Further studies by Bowen[64], Sekine and Yamaguchi[65] and ourselves[66] of the relationships between actomyosin-ATPase, the superprecipitation of actomyosin and the isodimensional shrinkage of an actomyosin thread suggest that the actomyosin-ATPase is not necessary for superprecipitation.

When a very low concentration of G-actin is polymerised, the polymer formed has half the molecular weight of normal F-actin and is easily cleaved by shear stress[67] (chapter 5, section 5). Studies of the reactions of such actin polymers with myosin and H-MM[68] showed that the final amount of superprecipitation of actomyosin with ATP measured by the turbidity change was unaltered from that of control actomyosin, but the ATPase activity was reduced by 80% or more. Thus the superprecipitation of actomyosin with ATP is not necessarily proportional to the ATPase activity of the actomyosin. This conclusion was confirmed by our study of the relationship between superprecipitation and deoxy-ATPase activity

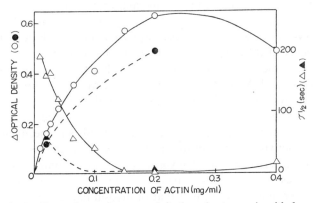

Fig. 7. Dependence of superprecipitation of actomyosin with deoxy-ATP upon F-actin concentration. 0.8mg/ml myosin, 0.1mM deoxy-ATP, 0.1M KCl, 1mM $MgCl_2$, 0.11mM $CaCl_2$, 0.1mM EGTA, 5mM Tris-maleate (pH 7.0), 25°C. ○, ●, optical density; △, ▲, $\tau_{1/2}$. ○ and △, control F-actin; ● and ▲, ADP-free F-actin prepared by the method of Kasai and his colleagues.

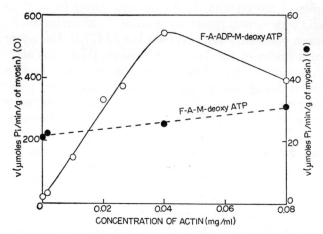

Fig. 8. Promotion of myosin-deoxy-ATPase by F-actin. 0.08mg/ml myosin, 0.1mM deoxy-ATP, 0.1M KCl, 1mM $MgCl_2$, 0.11mM $CaCl_2$, 0.10mM EGTA, 5mM Tris-maleate (pH 7.0), 25°C. ○, control F-actin (F-A-ADP); ●, ADP-free F-actin (F-A) prepared by the method of Kasai and his colleagues.

in the reaction of deoxy-ATP with actomyosin[66,69], using F-actin from which the bound ADP had been removed by the method of Kasai et al.[70] (chapter 5). Deoxy-ATP was used since, unlike ATP, it does not bind with actin, although it resembles ATP in that it liberates an initial burst of P_i by the formation of phosphoryl myosin and a myosin-phosphate-ADP complex[69]. Deoxy-ATP also contracts myofibrils (chapter 9). Figure 7 shows the dependence of the amount and rate of superprecipitation on F-actin concentration in the reaction of deoxy-ATP with actomyosin reconstituted from myosin and ADP-free F-actin prepared by the method of Kasai et al. The removal of ADP from the F-actin has almost no effect on superprecipitation. Figure 8 also shows the dependence of the deoxy-ATPase upon F-actin concentration for the same actomyosin specimen. The deoxy-ATPase activity of myosin to which control F-actin has been added is 26 times greater than that of myosin alone, but in contrast, the ADP free actin does not greatly enhance the deoxy-ATPase activity. Addition of the latter in an amount half that of myosin induces only 1% of the increase in the deoxy-ATPase activity found with the control actomyosin. The small degree of activation by the ADP-free F-actin was attributed to contamination by small amounts of ADP bound to F-actin. Similar results were obtained when CTP, which is difficult to bind to actin, replaced deoxy-ATP. These results clearly show that superprecipitation is independent of the actomyosin-ATPase.

5. Role of ADP Bound to F-Actin in Superprecipitation

Another problem in the mechanism of superprecipitation is the role of the nucleotide bound to F-actin. Straub and Feuer[71] discovered that ATP bound to G-actin is converted into ADP and P_i upon polymerisation (chapter 5). Since then, interest has developed in the physiological significance of the nucleotide bound to actin.

ATP bound to G-actin rapidly exchanges with external ATP in solution, while ADP complexed to F-actin does not exchange with external ADP[72]. However, ultrasonification of F-actin causes rapid exchange of bound ADP with other nucleotides[73]. The role of the nucleotide bound to actin in the reaction of actomyosin with ATP has not yet been elucidated. Recently, A. G. Szent-Györgyi and Prior[74] made the interesting observation that ADP bound to F-actin exchanges with external nucleotide during superprecipitation of actomyosin with ATP. This suggests the following: (1) ADP in F-actin accepts a phosphoryl group from phosphoryl myosin, thus being converted to ATP with subsequent change in the conformation of actin, followed by the regeneration of bound ADP; such a cycle can be repeated (*cf.* p. 205), (2) conformational changes of F-actin accompanying the superprecipitation of actomyosin result in the exchange of the nucleotide bound to F-actin.

We have already shown that ADP bound to F-actin is not directly related to superprecipitation. However, the measurements described in the above section used actin from which nucleotides had been removed with urea-EDTA by the method of Kasai *et al.*[70], and it is clear from experiments described later that this particular method damages the functions of actin. Therefore, nucleotides were removed from actin by the method of Mommaerts[75] in order to obtain intact, ADP-free actin, and the following two experiments were carried out to show that the ADP in actin does not participate in the actomyosin-ATPase reaction.

Our studies[68] of the nucleotide exchange which accompanies superprecipitation of F-actin-^{14}C-ADP-myosin and F-actin-^{14}C-IDP-myosin by ATP indicated that ^{14}C-IDP and ^{14}C-ADP are liberated, but not ^{14}C-ITP or ^{14}C-ATP. These observations, together with the fact that the binding between actin and ITP is not very strong (*cf.* chapter 5, section 2), can be easily explained by assuming that during superprecipitation of actomyosin

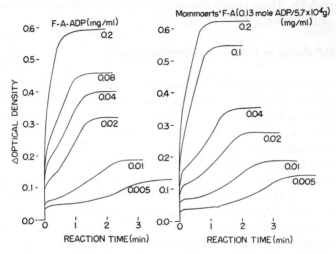

Fig. 9. Superprecipitation by deoxy-ATP of actomyosin reconstituted from myosin and ADP-free F-actin prepared by Mommaerts' method. 0.8mg/ml myosin, 0.1mM deoxy-ATP, 0.1M KCl, 1mM $MgCl_2$, 0.11mM $CaCl_2$, 0.1mM EGTA, 5mM Tris-maleate, pH 7.0, 25°C. Concentrations of control F-actin (F-A-ADP) and Mommaerts' F-actin (0.13 moles ADP/5.7×10^4 g) are indicated in the figure.

there is a loosening of the F-actin structure, which results in the exchange of the bound nucleotide with nucleotide in the medium, and that ADP or IDP bound to F-actin do not accept a phosphoryl group from phosphoryl myosin.

In order to further clarify the role of the nucleotide, we investigated the physiological functions of actin containing no nucleotide, prepared by Mommaerts' method[75], which is much milder than that used by Kasai and his colleagues. Figure 9 shows measurements of turbidity changes during superprecipitation by deoxy-ATP of actomyosin reconstituted from ADP-free F-actin and myosin. There was almost no difference from a control actin in the amount or rate of superprecipitation when the F-actin concentration was varied, keeping a constant concentration of myosin. Unlike the case with actin prepared by the method of Kasai and his colleagues, the deoxy-ATPase activity of this ADP-free actomyosin hardly differed from that of the control actomyosin, showing that there was considerable activation of the myosin-deoxy-ATPase by F-actin. In this case, deoxy-nucleotides were not incorporated into the actin. Bárány et al.[76]

also demonstrated that the ADP-free actomyosin obtained by sonification underwent superprecipitation with CTP or deoxy-ATP, and that there is an actomyosin-CTPase and deoxy-ATPase, but that there was no incorporation of nucleotide into actin.

These results all indicate that the ADP bound to F-actin is not directly related to superprecipitation or to the actomyosin-ATPase.* The result that TBS, which binds with a lysine residue at a position different from the binding site of actin, affects the myosin-ATPase similarly to actin[42], also suggests that F-actin is an allosteric effector for the reaction of myosin with ATP. Thus it is highly improbably that ADP bound to F-actin plays any important role in the reaction between myosin, actin and ATP. However, the above results do not preclude the possibility of a conformational change by F-actin during contraction. Moos and Eisenberg[77] recently reported that the rate of release of actin-bound nucleotide is greatly increased by myosin, not only in the presence of ATP, but also in the presence of ADP, where no superprecipitation occurs. It was suggested that aggregation of the myosin is essential to this effect of myosin since the effect is not observed at high ionic strength and it is not shown by H-meromyosin. Therefore, it was concluded that the increased exchangeability of the actin-bound nucleotide in actomyosin is unrelated to the ATPase activity or to the contraction of actomyosin.

6. Molecular Mechanism of Superprecipitation of Myosin B

The addition of high concentrations of ATP to myosin B at low ionic strength results first in a decrease in turbidity—the clearing response[4]—followed by a large increase in turbidity due to superprecipitation. Takahashi and Yasui[78] studied, by electron microscopy, morphological changes in myosin B filaments during superprecipitation. Even during the clearing phase an appreciable proportion of myosin filaments were bound to actin, but most had dissociated. This dissociation is thought to correspond to the relaxation of living muscle.** Negative staining and rotary shadowing techniques show that the lateral projections from myosin filaments bound to

---

* The nucleotide in actin is needed in order to maintain its conformation. Removal of nucleotide by the method of Kasai and his colleagues probably partially distorts the conformation, and this may diminish some functions.

** At low ionic strength, myosin forms a filament after dissociation, so that we cannot conclude from this result that at high ionic strength, when myosin is soluble, actomyosin is dissociated by ATP (cf. p.175).

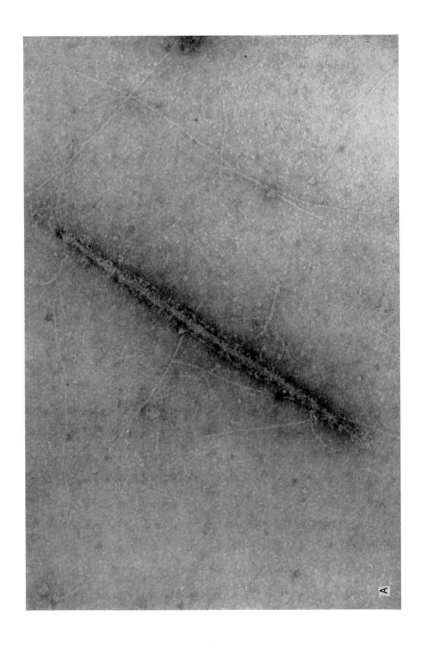

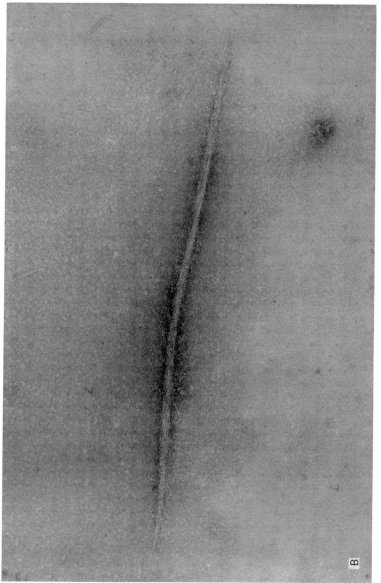

Fig. 10. Structure of the lateral projections from myosin filaments. (A) attached, and (B) detached from the actin filament. 1 mg/ml myosin B in 0.6M KCl, 2mM ATP, 10mM $MgCl_2$, and 10mM Tris-maleate at pH 7.0, was dialysed against 0.15M KCl, 1mM ATP, 5mM $MgCl_2$ and 6.7mM phosphate at pH 7.0. The solution inside the bag was diluted 10-fold with the dialysate and allowed to stand for 4 hr at room temperature. ×92,000. (Courtesy of Dr. K. Takahashi, Faculty of Agriculture, Hokkaido University.)

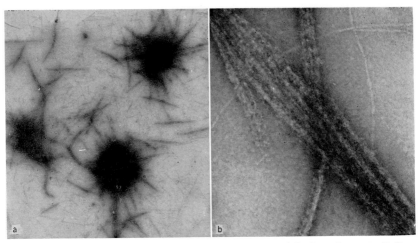

Fig. 11. Formation of unit aggregates during the superprecipitation of myosin B filaments. The myosin B solution (0.2mg/ml myosin B, 0.15M KCl, 0.2mM ATP, 5mM $MgCl_2$, 0.01mM $CaCl_2$, 10mM Tris-maleate, pH 7.0) was kept at 25°C. The electron micrograph shows the myosin B after addition of ATP, when the turbidity was 20% of its maximum value. Magnification: (a), ×8,600; (b), ×80,300. (Courtesy of Dr. K. Takahashi, Faculty of Agriculture, Hokkaido University.)

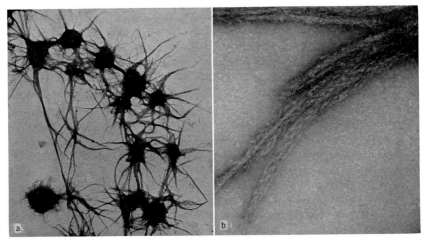

Fig. 12. Electron micrograph of a myosin B filament at the maximum turbidity of superprecipitation. Magnification: (a), ×8,600; (b), ×80,300. (Courtesy of Dr. K. Takahashi, Faculty of Agriculture, Hokkaido University.)

actin are at right angles to the filament axes. In contrast, the projections from myosin filaments dissociated from actin lie at an angle of 30–40° to the filament axes[78] (Fig. 10). When the turbidity of myosin B is at 20% of its maximum, some of the myosin and actin filaments aggregate to a loose complex filament (Fig. 11(a)), in which the lateral projections from the myosin filaments are again at right angles to the filament axes (Fig. 11(b)). At this stage free actin filaments rapidly complex myosin filaments to form aggregates with a configuration like that of a starfish. When the turbidity of myosin B has reached its maximum, many starfish-like structures are connected by actin filaments (Fig. 12 (a)). The actomyosin filament forms more compact arrow-head structures in which the myosin filament is only half of its original diameter (Fig. 12(b)), and which probably corresponds to living muscle in rigor. These observations show that changes in the filament structure of actomyosin during superprecipitation are more complex than the morphological changes during contraction of myofibrils (chapters 1 and 13).

The mechanism of superprecipitation is very complicated, but kinetic studies are still possible[65]. The rate of superprecipitation of myosin B slows as the myosin B concentration decreases, and there is no superpreci-

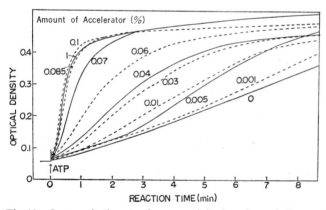

Fig. 13. Increase in the rate of superprecipitation of myosin B upon the initial addition of a trace of superprecipitated myosin B. 0.3mg/ml myosin B, 0.5mM ATP, 0.06M KCl, 2mM MgCl$_2$, 30μM EGTA, 20mM Tris-maleate, pH 7.0, 25°C. Superprecipitated myosin B, after sonification for 15 min, was added as promoter in the concentrations shown (% of total myosin B).

pitation at all at less than a critical concentration. The rate of superprecipitation of myosin B is accelerated by adding a minute amount of myosin B which has already been superprecipitated (Fig. 13). Adding a small amount of already superprecipitated myosin B, which had also undergone ultrasonic treatment, produces yet greater acceleration. Even the addition of $0.3 \mu g/ml$ of superprecipitated myosin B increases the rate of superprecipitation of myosin B at a concentration of $0.3 mg/ml$ by 20–30 times. However, as expected, there is no increase in ATPase activity. When superprecipitated myosin B is centrifuged for 30 min at $10^4 \times g$ into a supernatant and precipitate, which is then dissolved in 0.6M KCl, the acceleration effect disappears. Up to an amount of already superprecipitated myosin B (accelerator) of $0.45 \mu g/ml$, the reciprocal of $\lambda_{1/2}$ (the time to reach half the maximum turbidity change) for the superprecipitation of $0.3 mg/ml$ of myosin B with ATP increases linearly with increasing accelerator. With greater amounts of accelerator the rate is almost constant. However, the maximum change in turbidity is constant and independent of the addition of accelerator, hence the relationship between the amount and rate of superprecipitation of myosin B at various concentrations was studied in the presence of a large quantity of accelerator. When superprecipitated myosin B accelerator is not added, the rate of superprecipitation decreases considerably with decreasing myosin B concentration, but when accelerator is added, the precipitation is very fast and independent of myosin B concentration. Hence superprecipitation of myosin B is a nucleated growth process, and this casts considerable doubt on the validity of considering it a model for muscle contraction.

Since the mechanism of superprecipitation is so complicated, it will not in general be useful as a quantitative model of muscle contraction, and our particular emphasis in this chapter on the situation where actomyosin-ATPase activity is not proportional to the superprecipitation rate is a critical reflection on the lack of caution with which superprecipitation has been used as a muscle model. Nevertheless, it is convenient to use superprecipitation as a qualitative indicator of the 'activation' of myosin by ATP (section 2), and there are many cases when the two effects do show a good correlation, when superprecipitation may be usefully used as a contraction model. In fact several very interesting analyses have recently been made of the relation between the ATPase activity and the rate of superprecipitation of actomyosin. For example, by comparing these two rates, Levy and Ryan[79] concluded that contraction results only when the hydrolysis of ATP and the binding of ATP at the non-hydrolytic site occur in a concerted fashion. Tokiwa and Morales[80] partially modified a myosin solution with an ATP analogue, 6-mercapto-9-β-ribofuranosylpurine 5'-triphosphate, which is an affinity label for the ATPase sites of myosin. Two moles of this analogue attach to 1 mole of myosin, thereby eliminating the myosin-ATPase activity[81]. They compared the superprecipitation and ATPase reactions, in the presence of actin, of the partially modified myosin with those of a control, unmodified myosin. On the assumption that myosin has two identical globular heads, they concluded that only molecules with two native (unmodified) heads are effective in superprecipitation, while half-labelled molecules can participate in the hydrolysis of ATP. However, we believe that superprecipitation is too complex for a molecular mechanism of muscle contraction to be devised from these studies, although they are very interesting phenomena.

7. Effect of Various Factors on the Reaction of Actomyosin with ATP[3,4a]

Regulation of the contraction and relaxation of muscle depends on a complex mechanism involving Ca^{2+} and a regulatory protein factor (chapter 10). However, the ATPase activity of actomyosin and its colloidal properties are affected not only by the regulatory protein factor and Ca^{2+}, but also by the KCl and Mg-ATP concentrations and by temperature. In this section we shall summarise experimental results using myosin B which contains the regulatory protein factor and trace amounts of Ca^{2+}.

A) Effect of Mg-ATP concentration

At low ionic strengths and in the presence of low concentrations of Mg-ATP, myosin B solution exhibits a typical actomyosin-ATPase activity resulting in superprecipitation. However, when the Mg-ATP concentration is increased a very clear solution is obtained. This phenomenon was first observed by Spicer[4] and is called the clearing response. The ATPase activity gradually becomes inhibited by substrate, and at high ATP concentrations it is equal to the ATPase activity of myosin (Fig. 14)[3,82]. Maruyama and Gergely[83] measured the viscosity and the flow-birefringence of such clear solutions of actomyosin and showed that under these conditions the measured values were equal to the sum of those of F-actin and myosin. Ikemoto et al.[62] examined clear solutions of actomyosin by electron microscopy and found them to be dissociated into actin filaments and myosin filaments. This clearing response is thought to be identical in mechanism to the 'dissociation' of actomyosin by ATP at high ionic

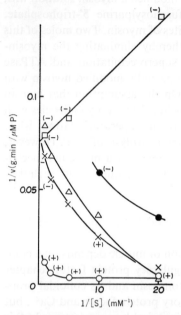

Fig. 14. Effect of KCl concentration on myosin B-ATPase. The relation of the rate to ATP concentration is presented as a Lineweaver-Burk plot. The $MgCl_2$ concentration is always higher than that of ATP by 0.1mM. pH 7.0, 25°C. +and −denote superprecipitation and the clearing response, respectively. KCl concentrations: ○, 0.03; ×, 0.075; △, 0.105; ●, 0.12; □, 0.15M.

strength. The clearing response probably corresponds to the resting state of living muscle. Evidence for the dissociation of actin from myosin in resting muscle has been obtained from measurements of the Young's modulus of glycerinated muscle[10,84], from measurements of the viscoelastic properties of glycerinated muscle[85] (chapter 8) and from the X-ray diffraction pattern of living muscle[86] (chapter 13). However, according to Pepe[87-89], H-MM antigenic sites in the region of overlap of the thin and thick filaments of glycerinated fibres are not exposed even in a relaxing medium. This suggests that at least the end of the myosin cross-bridge is always maintained in some fixed position relative to the actin filament.

B) Effect of KCl concentration

When the KCl concentration is gradually increased the Mg-ATP concentration necessary for the clearing response decreases, and at KCl concentrations of 0.2M or more superprecipitation no longer occurs, even at low Mg-ATP concentrations (Fig. 14)[55,82]. Thus at relatively high KCl concentrations the ATPase activity follows Michaelis-Menten kinetics over the ATP concentration range, as was also found with myosin-ATPase[82].

C) Effect of temperature
This is of interest because of its relationship to muscle contraction—muscle does not contract at very low temperatures. The ATPase activity of an actomyosin suspension at 0°C is low, regardless of the ATP concentration. Increasing the temperature causes superprecipitation even at high ATP concentrations[82]. The weakening of the bonding between actin and myosin as the result of increase in KCl concentration or decrease in temperature has already been described in chapter 6.

Higher concentrations of ATP rapidly transform $E_2^1 {\cdot\cdot\cdot ADP \atop \diagdown P}$ to $E_2^1 {\cdot\cdot\cdot ADP \atop \cdot\cdot\cdot P}$ (chapter 3, section 7), and the dissociation of actomyosin takes place when myosin is converted to this latter form. Another possible mechanism of dissociation of actomyosin is the increase in the concentration of E_{2S}^1 with increase in the ATP concentration (chapter 6, section 6). If the concentration of F-actin added to myosin is increased[82], the concentration of ATP at which substrate inhibition appears becomes higher. These results show that the clearing response with ATP occurs when the bonding between actin and myosin is weak. Hence factors such as KCl and Mg-ATP concentrations and temperature influence the interaction between actin and myosin, thereby regulating actomyosin-ATPase.

D) Effect of modification of SH-groups
According to the conclusions of section 3 (see p. 207), changing the rates of steps (3), (4) and (5) by some means would permit the regulation of ATPase activity and colloidal properties of actomyosin. Myosin contains about 2 moles of 'intrinsic' Ca^{2+} which is not removable by the usual purification methods, but can be irreversibly removed by treatment of myosin with PCMB, followed by cysteine, DTT or β-mercaptoethanol[90]. This last reagent completely removes the PCMB from myosin[91], so that its ATPase activity is completely recovered[90]. The ATPase activity of actomyosin reconstituted from this myosin and actin is very different from that containing Ca^{2+}[90]. Thus the control actomyosin exhibits the clearing response at high concentrations of Mg-ATP, the ATPase being similar to myosin-

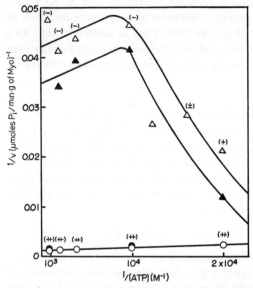

Fig. 15. Lineweaver-Burk plot of actomyosin-ATPase at steady-state. 0.075M KCl, pH 7.0, 24°C. The MgCl$_2$ concentration is higher than that of ATP by 1mM. Myosin: actin=2.5:1. ++ and + indicate superprecipitation immediately after and several minutes after addition of ATP; ——shows clearing response with ATP. △, control actomyosin; ▲, 0.2mM CaCl$_2$ added to the control; ○, actomyosin reconstituted from myosin treated with PCMB and cysteine; ●, the sample obtained by addition of 0.5mM EDTA to the above (○).

ATPase, while even at very high Mg-ATP concentrations actomyosin prepared from myosin treated with PCMB and β-mercaptoethanol shows typical actomyosin-ATPase activity which results in superprecipitation (Fig. 15). The optical rotatory dispersion, sedimentation coefficient and viscosity of myosin did not change appreciably upon treatment with PCMB and β-mercaptoethanol. However, when H-MM was treated with PCMB and DTT, the light chain, g_2', which is derived from the g_2 of myosin, decreased to about a half of that in the untreated control, while the contents of the other polypeptide chains were unchanged[92] (cf. chapter 2, section 6). Thus one of two g_2 chains in the myosin molecule seems to play an important role in the regulation by high concentrations of Mg-ATP of the actin-myosin system. However, there remains a possibility that one of the two g_2 is not a subunit necessary for the regulation, but is released from myosin as a

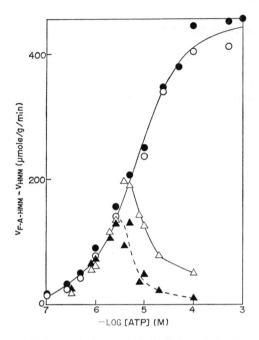

Fig. 16. Dependence of the ATPase activity of acto-H-meromyosin upon ATP concentration. 0.25mg/ml H-MM, 0.125mg/ml F-actin extracted at room temperature. 2mM $MgCl_2$, 50mM KCl, 20mM Tris-maleate, pH 7.0, 24°C. The ATP concentration was kept constant using a pyruvate kinase system. △, ▲, control H-MM; ○, ●, H-MM treated with PCMB and β-mercaptoethanol. ○, △, 0.1mM $CaCl_2$; ●, ▲, 0.1mM EGTA, Difference between the ATPase activity of acto-H-meromyosin ($v_{\text{F-A-HMM}}$) and that of H-meromyosin alone (v_{HMM}) is plotted against logarithm of ATP concentration.

result of the conformational change around the regulation site induced by the treatment with PCMB. The regulation system of myosin can be also selectively inactivated by heat and reactivated by dithiothreitol, a disulphide-reducing agent[93]. In order to find a reaction mechanism which could account for these marked changes in the properties of actomyosin-ATPase after treatment of myosin with PCMB and β-mercaptoethanol, we have investigated the reaction of acto-H-meromyosin with ATP[94]. At high ATP concentrations the rate of liberation of the initial stoichiometric burst of P_i is considerably greater than the rate of decrease in light-scattering intensity when acto-H-meromyosin reacts with ATP.

Hence the formation of $E_2^1 {\substack{\cdot\cdot \text{ADP} \\ \diagdown \text{P}}}$, as indicated by the initial burst, was faster than the formation of $E_2^1 {\substack{\cdot\cdot \text{ADP} \\ \cdot\cdot \text{P}}}$, as indicated by the dissociation of acto-H-meromyosin (see the reaction scheme in p. 207). As shown in Fig. 16, the actin extracted at room temperature is a complex of F-actin with a regulatory protein (chapter 10), and the ATPase of its complex with H-meromyosin is inhibited in the presence of Mg^{2+} and low concentrations of KCl if trace amounts of Ca^{2+} are removed by the addition of EGTA. There is also significant substrate inhibition by high concentrations of ATP. It is thought that step (5) of the reaction scheme given in p. 207 is much slower than step (3) when the concentration of F-actin is low, and that steps (3) and (4) are accelerated by Ca^{2+} and ATP respectively, although other mechanisms are also possible, as will be mentioned in chapter 10, section 5.

When H-meromyosin is treated with PCMB and β-mercaptoethanol the ATPase activity increases with increasing ATP concentration, independently of Ca^{2+} and with no substrate inhibition. The double reciprocal plot of rate of the acto-H-meromyosin-ATPase reaction against the concentration of F-actin indicated that the treatment of H-MM decreases markedly the 'apparent' dissociation constant of binding of H-MM with F-actin in the presence of ATP, while it does not affect the rate in the presence of a sufficient amount of F-actin[94]. When ATP was added to acto-H-meromyosin reconstituted from the PCMB-DTT-treated H-meromyosin and F-actin prepared by extraction at room temperature, H-meromyosin was not filtrated though a Millipore filter (pore size, 0.45μ), while the most of H-meromyosin of control acto-H-meromyosin was filtrated (Fig. 17)[92]. This was true even when the concentration of ATP was much higher than the value of K_m of acto-H-meromyosin-ATPase (cf. Figs. 16 and 17). The angular distribution of light-scattering intensity of acto-H-meromyosin reconstituted from the PCMB-DTT-treated H-meromyosin and F-actin showed that $\bar{M}_w$ was scarcely changed by adding ATP, but the radius of gyration increased to about twice as large as that in the absence of ATP (Fig. 18). It is worthy of our notice that the change in light-scattering intensity induced by ATP of PCMB-DTT-treated acto-H-meromyosin at low ionic strength is rather similar to that of myosin B at high ionic strength, which was described in chapter 6, section 4 (cf. Fig. 7A). Thus, the dissociation of atomyosin is not a step prerequisite to the actomyosin-type of ATPase. This idea has also been

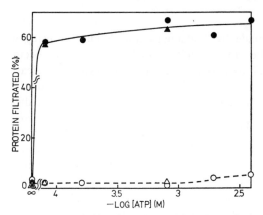

Fig. 17. Rapid filtration of acto-H-meromyosin in the presence of ATP though a Millipore filter. 0.71mg/ml H-MM, 0.40mg/ml F-actin, 2mM $MgCl_2$, 0.05M KCl, pH 7.0, 19°C. H-MM was prepared from untreated myosin (●, ▲) or from PCMB-DTT-treated myosin (○, △), and F-actin was prepared by extraction at room temperature. ○, ●, 0.1mM $CaCl_2$; △, ▲, 1mM EGTA. Pore size of the Millipore filter, 0.45μ.

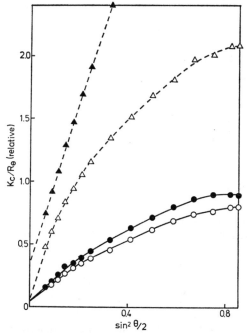

Fig. 18. The change in angular distribution of light-scattering of acto-H-meromyosin on adding ATP. 0.25mg/ml H-MM, 0.14mg/ml F-actin. Other conditions same as for Fig. 17. ●, ▲, control H-MM; ○, △, PCMB-DTT-treated H-MM. ○, ●, no ATP; △, ▲, 0.6mM ATP.

supported by the finding made by Matsunaga and Noda[94a], using a stopped flow method, that the superprecipitation of actomyosin can occur immediately after adding ATP to actomyosin, without showing any 'clearing response' as the initial step. On the other hand, Lymn and Taylor[94b] have recently proposed that the first step of actomyosin-ATPase is the dissociation of actomyosin by ATP: $\text{F-A-E} + \text{S} \rightarrow \text{F-A} + \text{ES} \rightarrow \text{F-A} + \text{E} \genfrac{}{}{0pt}{}{\cdot\text{ADP}}{\cdot\text{P}}$ $\rightarrow \text{F-A-E} + \text{ADP} + \text{P}_i$, from the results that the dissociation of acto-H-meromyosin by ATP occurs very rapidly, and that its rate increases almost linearly with increase in the concentration of ATP. However, these results can be explained by the mechanism discussed in chapter 6, section 6 that the dissociation of actomyosin occurs not only on the formation of myosin-phosphate-ADP complex but also on the formation of $^S\text{E}_{2S}{}^1$ at high concentrations of ATP. Another possibility is that ATP at high concentrations accelerates the dissociation of acto-H-meromyosin, since the dissociation in the presence of ATP is markedly promoted even by the increase in ionic strength due to a high concentration of ATP itself, as already shown by us[46]. Furthermore, Lymn and Taylor's mechanism cannot explain the inhibition of the ATPase activity of purified actomyosin by excess substrate (see, for example, Fig. 9, chapter 10).

These results show that the treatment of H-meromyosin with PCMB markedly decreases K'_D, so that almost all H-meromyosin-phosphate-ADP complex can bind with F-actin. In the case of control acto-H-meromyosin, K'_D is large and the decomposition of ATP occurs mainly through step (3), which is much faster than step (5), as shown above. Thus the ATPase activity is increased by the addition of a trace amount of Ca^{2+} and suppressed by increasing the ATP concentration. Conversely, in the case of H-meromyosin treated with PCMB and β-mercaptoethanol, K'_D is small and the decomposition of ATP occurs mainly through steps (4) and (5). As already described, step (4) is accelerated by increasing the ATP concentration but is not regulated by Ca^{2+}. Thus when H-meromyosin is treated with PCMB and β-mercaptoethanol the ATPase activity of the acto-H-meromyosin is proportional to the ATP concentration, but independent of Ca^{2+}. As would be expected from our molecular mechanism for actomyosin-ATPase, similar results have been obtained when step (5) was accelerated by increasing the F-actin concentration.

The lack of substrate inhibition of the actomyosin-ATPase after treatment with PCMB and β-mercaptoethanol is not due to inhibition of a contaminating regulatory protein factor, as Yasui et al.[95] speculated, since even with a purified H-MM-actin-ATP system, treatment of H-MM gives results similar to those found from the myosin-actin-ATP system (as described above) and the function of the regulatory protein factor is not inhibited by modification of the SH group[96] (chapter 10).

8. Comparative Studies on the Myosin-Actin-ATP System

So far we have described the properties of rabbit skeletal myosin and actin and their reactions with ATP. Now we will summarise the studies by other investigators on the function and properties of contractile proteins from other organs and species.

Contractile proteins have been isolated from cardiac muscle and studied by many investigators (cf. Ref. 97). Mammalian cardiac myosin was investigated by Olson et al.[98], Conway and Roberts[99] and others[100–102], and cardiac actin by Katz and Carsten[103] and McCubbin and Kay[104]. The interaction of cardiac actomyosin with ATP was studied by Tada[105] and Jacobson et al.[106] No fundamental differences have been observed between skeletal actomyosin and cardiac actomyosin.

Contractile proteins of smooth muscle have also been investigated by many researchers. Myosin was isolated from uterus[107,108], carotid[109,110,110a], gizzard[111] or esophagus[112]. F-Actin was also purified from carotid[110a]. Furthermore, Nonomura[113] and Kelly and Rice[114] demonstrated the presence of thick and thin filaments in smooth muscle. Lowy and Small[115], and Small and Squire[115a] have reported that myosin is present in vertebrate smooth muscle in the form of very long, ribbon-like structures, and that actin exists in the usual filament but is here arranged in small regular arrays between and around the myosin ribbons. On the other hand, Rice et al.[116] have reported that myosin occurs in cylindrical filaments of diameter 120–180Å, which are arranged on a regular lattice with a nearest neighbour spacing of 660Å or more. They[117] concluded that the myosin ribbons are formed by aggregation of round filaments in certain non-physiological conditions. S. Puszkin and Berl[118,119,119a] isolated actin-like and myosin-like proteins from brain, and called them neurin and stenin respectively. A colchicine-binding protein* was also purified from brain[119b], and its physical and chemical properties were shown to be similar to those of tublin (see below).

A contractile protein which has an arrow-head structure and undergoes superprecipitation has been isolated from equine leucocytes[120]. It shows

* A relevant property of microtubules is their high affinity for colchicine[119b].

an actomyosin-type ATPase activity with Ca^{2+}-sensitivity[120a]. The platelets contain an actomyosin-like protein called thrombosthenin. Thrombosthenin-M, which is like myosin, was isolated by Bettex-Galland and Lüscher[121]. The ATPase activity and the molecular structure of thrombosthenin-M are very similar to those of skeletal myosin[121a]. Puszkin et al.[122] extracted and purified a colchicine-binding protein, presumably derived from microtubules, from human platelets. This protein forms an arrow-head complex with H-meromyosin[123], and markedly altered the cation requirements for the ATPase activities of both muscle myosin and thrombosthenin-M; when added to either of these there was a rise in relative viscosity, which fell sharply upon addition of ATP.

The cilia and flagella of eukaryotes form a homologous group with a remarkable uniformity of structural organization: the nine-plus-two axial fibre complex. They should not be confused with bacterial flagella, which are quite different. Full accounts of their structural features have been given by Fawcett and Porter[124,125], Gibbons and Grimstone[126-128], Sleigh[129] and others. A structural protein, tublin[130], has been isolated from cilia and flagella, and this resembles muscle actin[131-136], although tublin contains GTP instead of ATP. However, Stephens[137] showed recently that the chemical structure of tublin is very different from that of actin, while Olmsted and his co-workers[137a] showed that tublin contains two separable proteins which differ in molecular weight (5.3 and 5.4×10^4) and in amino acid composition. An ATPase was first extracted by Engelhardt and Burnasheva[138-140] from flagella of spermatozoa and was called 'spermasin'. More recently, Gibbons and Rowe[141-144] isolated a soluble form of the enzyme from the cilia of Tetrahymena, and suggested that the ATPase is located on the arms of the nine outer doublets. He proposed the name 'dynein' for the ATPase. Dynein contains both 14S and 40S components, the latter being a polymer of the former. 14S dynein has about the same molecular weight as myosin from skeletal muscle, but myosin is a highly asymmetric molecule, while the dynein monomer is nearly globular. Myosin-ATPase is activated by Ca^{2+} but not by Mg^{2+}, and it shows little specificity for any particular nucleoside triphosphate. In contrast, dynein-ATPase is activated by both Ca^{2+} and Mg^{2+}, and it is relatively specific for ATP. Furthermore, Satir[145,146] proposed that sliding filaments are the source of the contractile process in the bending motion of flagella, and quite recently Summers and Gibbons[146a] have supported this hypothesis by observing that ATP induces active sliding between groups of the outer doublet tubes in trypsin-treated flagella of sperm.

ATP is not only required for mitosis[146b-d], but is even required for the maintenance of the structure of mitotic apparatus *in vivo*[146e]. A number of studies on the ATPase activity of isolated mitotic apparatus have appeared[146f,g]. Tublin[147] and a dynein-like ATPase[148] have been isolated as structural proteins from the microtubules[149-152] in mitotic spindles, and more recently Mazia and Petzelt have reported that an ATPase activity is concentrated in the mitotic apparatus isolated from the sea urchin egg by a new procedure[152a] and that the ATPase follows a cycle of activity in the egg during the mitotic cycle[152b]. The sliding of interpolar microtubules and chromosomal microtubules past each other has been proposed by McIntosh and his colleagues[153,154] as the mechanism of chromosomal movement. More recently, the presence of actin in sperm tails and mitotic spindles[154a,b] have been demonstrated by electron microscopy.

Actin and myosin have been isolated from mycomycetes[155-164], which exhibit a very active protoplasmic flow[165], and also from amoeba[166-169]. Nachmias *et al.*[164] have found that actin from mycomycetes can bind rabbit myosin S-1 in a manner almost similar to that in which S-1 is bound by rabbit actin. Similar observations on amoeba actin have been made by Pollard *et al.*[167]. A protein resembling actomyosin has also been prepared from vascular bundles of higher plants[167a,b]. Thus the myosin-actin-ATP system and other similar systems have been found in various motile apparatuses in various kinds of cells, and this strongly suggests that the myosin-actin-ATP system is the molecular basis of cell motility in general, although a different type of cell motility has been found in the case of the stalks of peritrich ciliates[170-173].

REFERENCES

1 A. Szent-Györgyi, "Chemistry of Muscular Contraction," 1st ed., Academic Press, New York (1947).
2 H. H. Weber and H. Portzehl, *Progr. Biophys. Biophys. Chem.*, **4**, 60 (1954).
3 N. A. Biró and A. G. Szent-Györgyi, *Hung. Acta Physiol.*, **2**, 120 (1949).
4 S. S. Spicer, *J. Biol. Chem.*, **199**, 289 (1952).
4a W. Hasselbach, *Z. Naturforsch.*, **7b**, 163 (1952).
5 T. Nihei and Y. Tonomura, *J. Biochem.*, **46**, 305 (1959).
6 M. C. Schaub and M. Ermini, *Biochem. J.*, **111**, 777 (1969).
7 I. Banga, *Stud. Inst. Med. Chem. Univ. Szeged*, **1**, 27 (1941).
8 K. Bailey, *Biochem. J.*, **36**, 121 (1942).
9 D. M. Needham, *in* "The Structure and Function of Muscle," ed. by G. H. Bourne, Academic Press, New York, Vol. 2, p. 55 (1960).

10 H. Portzehl, *Z. Naturforsch.*, **7b**, 1 (1952).
11 F. Turba and G. Kuschinsky, *Biochim. Biophys. Acta*, **8**, 76 (1952).
12 H. Portzehl, *Biochim. Biophys. Acta*, **14**, 195 (1954).
13 E. Bozler, *J. Gen. Physiol.*, **38**, 149 (1954).
14 S. Watanabe, *Arch. Biochem. Biophys.*, **54**, 559 (1954).
15 K. Takahashi, T. Mori, H. Nakamura and Y. Tonomura, *J. Biochem.*, **57**, 637 (1965).
16 H. Nakamura, T. Mori and Y. Tonomura, *J. Biochem.*, **58**, 582 (1965).
17 Y. Hayashi and Y. Tonomura, *J. Biochem.*, **63**, 101 (1968).
18 S. Kitagawa and Y. Tonomura, *J. Res. Inst. Catalysis, Hokkaido Univ.*, **8**, 91 (1960).
19 Y. Tonomura, K. Yagi, S. Kubo and S. Kitagawa, *J. Res. Inst. Catalysis, Hokkaido Univ.*, **9**, 256 (1961).
20 Y. Tonomura, S. Kitagawa and J. Yoshimura, *J. Biol. Chem.*, **237**, 3660 (1962).
21 Y. Tonomura, T. Kanazawa and K. Sekiya, in "Molecular Biology; Problems and Perspectives," ed. by A. E. Braunstein, Nauk U. S. S. R., Moscow, p. 213 (1964).
22 Y. Tonomura, T. Kanazawa and K. Sekiya, *Annu. Rep. Sci. Works, Fac. Sci., Osaka Univ.*, **12**, 1 (1964).
23 Y. Tonomura and S. Kitagawa, *Biochim. Biophys. Acta*, **26**, 15 (1957).
24 Y. Tonomura and S. Kitagawa, *Biochim. Biophys. Acta*, **40**, 135 (1960).
25 M. Ikehara, E. Ohtsuka, S. Kitagawa, K. Yagi and Y. Tonomura, *J. Amer. Chem. Soc.*, **83**, 2679 (1961).
26 R. G. Yount and D. E. Koshland, *J. Biol. Chem.*, **238**, 1708 (1963).
27 T. Kanazawa and Y. Tonomura, *J. Biochem.*, **57**, 604 (1965).
28 A. Inoue, K. Shibata-Sekiya and Y. Tonomura, *J. Biochem.*, **71**, 115 (1972).
29 D. R. Kominz, *Biochemistry*, **9**, 1792 (1970).
30 K. Imamura, T. Kanazawa, M. Tada and Y. Tonomura, *J. Biochem.*, **57**, 627 (1965).
31 Y. Tonomura and T. Kanazawa, *J. Biol. Chem.*, **240**, PC 4110 (1965).
32 N. Kinoshita, T. Kanazawa, H. Onishi and Y. Tonomura, *J. Biochem.*, **65**, 567 (1969).
33 Y. Yazawa and K. Yagi, *Biochim. Biophys. Acta*, **180**, 190 (1969).
34 K. Yagi, Y. Yazawa, F. Ohtani and Y. Okamoto, presented at a Japan - U.S. Seminar, Tokyo (1972).
35 H. Onishi, H. Nakamura and Y. Tonomura, *J. Biochem.*, **63**, 739 (1968).
36 A. Weber and W. Hasselbach, *Biochim. Biophys. Acta*, **15**, 237 (1954).
37 J. R. Bendall, *Biochem. J.*, **81**, 520 (1961).
38 W. J. Bowen, L. S. Stewart and H. L. Martin, *J. Biol. Chem.*, **238**, 2926 (1963).
39 Y. Tonomura, H. Nakamura, N. Kinoshita, H. Onishi and M. Shigekawa, *J. Biochem.*, **66**, 599 (1969).

40 T. Yasui and S. Watanabe, *J. Biol. Chem.*, **240**, 98 (1965).
41 S. Watanabe and T. Yasui, *J. Biol. Chem.*, **240**, 105 (1965).
41a J. M. Gillis and G. Maréchal, *J. Physiol.*, **214**, 41P (1971).
41b A. Inoue, unpublished.
42 H. Tokuyama and Y. Tonomura, *J. Biochem.*, **62**, 456 (1967).
43 S. Kubo, S. Tokura and Y. Tonomura, *J. Biol. Chem.*, **235**, 2835 (1960).
44 S. V. Perry and J. Cotterill, *Nature*, **206**, 161 (1965).
45 L. Leadbeater and S. V. Perry, *Biochem. J.*, **87**, 233 (1963).
46 K. Sekiya, K. Takeuchi and Y. Tonomura, *J. Biochem.*, **61**, 567 (1967).
47 S. V. Perry, J. Cotterill and D. Hayter, *Biochem. J.*, **100**, 289 (1966).
48 E. M. Szentkiralyi and A. Oplatha, *J. Mol. Biol.*, **43**, 551 (1969).
49 E. Eisenberg and C. Moos, *Biochemistry*, **7**, 1486 (1968).
50 E. Eisenberg and C. Moos, *J. Biol. Chem.*, **245**, 2451 (1970).
51 J. M. Jones and S. V. Perry, *Biochem. J.*, **100**, 120 (1966).
52 K. Yagi and Y. Yazawa, *J. Biochem.*, **60**, 450 (1966).
53 E. Eisenberg, C. R. Zobel and C. Moos, *Biochemistry*, **7**, 3186 (1968).
54 S. Ebashi, *J. Biochem.*, **50**, 236 (1961).
55 A. Weber, *J. Biol. Chem.*, **234**, 2764 (1959).
56 K. Maruyama and S. Watanabe, *J. Biol. Chem.*, **237**, 3437 (1962).
57 S. V. Perry, R. Reed, W. T. Astbury and L. C. Spark, *Biochim. Biophys. Acta*, **2**, 674 (1948).
58 K. Maruyama and D. R. Kominz, *J. Biochem.*, **65**, 465 (1969).
59 Y. Kozuki and Y. Tonomura, *J. Biochem.*, **62**, 726 (1967).
60 Y. Kozuki and Y. Tonomura, unpublished.
61 K. Takahashi and T. Yasui, *J. Biochem.*, **62**, 131 (1966).
62 N. Ikemoto, S. Kitagawa and J. Gergely, *Biochem. Z.*, **345**, 410 (1966).
63 T. Nihei and T. Yamamoto, *Biochim. Biophys. Acta*, **180**, 178 (1969).
64 W. J. Bowen, *Amer. J. Physiol.*, **165**, 10 (1951).
65 T. Sekine and M. Yamaguchi, *J. Biochem.*, **59**, 195 (1966).
66 M. Tada and Y. Tonomura, *J. Biochem.*, **61**, 123 (1967).
67 J. Yoshimura, H. Matsumiya and Y. Tonomura, *Annu. Rep. Sci. Works, Fac. Sci., Osaka Univ.*, **11**, 51 (1963).
68 H. Nakamura and Y. Tonomura, *J. Biochem.*, **61**, 242 (1967).
69 T. Tokiwa, T. Shimada and Y. Tonomura, *J. Biochem.*, **61**, 108 (1967).
70 M. Kasai, E. Nakano and F. Oosawa, *Biochim. Biophys. Acta*, **94**, 494 (1965).
71 F. B. Straub and G. Feuer, *Biochim. Biophys. Acta*, **4**, 455 (1950).
72 A. Martonosi, M. A. Gouvea and J. Gergely, *J. Biol. Chem.*, **235**, 1700 (1960).
73 S. Asakura, *Biochim. Biophys. Acta*, **52**, 65 (1961).
74 A. G. Szent-Györgyi and G. Prior, *J. Mol. Biol.*, **15**, 515 (1966).
75 W. F. H. M. Mommaerts, *J. Biol. Chem.*, **198**, 469 (1952).
76 M. Bárány, A. F. Tucci and T. E. Conover, *J. Mol. Biol.*, **19**, 483 (1966).
77 C. Moos and E. Eisenberg, *Biochim. Biophys. Acta*, **223**, 221 (1970).

78 K. Takahashi and T. Yasui, *J. Biochem.*, **60**, 231 (1966).
79 H. M. Levy and E. M. Ryan, *Biochem. Z.*, **345**, 132 (1966).
80 T. Tokiwa and M. F. Morales, *Biochemistry*, **10**, 1722 (1971).
81 A. J. Murphy and M. F. Morales, *Biochemistry*, **9**, 1528 (1970).
82 Y. Tonomura and J. Yoshimura, *Arch. Biochem. Biophys.*, **90**, 73 (1960).
83 K. Maruyama and J. Gergely, *J. Biol. Chem.*, **237**, 1095 (1962).
84 W. Hasselbach and H. H. Weber, *Biochim. Biophys. Acta*, **11**, 160 (1953).
85 H. Onishi, K. Miki, M. Kaneko and Y. Tonomura, unpublished.
86 H. E. Huxley and W. Brown, *J. Mol. Biol.*, **30**, 383 (1967).
87 F. A. Pepe, *J. Cell Biol.*, **28**, 505 (1966).
88 F. A. Pepe, *J. Mol. Biol.*, **27**, 227 (1967).
89 F. A. Pepe, *Int. Rev. Cytol.*, **24**, 193 (1968).
90 Y. Tonomura, J. Yoshimura and S. Kitagawa, *J. Biol. Chem.*, **236**, 1968 (1961).
91 Y. Tonomura, J. Yoshimura and T. Ohnishi, *Biochim. Biophys. Acta*, **78**, 698 (1963).
92 Y. Hayashi, H. Takenaka and Y. Tonomura, unpublished.
93 H. M. Levy and E. M. Ryan, *J. Gen. Physiol.*, **50**, 2421 (1967).
94 K. Sekiya and Y. Tonomura, *J. Biochem.*, **69**, 935 (1971).
94a T. Matsunaga and H. Noda, *J. Biochem.*, **60**, 674 (1966).
94b R. W. Lymn and E. W. Taylor, *Biochemistry*, **10**, 4617 (1971).
95 B. Yasui, F. Fuchs and F. N. Briggs, *J. Biol. Chem.*, **243**, 735 (1968).
96 D. J. Hartshorne and J. L. Daniel, *Biochim. Biophys. Acta*, **223**, 214 (1970).
97 A. M. Katz, *Physiol. Rev.*, **50**, 63 (1970).
98 R. E. Olson, E. Ellenbogen and R. Iyenger, *Circulation Res.*, **24**, 471 (1961).
99 G. E. Conway and J. L. Roberts, *Amer. J. Physiol.*, **208**, 243 (1965).
100 R. J. Luchi, E. M. Kritcher and P. T. Thyrum, *Circulation Res.*, **24**, 513 (1969).
101 P. T. Thyrum, E. M. Kritcher and R. J. Luchi, *Biochim. Biophys. Acta*, **197**, 335 (1970).
102 E. M. Kritcher, P. T. Thyrum and R. J. Luchi, *Biochim. Biophys. Acta*, **221**, 264 (1970).
103 A. Katz and M. E. Carsten, *Circulation Res.*, **13**, 474 (1963).
104 W. D. McCubbin and C. M. Kay, *Biochim. Biophys. Acta*, **214**, 272 (1970).
105 M. Tada, *J. Biochem.*, **62**, 658 (1967).
106 A. L. Jacobson, W. G. Cambell and J. M. Adams, *Biochemistry*, **10**, 1063 (1971).
107 D. M. Needham and J. M. Williams, *Biochem. J.*, **89**, 552 (1963).
108 C. Cohen, S. Lowey and J. Kucera, *J. Biol. Chem.*, **236**, PC 23 (1961).
109 G. Hamoir and L. Laszt, *Nature*, **193**, 682 (1963).
110 J. C. Rüegg, E. Strassner and R. H. Schirmer, *Biochem. Z.*, **343**, 70 (1965).
110a N. Shibata, T. Yamagami, H. Yoneda and H. Akagami, unpublished.

111 M. Bárány, K. Bárány, E. Gaetjens and G. Bailin, *Arch. Biochem. Biophys.*, **113**, 205 (1966).
112 M. Yamaguchi, Y. Miyazawa and T. Sekine, *Biochim. Biophys. Acta*, **216**, 411 (1970).
113 Y. Nonomura, *J. Cell Biol.*, **39**, 741 (1968).
114 R. E. Kelly and R. V. Rice, *J. Cell Biol.*, **42**, 683 (1969).
115 J. Lowy and J. V. Small, *Nature*, **227**, 46 (1970).
115a J. V. Small and J. M. Squire, *J. Mol. Biol.*, **67**, 117 (1972).
116 R. V. Rice, G. M. McManus, C. E. Devine and A. P. Somlyo, *Nature New Biology*, **231**, 242 (1971).
117 A. P. Somlyo, A. V. Somlyo, C. E. Devine and R. V. Rice, *Nature New Biology*, **231**, 243 (1971).
118 S. Puszkin, S. Berl, E. Puszkin and D. D. Clarke, *Science*, **161**, 170 (1968).
119 S. Berl and S. Puszkin, *Biochemistry*, **9**, 2058 (1970).
119a S. Puszkin and S. Berl, *Biochim. Biophys. Acta*, **256**, 695 (1972).
119b R. C. Weisenberg, G. G. Borisy and E. W. Taylor, *Biochemistry*, **7**, 4466 (1968).
120 N. Senda, N. Shibata, N. Tatsumi, K. Kondo and K. Hamada, *Biochim. Biophys. Acta*, **181**, 191 (1969).
120a N. Shibata, N. Tatsumi, K. Tanaka, Y. Okamura and N. Senda, *Biochim. Biophys. Acta*, **256**, 565 (1972).
121 L. Bettex-Galland and E. R. Lüscher, *Advan. Protein Chem.*, **20**, 1 (1965).
121a R. S. Adelstein, T. D. Pollard and W. M. Kuehl, *Proc. Natl. Acad. Sci. U. S.*, **68**, 2703 (1971).
122 E. Puszkin, S. Puszkin and L. M. Aledort, *J. Biol. Chem.*, **246**, 271 (1971).
123 M. Bettex-Galland, E. Probst and O. Behnke, *J. Mol. Biol.*, **68**, 533 (1972).
124 D. W. Fawcett, *in* "The Cell," ed. by J. Brachet and A. E. Mirsky, Academic Press, New York, Vol. 2, p. 217 (1961).
125 D. W. Fawcett and K. R. Porter, *J. Morphol.*, **94**, 221 (1954).
126 I. R. Gibbons and A. V. Grimstone, *J. Biophys. Biochem. Cytol.*, **7**, 697 (1960).
127 I. R. Gibbons, *J. Biophys. Biochem. Cytol.*, **11**, 179 (1961).
128 I. R. Gibbons, *in* "Molecular Organization and Biological Function," ed. by J. M. Allen, Harper & Row, Pub., New York, p. 211 (1967).
129 M. A. Sleigh, "The Biology of Cilia and Flagella," Pergamon Press, Oxford (1962).
130 H. Mohri, *Nature*, **217**, 1053 (1968).
131 F. L. Renaud, A. J. Rowe and I. R. Gibbons, *J. Cell Biol.*, **31**, 92A (1966).
132 F. L. Renaud, A. J. Rowe and I. R. Gibbons, *J. Cell Biol.*, **36**, 79 (1968).
133 M. L. Shelanski and E. W. Taylor, *J. Cell Biol.*, **34**, 549 (1967).
134 M. L. Shelanski and E. W. Taylor, *J. Cell Biol.*, **38**, 304 (1968).
135 R. E. Stephens, *J. Mol. Biol.*, **32**, 277 (1968).
136 R. E. Stephens, *J. Mol. Biol.*, **33**, 517 (1968).

137 R. E. Stephens, *Science*, **168**, 845 (1970).
137a J. B. Olmsted, G. B. Witman, K. Carlson, and J. L. Rosenbaum, *Proc. Natl. Acad. Sci. U. S.*, **68**, 2273 (1971).
138 V. A. Engelhardt, *Advan. Enzymol.*, **6**, 147 (1946).
139 V. A. Engelhardt and S. A. Burnasheva, *Biochimiya*, **22**, 513 (1957).
140 S. A. Burnasheva, *Biochimiya*, **23**, 558 (1958).
141 I. R. Gibbons, *Proc. Natl. Acad. Sci. U. S.*, **50**, 1002 (1963).
142 I. R. Gibbons, *Arch. Biol. Liége*, **76**, 317 (1965).
143 I. R. Gibbons and A. J. Rowe, *Science*, **149**, 424 (1965).
144 I. R. Gibbons, *J. Biol. Chem.*, **241**, 5590 (1966).
145 P. Satir, *J. Gen. Physiol.*, **50**, *Suppl.*, 241 (1967).
146 P. Satir, *J. Cell Biol.*, **39**, 77 (1968).
146a K. E. Summers and I. R. Gibbons, *Proc. Natl. Acad. Sci. U. S.*, **68**, 3092 (1971).
146b D. Epel, *J. Cell Biol.*, **17**, 315 (1963).
146c D. Mazia, *in* "The Cell," ed. by J. Brachet and A. E. Mirsky, Academic Press, New York, Vol. 3, p. 78 (1961).
146d A. I. Zotin, L. S. Milman and V. S. Faustov, *Exp. Cell Res.*, **39**, 567 (1965).
146e N. Sawada and L. I. Rebhun, *Exp. Cell Res.*, **55**, 33 (1969).
146f D. Mazia, R. R. Chaffee and R. M. Iverson, *Proc. Natl. Acad. Sci. U.S.*, **47**, 788 (1961).
146g T. Miki, *Exp. Cell Res.*, **29**, 92 (1963).
147 G. G. Borisy and E. W. Taylor, *J. Cell Biol.*, **34**, 535 (1967).
148 R. Weisenberg and E. W. Taylor, *Exp. Cell Res.*, **53**, 372 (1968).
149 S. Inoué and H. Sato, *J. Gen. Physiol.*, **50**, *Suppl.*, 259 (1967).
150 E. Robbins and N. K. Gonatas, *J. Cell Biol.*, **21**, 429 (1964).
151 A. Bajer, *Sym. Soc. Exp. Biol.*, **22**, 285 (1968).
152 L. I. Rebhun and G. Sander, *J. Cell Biol.*, **34**, 859 (1967).
152a D. Mazia, Ch. Petzelt, R. O. Williams and I. Meza, *Exp. Cell Res.*, **70**, 325 (1972).
152b Ch. Petzelt, *Exp. Cell Res.*, **70**, 333 (1972).
153 J. R. McIntosh, P. K. Hepler and D. G. Van Wie, *Nature*, **224**, 659 (1969).
154 J. R. McIntosh and S. Cleland, *J. Cell Biol.*, **43**, 89 A (1969).
154a O. Behnke, A. Forer and J. Emmersen, *Nature*, **234**, 408 (1971).
154b N. Gawadi, *Nature*, **234**, 409 (1971).
155 A. G. Loewy, *J. Cell. Comp. Physiol.*, **40**, 127 (1952).
156 P. O. P. Ts'o, L. Eggman and J. Vinograd, *J. Gen. Physiol.*, **39**, 801 (1956).
157 H. Nakajima, *Protoplasma*, **52**, 413 (1960).
158 S. Hatano and F. Oosawa, *J. Cell Physiol.*, **68**, 197 (1966).
159 S. Hatano and F. Oosawa, *Biochim. Biophys. Acta*, **127**, 488 (1966).
160 S. Hatano, T. Totsuka and F. Oosawa, *Biochim. Biophys. Acta*, **140**, 109 (1967).

161 S. Hatano and M. Tazawa, *Biochim. Biophys. Acta*, **154**, 507 (1968).
162 M. R. Adelman and E. W. Taylor, *Biochemistry*, **8**, 4964 (1969).
163 M. R. Adelman and E. W. Taylor, *Biochemistry*, **8**, 4976 (1969).
164 V. T. Nachmias, H. E. Huxley and D. Kessler, *J. Mol. Biol.*, **50**, 83 (1970).
165 N. Kamiya, *Protoplasmatologia*, **8**, 1 (1959).
166 D. E. Woolley, *J. Cell. Physiol.*, **76**, 185 (1970).
167 T. D. Pollard, E. Skelton, R. R. Weihing and E. D. Korn, *J. Mol. Biol.*, **50**, 91 (1970).
167a L.-F. Yen and T.-C. Shih, *Sci. Sinica*, **14**, 601 (1965).
167b L.-F. Yen, Y.-S. Han and T.-C. Shih, *Kexue Tong Bao*, **17**, 138 (1966).
168 R. R. Weihing and E. D. Korn, *Biochemistry*, **10**, 590 (1971).
169 D. T. Pollard, *Federation Meeting (Biochemistry)*, (1971).
170 L. Levine, *Biol. Bull.*, **111**, 319 (1956).
171 H. Hoffmann-Berling, *Biochim. Biophys. Acta*, **27**, 247 (1958).
172 W. B. Amos, *Nature*, **229**, 127 (1971).
173 T. Weis-Fogh and W. B. Amos, *Nature*, **236**, 301 (1972).

8

CONTRACTION OF MUSCLE MODELS BY ATP*

In vitro studies of structural proteins, particularly myosin and actin, isolated from muscle have been described in chapters 2 to 7. However, conclusions about *in vivo* phenomena such as muscle contraction should be drawn only with great caution from such *in vitro* studies alone. An important goal of the study of muscle contraction is to uncover the molecular mechanism of conversion of chemical to mechanical energy. In order for us to assess their physiological significance, the chemical processes found to occur with isolated structural proteins *in vitro*, when no mechanical work is performed, must ultimately be related to the dynamic properties of muscle. Since the protoplasmic membrane prevents free diffusion of the substrates in the reactions of which we are interested, and since the stages of excitation, contraction and relaxation overlap considerably, so that they cannot easily be studied separately, direct biochemical investigations of living

* Contributor: Yutaro Hayashi

muscle fibres are very difficult. Furthermore, the ranges over which external conditions such as, pH and ionic strength, can be varied are narrow. To bridge the gap between these two approaches, various muscle models have been used. Of all the proposed models, Natori's skinned muscle fibre[1,2], which is obtained by removing only the sarcolemma or protoplasmic membrane from living muscle, most closely resembles living muscle. Szent-Györgyi's glycerol-treatment of muscle[3] destroys the cell metabolic system as well as the sarcolemma and the T-system. The use of such specimens permits the study not only of the mechanical properties but also of the chemical properties of the contractile system itself. Perry and Corsi[4] have isolated myofibrils and we have used their system for studying the ability of ATP analogues to cause contraction (chapter 9). In this chapter we shall show that the smallest unit containing the complete contractile system of living muscle—the sarcomere[5]—can be used to investigate the biochemical phenomena related to contraction, although it is not suitable for measuring mechanical properties, such as tension.

There are also models obtained by combining and treating isolated contractile proteins, such as the synthetic threads made from myosin, actin and other structural proteins. These have been developed by Weber[6], Portzehl[7] and Hayashi et al.[8-10], and were used by this last research group to demonstrate the highly significant fact that contraction by ATP requires the presence of both myosin and actin, but of no other component. Further research in this direction is expected to develop a comprehensive muscle model with a regulatory system (chapter 10) like that in living muscle. This chapter will describe important work on the relationship between biochemical and mechanical phenomena using the above models.

1. Glycerol-treated Muscle Fibres

Szent-Györgyi[3] found that if muscle fibre bundles from rabbit psoas are kept in 50% glycerol-water solution at $-20°C$, they immediately contract when they are transferred to salt solutions containing ATP, even after several years' storage. Glycerol-treatment destroys the protoplasmic membrane and the T-system of the muscle, eliminates the metabolic system and removes interior ATP and CrP. However, X-ray diffraction and electron microscopy studies of its structure[11] indicate that the contractile system is practically unaltered from that in living muscle. Two important functions of ATP in the contractile system, contraction and relaxation, were investigated by Weber and Portzehl[12,13] using this muscle model. If a glycerol-treated muscle, fixed at a certain length, is immersed in a salt solution containing ATP, it decomposes the ATP and generates tension. If the load is quickly released to allow an immediate 5–10% shortening, the tension at once becomes zero, but then returns to the original value. This 'quick release' phenomenon is not observed in other elastic systems, and is characteristic of living muscle and glycerol-treated muscle in the presence of ATP. Removal of ATP from the muscle model results in rigor. The resistance of muscle fibre in rigor to being stretched is about 10,000 g·cm^{-2}·L·$\varDelta L^{-1}$, which is considerably greater than the resistance of a resting muscle containing ATP. This indicates that, in rigor, myosin and actin are strongly bound together in the myofibril. On the addition of ATP, resistance against stretch is approximately 5,000 g·cm^{-2}·L·$\varDelta L^{-1}$, so that there is only partial binding of myosin with actin under these conditions. After tension has been generated with ATP, the further addition of PP$_i$, inorganic triphosphate or ADP lowers the tension—this is called the 'plasticising effect.' Unlike ATP, these reagents are not decomposed, but they can cleave the binding between actin and myosin filaments and thus cause the muscle to relax. Under conditions when it is not decomposed, ATP shows a stronger plasticising effect than any of the above reagents. Thus when tension is induced by ATP, the addition of Salyrgan (an SH poison which inhibits the myosin-ATPase) considerably decreases the tension, and there is then only weak resistance to stretch, ca. 250 g·cm^{-2}·L·$\varDelta L^{-1}$. Reversal of the effect of Salyrgan by adding cysteine to restore the ATPase activity regenerates tension. Weber and Portzehl[12] concluded from these experiments that a high rate of ATP-decomposition is needed for muscle contraction, and that inhibition of ATP-decomposition causes relaxation of the muscle by ATP.

The current view of the mechanism of contraction and relaxation (chapters 10 and 13) is as follows: excitation of the plasma membrane of muscle induces discharge of Ca^{2+} from the sarcoplasmic reticulum. This then diffuses to the contractile system and is accepted by a regulatory protein present in the thin filaments. Myosin then binds with actin to form the actomyosin-ATPase, which then induces contraction. Following this, Ca^{2+} is removed from the contractile system by absorption back into the sarcoplasmic reticulum, cleaving the binding between myosin and actin by ATP, so that the ATPase activity returns to that of myosin and the muscle relaxes. Weber and Portzehl's conclusion about the coupling of the ATPase to contraction is still accepted as correct, and has been important in establishing the general mechanism of contraction and relaxation. In this chap-

240

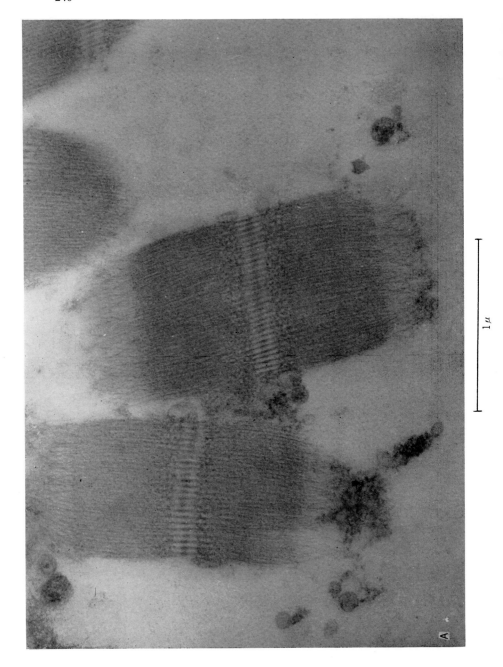

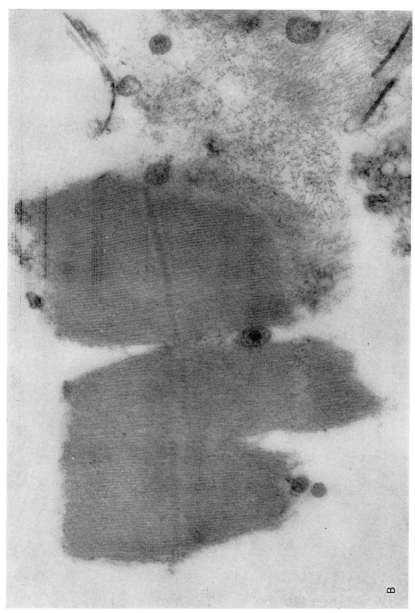

Fig. 1. Electron micrograph of an isolated single sarcomere. A: Before addition of ATP. B: After addition of ATP. (Courtesy of Professor T. Fukazawa, Faculty of Agriculture, Kyushu University.)

ter the mechanism of this coupling is described in detail, which is based mainly on our own work.

2. Contraction of Isolated Sarcomeres by ATP

The basic structural unit for the contraction of a myofibril is the sarcomere (chapter 1). We[5,14,15] have isolated single sarcomeres in substantial numbers from homogenates of chicken pectoral muscle by repeated differential centrifugation and a final filtration. Many of the sarcomeres clearly showed the A band and H-zone under phase contrast microscopy, but the I band was not very easy to see[5]. However, electron microscopy showed that the Z line had disappeared and the sample had been fragmented into sarcomeres (Fig. 1A)[16]. When ATP was added to a single sarcomere showing no Z line and in which both ends had cleaved at the ends of the I bands, contraction ceased when the H-zone and I band disappeared, so that the length was then equal to that of the original A band

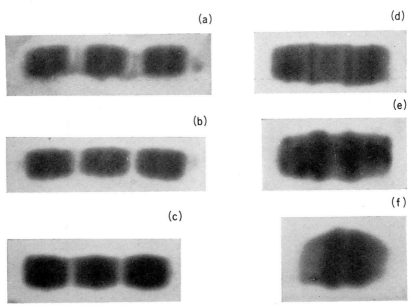

Fig. 2. Myofibrillar fragments at different stages of contraction. (a) Before and (b–f) after addition of ATP. 0.04M KCl, 5μM MgCl$_2$, 80μM ATP, 0.03M Tris-maleate, pH 7.0. Room temp. After initiation of contraction for 30 sec (b and c) and 1 min (d–f), shortening was blocked by adding 1mM EDTA.

(Fig. 1B). Double sarcomeres, or even larger groups which contained a central Z line, contracted further and a contraction band appeared, first at the M line—the centre of H-zone—as if the thin filaments collide together, then at the border of the A and I bands and finally at the position of the Z line (Fig. 2).

We have investigated the relationship between contraction of the sarcomere and decomposition of ATP under various conditions. Experiments using various concentrations of EDTA, EGTA, Mg^{2+}, Ca^{2+}, and ATP indicated that only Mg-ATP is needed for contraction of the sarcomere. However, under conditions where contraction was observed, the ATPase activity was also dependent on the concentration of Mg-ATP and not on that of free ATP, so that the ATP must be split entirely by the actomyosin-ATPase.

The relationship between the rate of splitting of ATP and the rate of contraction was also studied. Contraction was followed both by watching the shortening, and by the decrease in light-scattering intensity of the sarcomere suspension with the growth of the contraction band. The following relationship was obtained:

$$-dl/dt = -C\frac{d[S]/dt}{\left(1+\frac{K}{[S]}\right)^2},$$

where l is sarcomere length, [S] is [Mg-ATP], K is a constant of magnitude about 10^{-5}M at 0.04M KCl, pH 7.0 and 25°C, and C is a constant. Probably the term $\left(1+\frac{K}{[S]}\right)^2$ reflects the cleavage of actin from myosin by ATP, since the shortening due to movement of one cross-bridge will be inhibited when many cross-bridges are formed between the thick and thin filaments and they do not move synchronously. The above equation was valid over a broad range, the constant C being independent of the conditions. This indicates a fairly tight coupling between the actomyosin-ATPase reaction and the contraction. The measured value of C indicates that a contraction of 45% of the original length is equivalent to the decomposition of 7 moles of ATP per 10^6g of myofibrillar protein; a twitch involving a contraction of only 15% therefore corresponded to the splitting of 2.3 moles of ATP/10^6g[15]. Carlson et al.[17,18] and Mommaerts et al.[19] showed that the amount of the high energy phosphate compound split in a single twitch of living muscle was 2.0–2.3 moles/10^6g of myofibrillar protein, and the agreement of this value with ours shows that the contraction of isolated sarcomeres by ATP is physiologically significant.

3. Thread Models

Actomyosin is soluble at high ionic strengths, but precipitates at low ionic strengths (chapter 6). Hence one can prepare actomyosin threads by suddenly injecting the solution into distilled water, which causes coagulation [6]. Threads thus prepared show ATPase activity and shrink on the addition of ATP, but because of their poor orientation relative to the fibre axis of actomyosin, they contract isodimensionally, unlike living muscle[20]. Portzehl[7] and Hayashi[8] have prepared better oriented actomyosin threads. Hayashi[8] made a monomolecular layer of actomyosin, which was compressed in one direction to produce an oriented thread which contracted anisodimensionally on the addition of ATP at low ionic strengths (0.05M KCl), and which also showed ATPase activity. The shortened thread could be elongated by exposure to high ionic strengths (0.25M KCl). Alternation of these operations produced the same cycle of contraction and elongation observed with glycerinated muscle fibre[8]. Significantly, threads prepared using myosin only, after removal of the actin, split ATP but developed no tension. As the actin content of the actomyosin thread was increased, the tension generated also increased proportionately, up to a value of 27%[10]. These results indicate that muscle contraction is not due only to myosin and ATP, but involves the interaction of myosin, actin and ATP (chapter 1).

4. Relationship between Development of Tension, Splitting of ATP and Sarcomere Length

Ramsey and Street[21] were the first to measure the relationship between the length of a muscle fibre and its tension; they did so using an intact single fibre in isometric contraction. They showed that the tension was maximum at the resting length, and decreased upon shortening or lengthening (Fig. 3). Recently Gordon et al.[22,23] used uniform lengths of sarcomere to measure accurately the relationship between the sarcomere length and tension in living muscle fibre. The tension was at a constant maximum value when the sarcomere length ranged from 2.05 to 2.20μ. At shorter lengths, especially below 1.67μ, the tension decreased and became almost zero at 1.27μ. At 2.2μ or more the tension decreased linearly with increasing length, becoming zero again at 3.65μ (Fig. 3). A similar dependence of tension on sarcomere length (<2.0μ) was also reported by

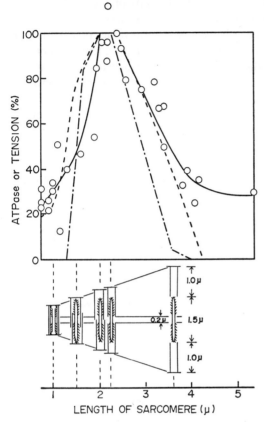

Fig. 3. Dependence of tension development and ATPase activity on sarcomere length. - - -: Ramsey and Street[21] and —·—·—: Gordon et al.[22, 23] show the relationship between isometric tetanus tension and sarcomere length. —— and ○, Hayashi and Tonomura's results on glycerol-treated muscle fibre[26, 27], showing the relationship between ATPase activity and sarcomere length during ATP-induced isometric contraction. The lower diagram represents the extent of overlapping of the thin filaments with the bridge from the thick filaments, based on the 'sliding theory' of muscle contraction and the double filament structure of muscle fibres (*cf.* chapter 1).

Hellam and Podolsky[24], who used skinned muscle fibres. These results agree qualitatively with those of Ramsey and Street. The double filament structure of the myofibrils and the 'sliding theory' of contraction were described in chapter 1, while Fig. 3 shows for various lengths of sarcomere the size of the region a_0 in which the projections of the thick filaments overlap the thin filaments. Gordon *et al.*'s relationship between tension and sarcomere length agrees excellently with theoretical values predicted on the assumption that at sarcomere lengths of 2.0μ or greater, tension is proportional to a_0, while at lengths of 3.6μ or more, when $a_0=0$, the tension decreases to zero. Hence tension is generated by the interaction between the thin filaments and the projections from the thick filaments, which correspond to the H-MM portion of the myosin molecule (*cf.* chapter 2, section 7). The volume of muscle is kept constant during shortening, so that the distance between the thick and thin filaments increases with shortening[25]. In spite of such a change in the interfilament distance, the tension developed is proportional to the length of the overlapping region of the two filaments, as mentioned above. This has been considered to be due to the action of the flexible part of the cross-bridge, *i.e.*, the subfragment-2 of myosin[26] (*cf.* chapter 2, section 5).

Sandberg and Carlson[27] studied the relationship between the rate of splitting of CrP and the sarcomere length in an isometric tetanus of living muscle under nitrogen, under conditions in which ATP synthesis by glycolysis had been inhibited by treatment with iodoacetamide. There are two components of the rate of decomposition of CrP—one proportional to the tension generated and dependent on sarcomere length, and the other

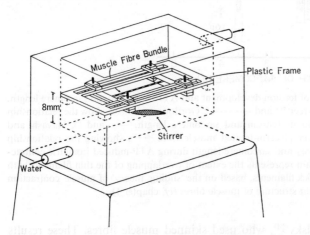

Fig. 4. Apparatus for measurement of the ATPase activity of fibre bundles at fixed sarcomere length. The two ends of a fibre bundle were clamped between pairs of acrylic plastic plates with stainless steel screws. The temperature of the bath was maintained constant by running water.

independent of these quantities. The latter may correspond to consumption of energy in the calcium pump of the sarcoplasmic reticulum (cf. chapter 11).

In order to understand better the relationship between sarcomere length and ATPase activity[28,29], the ATPase activity was measured during an ATP-induced 'contraction' of a glycerinated fibre bundle at fixed fibre length. The ATPase activity was determined by coupling a pyruvate kinase system to the myofibrillar ATPase and measuring the pyruvate liberated. The diffusion of ATP into the muscle fibre was not rate-determining under the conditions used. The ends of the fibre bundles were clamped to plastic plates (Fig. 4) and the activity corresponding to the clamped part was subtracted from the total activity in order to derive the specific activity of the portion of relatively constant sarcomere length. For lengths of 2.0–2.5μ the activity was almost constant, but when the sarcomere length was increased the activity decreased linearly until, at about 4μ, it was between a third and a quarter of the maximum value, after which it remained constant (Fig. 3). This result agrees with those of Ward et al.[30] However, at lengths of 2μ or less, the ATPase activity drastically decreased again, reaching a quarter of the maximum value at about 1μ. This relationship between ATPase activity and sarcomere length agrees well with the dependence of tension on sarcomere length found by Ramsey and Street.

There is disagreement between our measurements and Gordon and colleagues' result at short and long sarcomere lengths, but this may be due to non-uniformity of sarcomere length during the measurement of the ATPase activity in our studies, and also to a contribution from myosin-ATPase which is not directly related to contraction. Thus, these measurements indicate that development of tension and splitting of ATP both result from the interaction of the thin filaments with the projections from the thick filaments. Recently, Jagendorf-Elfvin[31] has determined the position of deposition of a lead phosphate precipitate in glycerinated muscle during ATP-induced contraction, in an attempt to locate the primary site of contractile ATPase activity. She found that the precipitate was most concentrated in those regions where there is most contact between the thick and thin filaments.

If our above conclusion is true, we must find the reason for the decrease in ATPase activity and tension at sarcomere lengths of 2μ or less. During contraction of the scutal depressor of a barnacle and the ventral longitudinal muscle of a blowfly larva, the thick filament is reported to pass through the Z line instead of stopping there, and to reach the next sarcomere. This 'supercontraction' was described by Hoyle et al.[32] and Osborne[33], and suggests that even when the sarcomere length has been sufficiently shortened for the thick filament to reach the Z line, there is still a driving force for contraction. Thus at a length of 1μ there is still an interaction between the thin filaments and the projections from the thick filaments. However, in the myofibrils of ordinary striated muscle, when the length of the sarcomere becomes about 1μ the thick filaments fold at both ends and touch the Z line, while the thin filaments also touch the Z line; this prevents cross-bridge movement, as shown in Fig. 3.* Thus if we postulate that movement of the cross-bridges from the thick filaments which contact with the thin filaments is coupled with the actomyosin-

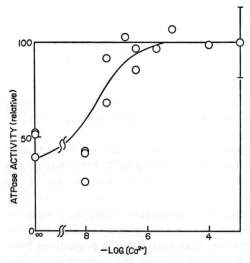

Fig. 5. Dependence of the ATPase activity of a fibre bundle at resting length on Ca^{2+} concentration. Muscle fibre bundles were used after storage in 50% glycerol for 3 months. 2.4mM ATP, 1mM PEP, 20–60μg/ml PK, 50mM KCl, 3.4mM $MgCl_2$, pH 7.0, 25°C.

* Hagopian's[34] electron microscope observations of contraction bands in glycerinated chick pectoral muscle indicate that at a sarcomere length of 1.3μ the tapered ends of the thick filaments are able to penetrate the Z line in an orderly arrangement, and are able to form a dense zone in the adjacent sarcomere. Therefore, another possible cause of the low ATPase activity at short sarcomere lengths is that this new arrangement of the thick and thin filaments prevents good contact between the thin filaments and the projections.

ATPase, and that the ATPase is coupled with development of tension, we can easily explain the decrease in both tension and ATPase activity at sarcomere lengths of about 1μ by the two filaments touching the Z line and suppressing the cross-bridge movement.

The ATPase activity of a fibre bundle of sarcomere length $2.00-2.65\mu$ stored in glycerol-water for about 3 months stays constant at free Ca^{2+} concentrations of $1\mu M$ or more, but decreases to about 40% of the maximum value at lower Ca^{2+} concentrations, although it is not reduced to that of myosin even in the presence of EGTA (Fig. 5). The activity in the presence of EGTA depended on the time of storage of the fibre, so that a sample kept 3–13 days showed only 15% of the value obtained in the presence of Ca^{2+}, but a specimen stored for 9 months showed full activity even when sufficient EGTA was present. Bozler[35,36] has already demonstrated that the Ca^{2+}-sensitivity of tension development in a glycerinated muscle fibre decreases with storage time, indicating that tension development and ATPase activity in actomyosin are based on the same elementary process, which is probably regulated by Ca^{2+}. Two explanations of the decrease in Ca^{2+}-sensitivity of the ATPase activity of muscle fibres stored in glycerol-water are feasible: (1) that the action of the regulatory protein (chapter 10) disappears during storage and (2) that myosin is converted to a state similar to that produced by treatment with PCMB and β-mercaptoethanol (chapter 7, section 7). It is not known which mechanism is correct.

5. Kinetic and Viscoelastic Properties of Cross-Bridges

According to Hanson and H. E. Huxley's sliding theory of muscle contraction (chapters 1 and 13), the development of tension by muscle is ultimately attributable to movement of the cross-bridges from the thick filament. Therefore, in order to elucidate the molecular mechanism of muscle contraction, we must investigate kinetic and viscoelastic properties of the cross-bridges and the interrelationships between the movements of the many cross-bridges attached to one thick filament.

According to Podolsky and Teichholz[37], the velocity of shortening of skinned muscle fibres under any relative load P/P_o is not altered by changes in the Ca^{2+} concentration that markedly affect the contractile force. Thus the force is proportional to the number of cross-bridges, but although calcium ions appear to control the number of sites at which cross-bridges can be formed, they have no significant effect on the kinetic properties of an individual bridge. Furthermore, the same relationship between tension and velocity was obtained during relaxation by removal of Ca^{2+} with EGTA. Therefore, it is concluded that there is no time-lag between the inactivation of the site and the decrease in force, and that the kinetic properties of each cross-bridge (site) does not change during the relaxation process. This is because the number of cross-bridge contacts is probably small (less than 20% of those in rigor) during maximal activation of muscle, as found by H. E. Huxley and Brown[38] from their X-ray analysis of living frog muscle. Thus, it appears that we can construct a molecular model of muscle contraction, as was done by A. F. Huxley (chapter 13), which can interpret the various aspects of contraction of muscle cells even though it considers apparently only one cross-bridge.

However, we deduced the interdependence of cross-bridges in contracted muscle fibres from our own studies[39] of the viscoelasticity of glycerinated muscle fibres. The viscoelastic properties of fibres were measured by the method of forced oscillation, using an apparatus (Fig. 6) similar to that described by Miki et al.[40] At various frequencies (ν Hz; $\omega=2\pi\nu$ rad/sec) of an oscillator of amplitude F, the amplitude A of oscillation of the fibre and the time-interval Δt between these two oscillations were measured. The tensile storage modulus[41] then approximates to

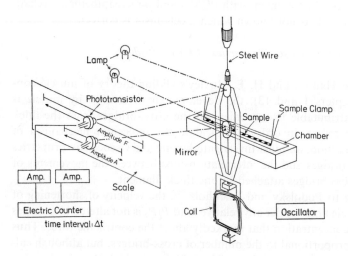

Fig. 6. Apparatus for the measurement of viscoelasticity of glycerol-treated muscle fibre. The bath temperature was kept constant by flowing an ATP solution of constant temperature.

$$E'(\omega) = \frac{lc'}{2ar^2}\left(\frac{F}{A} - 1\right),$$

where l is the length of muscle fibre, c' the torsional constant of the wire, a the cross-sectional area of muscle fibre and r the distance between wire and fibre. The tensile loss modulus is given by

$$E''(\omega) = \omega\eta'_\iota = \frac{lc'}{2ar^2}\left(2\pi \cdot \Delta t \cdot \nu \cdot \frac{F}{A}\right),$$

where η'_ι is the real part of the tensile viscosity.

Figure 7 shows the dependences on ν of the relative values of the tensile loss modulus $(2\pi \cdot \Delta t \cdot \nu F/A)/a$ and the tensile storage modulus $(F/A-1)/a$ in 0.05M KCl, 2mM MgCl$_2$, 10μM CaCl$_2$, 10% glycerol and 2mM ATP at pH 7.0 and 19°C. Over the range of ν used, both values decreased when the sarcomere length was increased from 2.38 to 3.66μ. Furthermore, the dependence of $tan\,\delta (=E''(\omega)/E'(\omega))$ on ν was independent of the sarcomere length. Therefore, in Fig. 8 only the tensile storage modulus is plotted

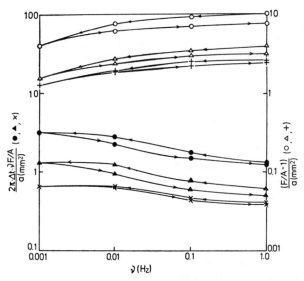

Fig. 7. Dependence on frequency, ν, of the relative values of the tensile loss modulus, $(2\pi \cdot \Delta t \cdot \nu \cdot F/A)/a$, and the tensile storage modulus, $(F/A-1)/a$, of glycerol-treated muscle fibre bundles. 2mM ATP, 50mM KCl, 2mM MgCl$_2$, 10μM CaCl$_2$, 10% glycerol, pH 7.0, 19°C. The arrow indicates the direction of measurement. Sarcomere length: ○, ●, 2.38; △, ▲, 3.17; +, ×, 3.66μ.

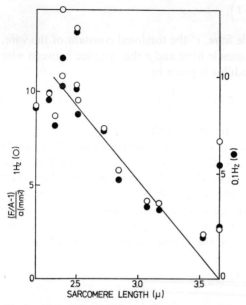

Fig. 8. Dependence of the tensile storage modulus on sarcomere length. Experimental conditions as in Fig. 7. The relative values of the tensile storage modulus at $\nu = 1$Hz (○) and 0.1Hz (●) are plotted against sarcomere length.

against the sarcomere length. It decreases linearly with increase in the sarcomere length from 2.2 to 3.5μ, and the line intercepts the abscissa at a sarcomere length of 3.6μ. This dependence of tensile storage modulus on sarcomere length agrees well with that of tension development described in section 4. Therefore, we can conclude that the viscoelasticity of contracted muscle fibre is proportional to the area of the over-lapping region of the thick and thin filaments. At sarcomere lengths longer than 3.5u, the tensile storage modulus increases again with increase in the sarcomere length. This may probably be attributed to components other than cross-bridges, *e.g.* parallel elastic components.

Figure 9 shows the dependences on ν of the relative values of the tensile loss modulus $(2\pi \cdot \Delta t \cdot \nu F/A)/a$ and the tensile storage modulus $(F/A - 1)/a$ at a fixed sarcomere length of 2.4μ in the presence of three different ATP concentrations. The ranges of frequency of dispersion for both these two moduli shift to higher frequencies with increase in the ATP concentration, and the curves at different ATP concentrations coincide with each other

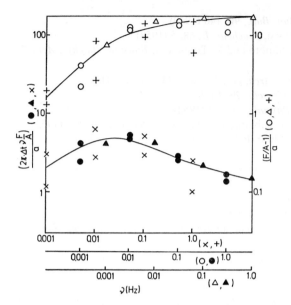

Fig. 9. Dependence on frequency, ν, of the relative values of the tensile loss modulus, $(2\pi \cdot \Delta t \cdot \nu \cdot F/A)/a$, and the tensile storage modulus, $(F/A-1)/a$, of muscle fibre bundles at different ATP concentrations. 50mM KCl, 2mM $MgCl_2$, in excess of ATP concentrations. 3.2μM Ca^{2+}, 10% glycerol, pH 7.0, 19°C. ATP concentrations: △, ▲, 1; ○, ●, 4; +, ×, 10mM.

when only the abscissa (log ν) is shifted. Similar shift in the range of frequency of dispersion is also observed with decrease in the concentration of Ca^{2+} ions on adding EGTA or with decrease in temperature of the bath. It must be noted that all these procedures control the number of sites at which cross-bridge contacts can be formed. Thus it is concluded that the elasticity of glycerol-treated muscle fibres at contracted state is independent of the number of cross-bridge contacts per unit length of the overlapping region, while the viscosity decreases with decrease in the number of cross-bridge contacts between the thick and thin filaments.

REFERENCES

1 R. Natori, *Jikeikai Med. J.*, **1**, 119 (1954).
2 R. Natori, *Jikeikai Med. J.*, **2**, 1 (1955).

3   A. Szent-Györgyi, *Biol. Bull.*, **96**, 140 (1949).
4   S. V. Perry and A. Corsi, *Biochem. J.*, **68**, 5 (1958).
5   T. Fukazawa, Y. Hashimoto and Y. Tonomura, *Biochim. Biophys. Acta*, **75**, 234 (1963).
6   H. H. Weber, *Pflüger's Arch. ges. Physiol.*, **235**, 205 (1934).
7   H. Portzehl, *Z. Naturforsch.*, **7b**, 1 (1952).
8   T. Hayashi, *J. Gen. Physiol.*, **36**, 139 (1952).
9   T. Hayashi and R. Rosenbluth, *J. Cell. Comp. Physiol.*, **40**, 495 (1952).
10 T. Hayashi, R. Rosenbluth, P. Satir and M. Vozick, *Biochim. Biophys. Acta*, **28**, 1 (1958).
11 H. E. Huxley, *J. Biophys. Biochem. Cytol.*, **3**, 631 (1957).
12 H. H. Weber and H. Portzehl, *Progr. Biophys. Biophys. Chem.*, **4**, 60 (1954).
13 H. H. Weber, *in* "Molecular Biology," ed. by D. Nachmansohn, Academic Press, New York, p. 25 (1960).
14 K. Takahashi, T. Mori, H. Nakamura and Y. Tonomura, *J. Biochem.*, **57**, 637 (1965).
15 H. Nakamura, T. Mori and Y. Tonomura, *J. Biochem.*, **58**, 582 (1965).
16 T. Fukazawa and T. Yasui, personal communication.
17 F. D. Carlson and A. Siger, *J. Gen. Physiol.*, **44**, 33 (1960).
18 F. D. Carlson, D. J. Hardy and D. R. Wilkie, *J. Gen. Physiol.*, **46**, 851 (1963).
19 W. F. H. M. Mommaerts, K. Seraydarian and G. Marechal, *Biochim. Biophys. Acta*, **57**, 1 (1962).
20 A. Szent-Györgyi, "Chemistry of Muscular Contraction," 1st ed., Academic Press, New York (1947).
21 R. W. Ramsey and S. F. Street, *J. Cell. Comp. Physiol.*, **15**, 11 (1940).
22 A. M. Gordon, A. F. Huxley and F. J. Julian, *J. Physiol.*, **184**, 143 (1966).
23 A. M. Gordon, A. F. Huxley and F. J. Julian, *J. Physiol.*, **184**, 170 (1966).
24 D. C. Hellam and R. J. Podolsky, *J. Physiol.*, **200**, 807 (1969).
25 E. Rome, *J. Mol. Biol.*, **27**, 591 (1967).
26 H. E. Huxley, *Science*, **164**, 1356 (1969).
27 J. A. Sandberg and F. D. Carlson, *Biochem. Z.*, **345**, 212 (1966).
28 Y. Hayashi and Y. Tonomura, *J. Biochem.*, **60**, 484 (1966).
29 Y. Hayashi and Y. Tonomura, *J. Biochem.*, **63**, 101 (1968).
30 P. C. J. Ward, C. Edwards and E. S. Benson, *Proc. Natl. Acad. Sci. U. S.*, **53**, 1377 (1965).
31 M. Jagendorf-Elfvin, *Tissue and Cell*, **2**, 311 (1970).
32 G. Hoyle, J. H. McAlear and A. Selverston, *J. Cell Biol.*, **26**, 621 (1965).
33 M. P. Osborne, *J. Ins. Physiol.*, **13**, 1471 (1967).
34 M. Hagopian, *J. Cell Biol.*, **47**, 790 (1970).
35 E. Bozler, *Amer. J. Physiol.*, **167**, 276 (1951).
36 E. Bozler, *Amer. J. Physiol.*, **168**, 760 (1952).
37 R. J. Podolsky and L. E. Teichholz, *J. Physiol.*, **211**, 19 (1970).

38 H. E. Huxley and W. Brown, *J. Mol. Biol.*, **30**, 383 (1967).
39 H. Onishi, K. Miki, M. Kaneko and Y. Tonomura, unpublished.
40 K. Miki, K. Hikichi and M. Kaneko, *Jap. J. Phys.*, **6**, 931 (1967).
41 J. D. Ferry, "Viscoelastic Properties of Polymers," John Wiley & Sons, New York (1960).

9
ATP ANALOGUES*

The reaction of myosin, actin and ATP was discussed in previous chapters, mainly with regard to the structures of myosin and actin and the changes which occur in the protein structure upon interaction with ATP. We shall now examine the influence of the nucleotide structure. The use of ATP analogues to help elucidate the mechanism of reaction of myosin with ATP and the role of the nucleotide in actin were described briefly in chapters 3 and 5. Since ATP is essential to biological energy transformation, it is very important to know in detail the significance of its structure towards its reactions *in vitro* and *in vivo*. We shall now try to relate the structure of ATP to its reactions with myosin and actin.

Natural products such as ITP, CTP, GTP and UTP were the first ATP analogues used, and their reactions with actomyosin have been studied intensively[1-6], particularly by Blum[1] and Hasselbach[2]. However, since

* Contributor: Kazuko Shibata-Sekiya

these compounds can provide only limited information, their structures all being very similar, a variety of synthetic ATP analogues have been prepared and their reactions with myosin and actin studied. Fortunately, standard methods for the synthesis of organic triphosphates have been developed by Todd and his co-workers[7,8] and by Smith and Khorana[9,10], and it is now fairly easy to synthesise ATP analogues with many different structures. The ATP molecule contains three structural groups—the adenine base, ribose and triphosphate. We have synthesised 21 analogues with modifications in each group and studied their reactions, together with those of deoxy-ATP, deacetyl-ATP[2] and adenylylmethylene DP[11], with actomyosin, in order to find the physiological significance of the structure of ATP[12-15]. The reactions of adenosine tetraphosphate and 6-mercaptopurine ribose TP with actomyosin have also been investigated by Winand-Devigne et al.[16] and Murphy and Morales[17] respectively. More recently, Yount et al.[18] have investigated quantitatively the reactions of adenylylimido DP and adenylylmethylene DP with actomyosin. The syntheses of adenosine 5'-O-(1-thiotriphosphate)[18a] and adenosine 5'-bis (dihydroxyphosphinylmethyl)-phosphinate[18b] have also been reported. However, the interaction between these ATP analogues with modified chains and actomyosin has not yet been investigated.

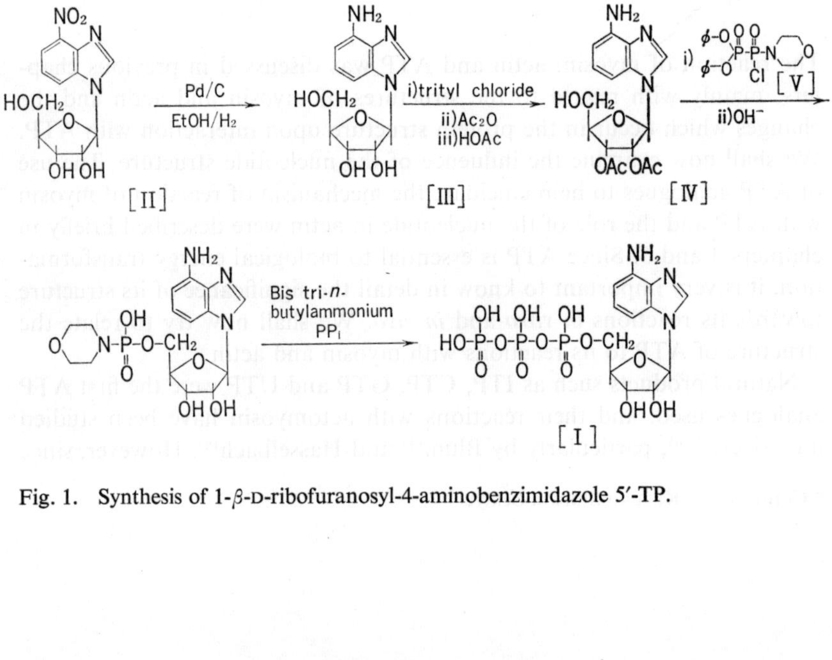

Fig. 1. Synthesis of 1-β-D-ribofuranosyl-4-aminobenzimidazole 5'-TP.

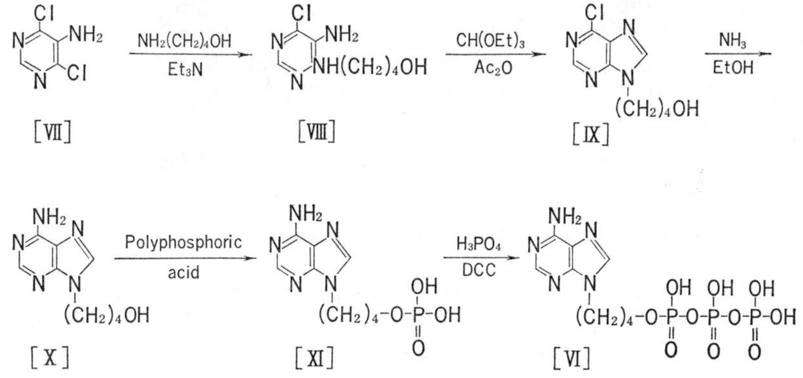

Fig. 2. Synthesis of 9-(4'-hydroxybutyl)-6-aminopurine 4'-TP.

1. Synthesis of ATP Analogues

Typical compounds obtained by modification of each of the three structural groups in ATP are 1-β-D-ribofuranosyl-4-aminobenzimidazole 5'-triphosphate[13] (abb. 4-aminobenzimidazole ribose TP), 9-(4'-hydroxybutyl)-6-aminopurine 4'-triphosphate[12] (abb. 9-(4'-hydroxybutyl)-adenine 4'-TP) and adenosine 5'-sulphatopyrophosphate[12]. Their syntheses are shown in Figs. 1, 2 and 3 respectively.

A) 4-Aminobenzimidazole ribose TP (I)
1-β-D-Ribofuranosyl-4-nitrobenzimidazole (II)[19] was hydrogenated in ethanol in the presence of palladised charcoal, giving 1-β-D-ribofuranosyl-4-aminobenzimidazole (III), which upon acetylation formed 1-(2',3'-di-O-acetyl)-β-D-ribofuranosyl-4-aminobenzimidazole (IV). IV was treated with P-diphenyl P'-morpholinopyrophosphorochloridate (V), deacetylated at the 2',3' positions with alkali and reacted with bis-tri-n-butylammonium pyrophosphate. The triphosphate (I) obtained was purified and isolated by column chromatography on Dowex I.

B) 9-(4'-Hydroxybutyl)-6-aminopurine 4'-triphosphate (VI)
Treatment of 4,6-dichloro-5-aminopyridine (VII)[20] with 4-amino-1-butanol and triethylamine gave 4-(4'-hydroxybutyl)-5-amino-6-chloropyridimide (VIII), which cyclised, with acetic anhydride and ethyl orthoformate,

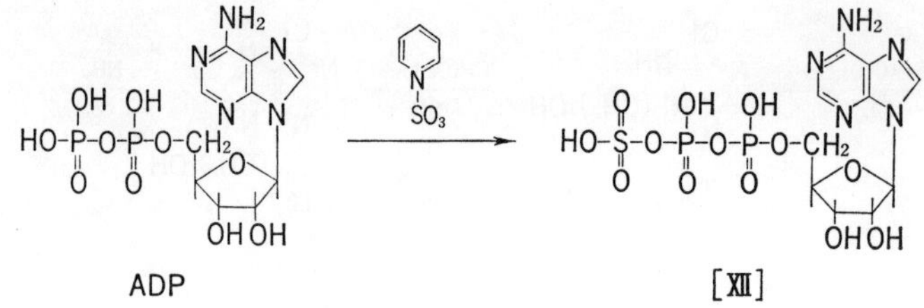

Fig. 3. Synthesis of adenosine 5'-sulphatopyrophosphate.

to 9-(4'-hydroxybutyl)-6-chloropurine (IX). Treatment with ethanol saturated with ammonia replaced the 6-Cl by NH_2, giving 9-(4'-hydroxybutyl)-6-aminopurine (X), which was reacted with phosphoric acid and phosphorous pentoxide to yield 9-(4'-hydroxybutyl)-6-aminopurine 4'-monophosphate (XI). This was converted to the triphosphate (VI) with phosphoric acid in the presence of dicyclohexyl carbodiimide, and purified with charcoal and ion exchange resin.

C) Adenosine 5'-sulphatopyrophosphate (XII)
Adenosine 5'-diphosphate was reacted with pyridine sulphur trioxide[21] in aqueous bicarbonate solution. The compound was purified with charcoal, celite and Dowex I.

The purity at each step of these syntheses was checked by elementary analysis, absorption maximum and R_f.

The chemical structures of analogues modified in the triphosphate, adenine and ribose groups are shown in Tables I, II and III respectively, together with results of studies of their reactions with actomyosin. Results obtained with several natural analogues and studies by other investigators are also included. The phenomena measured were the decrease in light-scattering intensity of myosin B in 0.6M KCl solution, the steady-state rates of hydrolysis of the analogues by myosin B at low and high ionic strengths and whether or not there was contraction of myofibrils obtained from rabbit skeletal muscle by the method of Perry and Corsi[22]. Although not shown in the Tables, the initial burst of P_i from myosin-ATPase and the clearing response of actomyosin was also measured with some of the analogues[12].

2. Decrease in Light-scattering Intensity of Myosin B Solution

Almost all of the analogues, except for a few in Table I in which the

TABLE I. Interaction between Actomyosin and ATP Analogues with Modified Phosphate Chains

Compound	Structure	Decrease in light-scattering of myosin B[a]	Rate of hydrolysis at high ionic strength in the presence of Ca^{2+} [b]	Rate of hydrolysis at low ionic strength in the presence of Mg^{2+} [c]	Contraction of myofibrils[d]
ATP		100	100	100	++
Adenosine tetraphosphate		100	16	2.5	
P^3-Methyl ATP		72	0	0	—
Adenosine sulphatopyrophosphate		0 (?)	0	0	—
Adenylylmethylene DP			0	0	—
Adenylylimido DP		100	0	0	—
ADP		0	0	0	—

The values given are relative to ATP=100. [a]Decrease in intensity of light-scattering of a 0.6M KCl solution of myosin B after the addition of an ATP analogue was measured in the presence of 1–10mM $MgCl_2$ at pH 7 and room temperature. [b]Rate of hydrolysis by myosin B in 0.6M KCl–7mM $CaCl_2$ at pH 7 and 20°C. [c]Rate of hydrolysis by myosin B in 0.075M KCl–2mM $MgCl_2$ at pH 7 and 20°C. [d]Contraction of isolated myofibrils in 0.05–0.075M KCl and 2–5mM $MgCl_2$ at pH 7 and room temperature.

triphosphate chain had been modified, decreased the light-scattering intensity of myosin B solution in 0.6M KCl and Mg^{2+}. The maximum decrease in light-scattering intensity equalled that found with ATP. Neither adenosine sulphatopyrophosphate nor ADP decreased the scattering intensity, showing that γ-phosphate is necessary for this. However, P^3-methyl ATP, which is methylated on the OH of the γ-phosphate, induced 72% of the decrease in light-scattering intensity found with ATP, indicating that the dissociation of protons from the γ-phosphate is not essential for changing the light-scattering intensity of actomyosin. This is not true for the hydrolysis of the analogues by myosin, or for contraction of the myofibrils.

While PP_i, like ATP, decreases the light-scattering intensity[23], ADP does not. The change in light-scattering intensity is caused by alteration of the size and shape of actomyosin, probably because of interaction between ATP and its analogues and the site 2 of myosin (chapter 6, section 6). Evidence for this comes from kinetic studies of the initial reaction of myosin with ATP, and from the dissociation of actomyosin by ATP[24,25]. The difference between PP_i and ADP could be caused by differences in binding. While PP_i may be able to bind to site 2 of myosin in a manner similar to the γ-phosphate of ATP, ADP may, in contrast, bind with myosin through its adenine 6-N and ribose 3'-O. This could fix the position of the pyrophosphate chain of ADP and prevent it binding to site 2 of myosin, so that no change in light-scattering intensity occurs. Since either Mg^{2+} or Ca^{2+} is essential for the maximum change in light-scattering intensity to be observed, simultaneous chelation of the divalent metal ion by phosphate and myosin may be necessary. 6-Deoxy-UTP and 3'-deoxythymidine TP (Tables II and III) respectively reduce the light-scattering intensity by only 15% and 17% of the maximum decrease found with ATP, but the similar

TABLE II. Interaction between Actomyosin and ATP Analogues with Modified Base Parts

Compound	Structure	Decrease in light-scattering of myosin B	Rate of hydrolysis at high ionic strength in the presence of Ca^{2+}	Rate of hydrolysis at low ionic strength in the presence of Mg^{2+}	Contraction of myofibrils
Monomethyl ATP		89	180	100	++

ATP ANALOGUES

Compound	Structure				
Dimethyl ATP		100	70	10	++
2′, 3′-O-Isopropylidene-6-morpholinopurine riboside TP		71.5	6	16	—
4-Aminobenzimidazole riboside TP		100	69	30	+
Tubercidin TP			49	270	±
ITP			300	50	++
2′, 3′-O-Isopropylidene-6-chloropurine ribose TP		93.5	28	220	+
6-Mercaptopurine ribose TP			140		+
2′, 3′-O-Isopropylidene-6-mercaptopurine ribose TP		100	80–90	190–270	—
2, 6-Dimethylmercaptopurine ribose TP			17	5	—

Compound	Structure				
UTP		200		85	++
Monomethyl CTP		100	158	75	+
6-Deoxy-UTP		15	0	5	−
Ribose TP		5			
TP$_i$		8			
PP$_i$		100	0	0	−

The values given are relative to ATP=100.

TABLE III. Interaction between Actomyosin and ATP Analogues with Modified Ribose Groups

Compound	Structure	Decrease in light-scattering of myosin B	Rate of hydrolysis at high ionic strength in the presence of Ca^{2+}	Rate of hydrolysis at low ionic strength in the presence of Mg^{2+}	Contraction of myofibrils
Deoxy-ATP		100	110	110	++
2′,3′-O-Diacetyl ATP				120	++
2′,3′-O-Isopropylidene ATP		71	40	53	++

ATP ANALOGUES

Compound		Col1	Col2	Col3	Col4
Glucosyl-adenine 6'-TP		100	56	62	—
Thymidine TP		99	40	152	++
3'-Deoxy-thymidine TP		17	7	22	—
9-(2', 3'-Di-O-acetyl)-D-erthrityl-adenine 4'-TP		100	78	52	+
9-(4'-Hydroxy-butyl)-adenine 4'-TP		100	75	45	—
9-(3'-Hydroxy-propyl)-adenine 3'-TP		100	100	30	—
9-(2'-Hydroxy-ethyl)-adenine 2'-TP		97	65	40	—

The values given are relative to ATP=100.

compounds, monomethyl CTP and thymidine TP, are about as active as ATP. This indicates that the 6-N or O atom of the base and the 3'-O atom of ribose are necessary for the correct fitting of the phosphate chain to site 2 of myosin.

The light-scattering results indicate the relative strengths of binding to actomyosin to be deoxy-ATP⩾ATP⩾dimethyl ATP>9-(2'-hydroxyethyl) adenine TP>UTP⩾9-(4'-hydroxybutyl) adenine TP>9-(3'-hydroxypropyl) adenine TP>ITP[13]; ATP⩾methyl CTP>glucosyladenine TP> 4-aminobenzimidazole ribose TP>erythrityladenine TP[15]. However, the abilities of the nucleotides to contract myofibrils do not follow the same order. Yount et al.[18] have observed that adenylylimido DP is a potent inhibitor of ATPase and its binding constants deduced by a kinetic method are near the reciprocals of K_m values, and that it effectively replaces ATP in light-scattering decrease of actomyosin, but does not induce contraction of muscle models. These results seem to rule out any theory of contraction in which the motivating force is based simply on the binding of a negatively charged substrate to a positively charged actomyosin (cf. chapter 13, section 1).

3. Hydrolysis by Myosin

The myosin-ATPase reaction was measured using myosin or myosin B in the presence of 0.6M KCl and 7mM $CaCl_2$. Only triphosphate chains are hydrolysed by myosin, while PP_i and ADP do not react; neither is P^3-methyl ATP hydrolysed, indicating that the dissociation of OH in the γ-phosphate group is essential for hydrolysis.

Since ribose TP reacts only 5% as rapidly as ATP (Table II), a base is a prerequisite for fast hydrolysis. Relative to ATP, the rates of hydrolysis of ITP, UTP, monomethyl ATP, monomethyl CTP, 2′, 3′-O-isopropylidene-6-mercaptopurine ribose TP, 2′, 3′-O-isopropylidene-6-chloropurine ribose TP, ribose TP and 6-deoxy UTP are, respectively, 3.0, 2.0, 1.8, 1.58, 0.9, 0.28, 0.05 and zero. Thus hydrogen bonding between the 6-N or O of the purine or pyrimidine base and a hydrogen donor in myosin facilitates decomposition; rapid decomposition occurs only when the strength of the hydrogen bonds is appropriate. The slow rates of reaction of dimethyl ATP, 2′,3′-O-isopropylidene-6-morpholinopurine riboside TP and 2,6-dimethylmercaptopurine ribose TP, respectively 0.7, 0.06 and 0.17 relative to ATP, are probably due to steric hindrance by the dimethyl and morpholino groups. 4-Aminobenzimidazole ribose TP and Tubercidin TP are hydrolysed respectively at 0.69 and 0.49 of the rate of ATP, and the basic branches of these molecules are weaker π-electron donors than the adenine in ATP. As described in chapter 4, hydrolysis of ATP by myosin is inhibited by specific chemical modification of a tyrosine residue[26], and the UV absorption spectrum suggests that complexing of H-meromyosin by ATP involves a tyrosine residue[27]. Also, TNP-myosin, formed by trinitrophenylation of the ε-amino group of lysine by TBS, forms a very tight Michaelis complex with ATP, indicating that there is conjugation between the adenine base of ATP and the TNP group[28]. Taken all together, these results indicate that the decomposition of ATP by myosin is accelerated by π-electron donation from the adenine base to tyrosine.

The ribose ring can be drastically modified without affecting markedly the rate of hydrolysis by myosin (Table III); thus deoxy-ATP and thymidine 5'-TP react 110 and 40% respectively as rapidly as ATP with myosin. However, 3'-deoxythymidine TP reacts only 7% as rapidly as ATP, whereas 9-(4'-hydroxybutyl)-adenine 4'-TP, which contains a flexible group in place of ribose, is hydrolysed at 75% the speed of ATP. These results show that when the ribose portion is rigid, it is necessary for it to include a 3'-O atom for fast hydrolysis to occur. Since 2',3'-O-isopropylidene ATP is also hydrolysed by myosin, its 3'-O atom is thought to form a hydrogen bond with the protein. 9-(2',3'-Di-O-acetyl)-D-erthrityl-adenine 4'-TP, 9-(4'-hydroxybutyl)-adenine 4'-TP, 9-(3'-hydroxypropyl)-adenine 3'-TP and 9-(2'-hydroxyethyl)-adenine 2'-TP, which all lack ribose rings, react at rates relative to ATP of 0.78, 0.75, 1.0 and 0.65, respectively. Thus there can be rapid hydrolysis, even when there is no 3'-O atom to form a hydrogen bond, so long as the ribose is replaced by a flexible chain.

Since myosin B-ATPase contains both myosin- and actomyosin-ATPases, its hydrolysis reactions at low ionic strengths in the presence of Mg^{2+} (Tables I, II and III) cannot be used to determine the role of each part of the nucleotide structure in hydrolysis by actomyosin-ATPase. In chapter 8 it was shown that contraction of a muscle model is coupled to its actomyosin-ATPase, and there is a similar relationship between the hydrolysis of 22 analogues of ATP by actomyosin and the contraction of muscle fibres. Those compounds which, under the above conditions, were hydrolysed by myosin B less than half as rapidly as ATP usually did not contract muscle fibres, while all but five of the others did.

The initial burst of P_i liberated by the reaction with myosin-ATPase was measured for only a small number of analogues. The steady-state rate of hydrolysis of 9-(2'-hydroxyethyl)-adenine 2'-TP by myosin B at high ionic strength is 65% of that of ATP, but there is no initial burst. Furthermore, at low ionic strength it is hydrolysed by myosin B-ATPase 40% as rapidly as ATP, but it does not contract myofibrils. This supports the conclusion of chapter 7 that the phosphorylation of myosin, which is observable as the initial burst, is the primary reaction of the actomyosin-ATPase and is coupled to muscle contraction. The hydrolyses of ATP, deoxy-ATP and diacetyl ATP by ordinary myosin B, which is thought to contain a protein relaxation factor (chapter 10), all show substrate inhibition at low ionic strength in the presence of Mg^{2+}. Recently, Weber[29] showed that

ITP, GTP and UTP, as well as ATP are capable of causing complete relaxation of myofibrils. For ITP and GTP the required concentrations are about 100 times higher than those for ATP, whereas UTP is maximally effective also in low concentrations. However, substrate inhibition is not observed with monomethyl ATP, which is rapidly hydrolysed and is able to contract myofibrils, or with analogues in which the ribose ring has been replaced by a flexible chain. Thus it seems that the restrictions imposed by the structures of the analogues upon their reactions increase in the order: hydrolysis by myosin, phosphorylation of myosin, contraction and relaxation.

4. Contraction of Myofibrils

Adenosine 5'-sulphatopyrophosphate, adenylylmethylene DP and ADP do not contract myofibrils (Table I), and, as in the case of hydrolysis, a triphosphate chain is needed for contraction. P^3-Methyl ATP does not induce the contraction of myofibrils, showing that the dissociation of an OH group in the terminal phosphate is essential to contraction; this agrees with the conclusion of chapters 3 and 7 that the phosphorylation of myosin is basic to contraction.

The order of ability of purine nucleotides to cause contraction is ATP > monomethyl ATP > ITP ≫ 2',3'-O-isopropylidene-6-chloropurine ribose TP ≫ 2', 3'-O-isopropylidene-6-mercaptopurine ribose TP > 2,6-dimethylmercaptopurine ribose TP ≈ 2', 3'-O-isopropylidene-6-morpholinopurine ribose TP. It appears to depend on hydrogen bonding between a 6 N or O atom of the base and myosin, and, unlike the hydrolysis by myosin-ATPase described above, increases with increase in the strength of hydrogen bonding. 4-Aminobenzimidazole ribose TP can induce weak contraction of myofibrils, but Tubercidin TP cannot, showing that the donation of π-electrons from the nucleotide base to the tyrosine residue of myosin is more necessary for myofibril contraction than for hydrolysis. With pyrimidine nucleotides, hydrogen bonding between the 6 atom of the base and myosin is necessary for contraction, while the ability to cause contraction is in the order: UTP ≫ monomethyl CTP ≫ 6-deoxy-UTP.

The ribose ring plays three roles in the contraction of myofibrils. Firstly, it maintains a suitable distance between the base and the triphosphate chain. Thus 9-(4'-hydroxybutyl)-adenine 4'-TP, 9-(3'-hydroxypropyl)-adenine 3'-TP and 9-(2'-hydroxyethyl)-adenine 2'-TP do not contract myofibrils, although they are hydrolysed fairly rapidly by myosin B; also glucosyl-adenine TP, in which the distance between the 3-O atom bound to the phosphate and the 6-N is 0.77 Å longer than in ATP, is incapable of contracting myofibrils. Secondly, rigidity of the ribose portion is essential for contraction. If the ribose ring, with its attendant rigidities, is replaced by a flexible chain which can rotate freely, binding to the active site of myosin is accompanied by a loss of configurational entropy. Compounds containing such flexible chains do not contract myofibrils. Thus 9-(2′,3′-di-O-acetyl)-D-erythrityl adenine TP is less able than ATP to cause contraction even though the distances between their bases and triphosphate chains are equal. Thirdly, the ribose 3′-O atom forms a hydrogen bond with myosin. This is also necessary for the hydrolysis reaction. Thus 3′-deoxythymidine TP cannot cause contraction of myofibrils, while deoxy-ATP, thymidine TP, 2′, 3′-O-diacetyl ATP and 2′, 3′-O-isopropylidene ATP can.

A model of the ES complex[30] was described in chapter 4, section 6, on the basis of the reactions of myosin with ATP analogues discussed in this chapter, and the results of chemical modification experiments in chapter 4. This model involves hydrogen bonding between myosin, the 6-N atom of the adenine base and the 3′-O atom of the ribose ring of ATP. The binding site on myosin is thought to be an NH group of the peptide Pro·Lys at the TBS binding site. It was suggested that the triphosphate chain conjugates through divalent metal ions with the Asp of Asp·Pro·Pro·Lys at the active site[31], and that the terminal pyrophosphate group binds to the SH group of cysteine and the guanidinium group of arginine in the peptide Ileu·Cys-SH·Arg which was discovered by Yamashita et al.[32]. This model explains the interaction of trinitrophenyl myosin with ATP as conjugation between the adenine base and the trinitrophenyl group. As described in chapters 3 and 6, various steps of the reaction between myosin and ATP were elucidated by kinetic studies, and we expect that further such studies with ATP analogues will lead to a more certain chemical model involving several reaction intermediates. Future studies on ATP analogues which contain 'reporter groups' may be also very beneficial to clarification of the molecular mechanism of the interaction of ATP with myosin. Recent studies on a fluorescent analogue[33,34] and on a spin-label analogue[35] are indications of future developments in this field. In particular the use of 1, N^6-ethenoadenosine triphosphate[34] is very promising, since the fluorescence emission spectrum of the analogue shows a maximum near 410 nm and can be easily measured without interference from protein fluorescence[34], and since we[36] have recently shown that the analogue binds strongly to myosin and the intensity of fluorescence of the analogue decreases considerably on its binding with myosin.

REFERENCES

1 J. J. Blum, *Arch. Biochem. Biophys.*, **55**, 486 (1955).
2 W. Hasselbach, *Biochim. Biophys. Acta*, **20**, 355 (1956).
3 R. Bergkvist and A. Deutsch, *Acta Chem. Scand.*, **8**, 1105 (1954).
4 R. E. Ranney, *Amer. J. Physiol.*, **183**, 197 (1955).
5 H. Portzehl, *Biochim. Biophys. Acta*, **14**, 195 (1954).
6 W. W. Kielley, H. M. Kalckar and L. B. Bradley, *J. Biol. Chem.*, **219**, 95 (1956).
7 J. Baddiley, A. M. Michelson and A. R. Todd, *J. Chem. Soc.*, 582 (1949).
8 V. M. Clark, G. W. Kirby and A. R. Todd, *J. Chem. Soc.*, 1497 (1957).
9 M. Smith and H. G. Khorana, *J. Amer. Chem. Soc.*, **80**, 1141 (1958).
10 H. G. Khorana, "Some Recent Developments in the Chemistry of Phosphate Esters of Biological Interest," John Wiley & Sons, New York (1961).
11 C. Moos, N. R. Alpert and T. C. Myers, *Arch. Biochem. Biophys.*, **88**, 183 (1960).
12 M. Ikehara, E. Ohtsuka, S. Kitagawa, K. Yagi and Y. Tonomura, *J. Amer. Chem. Soc.*, **83**, 2679 (1961).
13 M. Ikehara, E. Ohtsuka, S. Kitagawa and Y. Tonomura, *Biochim. Biophys. Acta*, **82**, 74 (1964).
14 M. Ikehara, E. Ohtsuka, H. Uno, K. Imamura and Y. Tonomura, *Biochim. Biophys. Acta*, **100**, 471 (1965).
15 Y. Tonomura, K. Imamura, M. Ikehara, H. Uno and F. Harada, *J. Biochem.*, **61**, 460 (1967).
16 J. Winand-Devigne, G. Hamoir and C. Liébecq, *Eur. J. Biochem.*, **1**, 29 (1967).
17 A. J. Murphy and M. F. Morales, *Biochemistry*, **9**, 1528 (1970).
18 R. G. Yount, D. Ojala and D. Babcock, *Biochemistry*, **10**, 2490 (1971).
18a F. Eckstein and H. Gindl, *Eur. J. Biochem.*, **13**, 558 (1970).
18b D. B. Trowbridge and G. L. Kenyon, *J. Amer. Chem. Soc.*, **92**, 2181 (1970).
19 Y. Mizuno, M. Ikehara and F. Ishikawa, *Chem. Pharm. Bull. Tokyo*, **10**, 761 (1962).
20 R. W. Chambers and H. G. Khorana, *J. Amer. Chem. Soc.*, **80**, 3749 (1958).
21 P. Baumgarten, *Ber.*, **59**, 1166 (1926).
22 S. V. Perry and A. Corsi, *Biochem. J.*, **68**, 5 (1958).
23 F. Morita and Y. Tonomura, *J. Amer. Chem. Soc.*, **82**, 5172 (1960).
24 H. Onishi, H. Nakamura and Y. Tonomura, *J. Biochem.*, **64**, 769 (1968).
25 N. Kinoshita, T. Kanazawa, H. Onishi and Y. Tonomura, *J. Biochem.*, **65**, 567 (1969).

26 T. Shimada, *J. Biochem.*, **67**, 185 (1970).
27 K. Sekiya and Y. Tonomura, *J. Biochem.*, **61**, 787 (1967).
28 S. Kubo, S. Tokura and Y. Tonomura, *J. Biol. Chem.*, **235**, 2835 (1960).
29 A. Weber, *J. Gen. Physiol.*, **53**, 781 (1969).
30 Y. Tonomura, S. Kubo and K. Imamura, *in* "Molecular Biology of Muscular Contraction," ed. by S. Ebashi, F. Oosawa, T. Sekine and Y. Tonomura, Igaku Shoin, Tokyo, p. 11 (1965).
31 N. Azuma, M. Ikehara, E. Ohtsuka and Y. Tonomura, *Biochim. Biophys. Acta*, **60**, 104 (1962).
32 T. Yamashita, Y. Soma, S. Kobayashi, T. Sekine, K. Titani and K. Narita, *J. Biochem.*, **55**, 576 (1964).
33 M. Onodera and K. Yagi, *Biochim. Biophys. Acta*, **253**, 254 (1971).
34 J. A. Secrist III, J. R. Barrio and N. J. Leonard, *Science*, **175**, 646 (1972).
35 R. Cooke and J. Duke, *J. Biol. Chem.*, **246**, 6360 (1971).
36 H. Onishi and Y. Tonomura, unpublished.

10

THE REGULATION OF MUSCLE CONTRACTION BY CALCIUM IONS AND PROTEIN FACTORS*

The excitation of a motor nerve releases acetylcholine from the nerve ending, inducing an end plate potential which then generates an impulse in the protoplasmic membrane of the muscle fibre[1,2]. Hodgkin and Huxley showed that the mechanism of initiation and propagation of this impulse can be explained by assuming that there is a large transient change in the permeability of the membrane to Na^+ and $K^{+[3]}$. This is a major problem in physiology, and is discussed in detail in several excellent books[3-5a]. Consequently, the mechanism of membrane excitation will not be considered in this chapter, since it is not directly related to the main subject of this monograph. According to Hodgkin and Huxley's theory, the non-uniform distribution of Na^+ and K^+ inside and outside the cell is the basis of excitation, and is due to active transport of Na^+ through the membrane.**

---

* Contributor: Yasuyuki Kozuki
** Ref. 6 discusses the distribution of ions in muscle fibre, and its relationship to the ion flux and membrane potential.

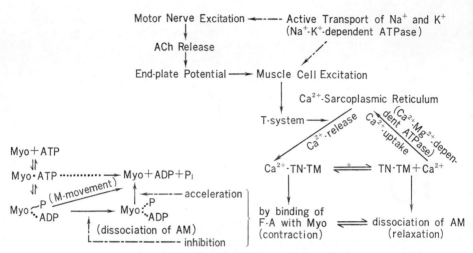

Fig. 1. General scheme for coupling of excitation to contraction.

As described in chapter 1, an excitation of the plasma membrane of a muscle fibre is relayed through the T-system to the sarcoplasmic reticulum (SR), which then releases Ca^{2+}. The Ca^{2+} is a controlling factor for the actin-myosin-ATP system: when the Ca^{2+} is released from the SR the system assumes the contracted state; when the Ca^{2+} is reabsorbed by the SR, actomyosin reverts to its original relaxed state.

Muscle proteins and those reactions which are directly related to contraction were described in previous chapters. There are many problems concerning the coupling of excitation with contraction, and the following chapters will discuss those relevant to the main theme of this book. This chapter describes regulation of muscle contraction by Ca^{2+} and protein factors, chapter 11 discusses the molecular mechanism of uptake of Ca^{2+} by the sarcoplasmic reticulum, and chapter 12 considers the mechanism of the Na^+-K^+-dependent membrane ATPase. Figure 1 summarises the details of the contractile system described in previous chapters and the relationship between the regulatory system, excitation and contraction, which will be described in this and subsequent chapters.

1. The Role of Ca^{2+} in the Regulation of Contraction and Relaxation

The work of Ringer[7], Mines[8] and, more recently, Ware et al.[9] indicated that in the absence of Ca^{2+} the mechanical activity of cardiac muscle is eliminated, despite the continuing electrical activity of the cell membrane. Heilbrunn and Wiercinski[10] were among the first to suggest that Ca^{2+} is required for the activation of skeletal muscle. Of the major ions which occur naturally in protoplasm, only Ca^{2+} is effective when injected directly into the sarcoplasm of single muscle fibres from frogs. However, our knowledge of the regulation of contraction and relaxation of living muscle has advanced significantly only in the last few years[11–14]. While we cannot yet explain all the details, the main features are now understood.

In 1951, Marsh[15] discovered a factor (the Marsh factor) in muscle homogenates which relaxes an actomyosin thread in the presence of Mg-ATP. However, at that time the nature of this factor was not completely understood. Ebashi and Lipmann[16] and Portzehl[17] showed that the Marsh factor is simply a microsomal vesicle of muscle. Subsequently the mechanism of relaxation by microsomal vesicles (described in detail in chapter 11) has received considerable attention. Hasselbach and Makinose[18] and Ebashi and Lipmann[16] found that the microsomal vesicles of muscle take up large amounts of Ca^{2+} in the presence of Mg-ATP, which indicates that Ca^{2+} plays an important role in muscle relaxation.

Aside from the work on muscle vesicles, we[19] had already observed in 1953 that traces of Ca^{2+} activate actomyosin-ATPase under conditions similar to those occurring naturally, *i.e.*, in low concentrations of KCl in the presence of Mg^{2+} (Fig. 2). Hence, we proposed that the reaction of

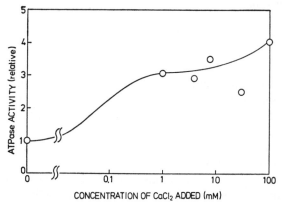

Fig. 2. Activation of actomyosin-ATPase by Ca^{2+} in the presence of Mg^{2+}. The ATPase activity is given as that relative to the value in the absence of $CaCl_2$ added. About 0.1M of $K^+ + Na^+$, 10mM $MgCl_2$, pH 6.5, 22°C.

actomyosin with ATP is regulated by Ca^{2+}, but our results attracted very little attention. Bozler[20] and Watanabe[21] showed that in the presence of Mg-ATP glycerinated muscle fibres and actomyosin suspensions undergo relaxation and dissolution respectively upon the addition of EDTA. Subsequent addition of Mg^{2+} led to no further change, but a trace of Ca^{2+} caused either contraction or superprecipitation, suggesting that EDTA induces relaxation by removing Ca^{2+}. Weber[22] and Ebashi[23] also found that traces of Ca^{2+} are essential to the actomyosin-ATPase and to the superprecipitation of actomyosin, while Ebashi[23] demonstrated that the relaxing effect of various metal-chelating agents parallels their ability to chelate Ca^{2+}. This confirms that Ca^{2+}, together with the microsomal vesicles (chapter 11), is the critical factor directly controlling the contraction and relaxation of muscle, both biochemically and physiologically.

Perry and Grey[24] noticed that the ATPase activity of natural actomyosin is sensitive to EDTA and EGTA, while the ATPase of actomyosin reconstituted from purified actin and myosin is insensitive. Ebashi and Ebashi[25] reported similar differences in the effect of EGTA on the superprecipitation of both types of actomyosin. They also investigated the nature of the protein factor responsible for the recovery of Ca^{2+}-sensitivity in natural actomyosin which had been made insensitive to Ca^{2+} by treatment with trypsin, and the relationship of this factor to the role of Ca^{2+} in the regulation of contraction and relaxation.

2. Properties of Tropomyosin*

In 1946 Bailey[27,28] extracted and crystallised a new structural protein, tropomyosin, from rabbit skeletal muscle, and examined its physicochemical properties in detail.** However, the physiological function of tropomyosin remained unknown until recently, although it constitutes 12–15% of muscle structural protein. Tropomyosin is very stable towards organic solvents, acids, alkalis and heat, and so is more easily purified and crystallised than other muscle structure proteins. Tsao et al.[34] obtained for it a molecular weight of 53,000. However, the molecular weight of tropomyosin was found more recently to be about 70,000 by hydrodynamic and light-scattering methods[35,36]. In the presence of reducing agents and denaturing solvents, tropomyosin dissociates into two similar subunits,

* Ref. 26 gives excellent description of general properties of vertebrate tropomyosin.
** In this section we will discuss the properties of vertebrate tropomyosin only. For invertebrate tropomyosins see Refs. 29–33.

TABLE I. Amino Acid Compositions of Tropomyosin, Native Tropomyosin and Troponin

	Tropomyosin[38] (a)	Troponin[39] (b)	Native tropomyosin[39]	0.6(b)+(a) / 1.6
Asp	89	83	83	85
Thr	26	22	20	24
Ser	40	31	34	37
Glu	212	159	197	192
Pro	2	26	11	11
Gly	11	43	23	23
Ala	108	74	92	95
Val	27	37	31	32
Met	16	27	23	20
Ileu	29	33	36	31
Leu	95	65	82	84
Tyr	15	12	15	14
Phe	4	23	12	11
His	5	17	10	10
Lys	113	100	106	108
Arg	41	66	50	50
(NH_3)	(64)	(60)	(66)	
Total	833	818	825	827

each of molecular weight about 34,000[36]. The two polypeptide chains of this protein are not identical as is indicated by its separation into two bands upon SDS-gel electrophoresis[37].

The amino acid composition of tropomyosin[38] (Table I) is characterised by a very large proportion, 62–66%, of residues with polar side chains. Tropomyosin has no tryptophane. Two C-terminal residues, isoleucine and serine, have been reported[38,40]. The determination of the number of unique sequences about the histidyl, methionyl and cysteinyl residues of tropomyosin has confirmed the distinction between its two polypeptide chains[41].

Tropomyosin fibres show the typical α-pattern[42], which is indicative of a coiled-coil structure[43]. Moreover, optical rotatory dispersion measurements and wide-angle X-ray diffraction show that tropomyosin is more than 90% α-helical[44]. These data are consistent with a two-chain α-helical coiled-coil molecule about 450Å long. Tropomyosin polymerises end to end when the ionic strength is decreased[27,28,34], and is notable for the range of polymorphic aggregates which it can build[45,46].

As will be described in section 5, the conformation of F-actin changes when Ca^{2+} binds to a Ca^{2+}-receptor protein—troponin—and this transfer of information is mediated by tropomyosin. For this purpose, tropomyosin should bind not only to troponin (section 4) but also to F-actin. The interaction of F-actin with tropomyosin was recently demonstrated by viscometric[47-49], ultracentrifugal[47-50] and flow-birefringence[51] methods.

3. The Discovery of the Relaxing Protein Factor

In 1964 Ebashi[25] showed that a protein other than actin and myosin is important in muscle contraction and relaxation. The properties of this protein resembled those of tropomyosin, and so it was called 'native' tropomyosin. This nomenclature implies that 'native' tropomyosin is tropomyosin extracted without impairment of its physiological functions, in contrast to Bailey's tropomyosin which is extracted in an inactivated state. The physiological function of 'native' tropomyosin is to restore Ca^{2+}-sensitivity to actomyosin which has been desensitised to the removal of Ca^{2+} with EGTA by treatment with trypsin. Hence, 'native' tropomyosin is sometimes called the 'EGTA-sensitising factor,' but we shall use the more common name of 'relaxing protein,' and will apply the general designation 'regulatory protein' to all those proteins which help regulate the actin-myosin-ATP system. At about the same time as Ebashi's work, Szent-Györgyi and Kaminer[52] heated a thoroughly washed rabbit muscle to 65°C, and isolated a protein factor called 'metin,' which induces a metachromatic property in actomyosin, suggesting that it binds to Ca^{2+}. Azuma and Watanabe[53,54] fractionated metin with ammonium sulphate and obtained a minor component which is an EGTA-sensitising factor. An EGTA-sensitising factor was also obtained by Perry et al.[55] from the ammonium sulphate fractionation of aqueous extracts of myofibrils, and also by Katz[56] from the ammonium sulphate fractionation of the tropomyosin-rich supernatant from partially polymerised actin solution.

The EGTA-sensitising factor obtained by these various methods is always rich in tropomyosin, but it was unclear at first whether it is just tropomyosin or a complex of tropomyosin with another protein. Since the amino acid composition of their 'native' tropomyosin differed from that of Bailey's tropomyosin (Table I), Ebashi and Ebashi[25] searched for a second component, and isolated a new structural protein, 'troponin'[39,57]. The so-called 'native' tropomyosin is a complex of troponin with tropomyosin[57,58,58a]. According to Ebashi and Endo[13], the weight ratio of tropomyosin to troponin in skeletal muscle is 2:1. These results have been confirmed more recently by Arai and Watanabe[59,60] and Hartshorne and Mueller[61,62].

In order to elucidate the action of the relaxing protein, it is necessary to understand the properties of troponin. The fractionation of troponin into two distinct proteins at low pH and high ionic strength was achieved by Hartshorne et al.[63-65], and later by Schaub and Perry[66]. One of the fractions, termed troponin B, effected a Ca^{2+}-insensitive inhibition of reconstituted actomyosin-ATPase. This inhibition was enhanced by the addition of tropomyosin. The other fraction, troponin A, conferred Ca^{2+}-sensitivity to the troponin B-tropomyosin system. All the properties of the original troponin were regained upon mixing troponin A and troponin B. Schaub and Perry[66] dissociated the troponin into the inhibitory component of molecular weight 2.3×10^{4}[67] and the calcium-sensitising component of molecular weight 1.8×10^{4}[67], by chromatography on SE-Sephadex in urea. The inhibitory component is specific for the actomyosin-type of ATPase, is remarkably stable to treatment with dissociating agents, heat, acids, alkalis and carboxymethylation and is neutralised by increasing amounts of actin, but not of myosin or tropomyosin[67]. Hartshorne and Pyun[68] have purified troponin A further by DEAE-cellulose column chromatography. Troponin A has a strong affinity to Ca^{2+}, and its molecular weight estimated by SDS-gel electrophoresis is 18,500. More recently, Greaser and Gergely[69] separated purified troponin into four major protein fractions by chromatography on DEAE-Sephadex in 6M urea, and, using SDS-gel electrophoresis, determined their molecular weights to be 14,000, 24,000, 35,000 and 21,000 respectively. The component of molecular weight 24,000 has a very potent inhibitory effect on the actomyosin-ATPase activity, both in the presence and absence of Ca^{2+}. Greaser and Gergely[69a] and Schaub et al.[69b] also reported that a component with a molecular weight of 21,000 is a Ca^{2+}-receptive protein[69c]. Reconstitution experiments show that the conponents of molecular weights 21,000, 24,000 and 35,000 restore the Ca^{2+}-requirement for the actomyosin-ATPase in the presence of tropomyosin. Murray and Kay[70] have also reported similar results on the inhibitory component. On the other hand, Wilkinson et al.[70a] reported that in the three components of troponin only the inhibitory component and the Ca^{2+}-receptive component are necessary to restore EGTA-sensitivity to the ATPase activity of desensitised actomyosin. Ebashi et al.[71] claimed that troponin consists of two components, a tropomyosin-binding protein (troponin-1) of molecular weight about 40,000 and a Ca^{2+}-receptive protein (troponin-2) of molecular weight about 22,000. They reported that recombination of these separated components restores the original activity. However it seems probable that the protein of molecular weight about 22,000 of Ebashi et al. contains both the components of molecular weights 21,000 and 24,000, because of the inefficiency of their method for separation of the components in troponin.

Thus it is established that troponin is a complex protein composed of several components, although it remains controversial which components are essential for its function. Therefore, we[71a] re-examined the composition of troponin prepared by the method of Ebashi et al.[71] and purified by DEAE-cellulose column chromatography[72]. As shown in Fig. 3, purified troponin showed in SDS-gel electrophoresis only three components. Their molecular weights were estimated to be 37,000, 23,000 and 19,000 respectively, and their relative amounts were estimated to be 48, 34 and 18% in weight basis, or 1.0, 1.1 and 0.8 in molar basis respectively. The individual protein components of troponin were separated by chromatography on DEAE-Sephadex A-50 in 50mM Tris-HCl (pH 8.0), 6M urea and 1mM DTT[69]. As shown in Fig. 4, the original troponin activity was only recovered on combination of the three components of troponin before removal of urea. However, all the mixtures of two components in the three components of troponin at a weight ratio of 1 : 1 or 1 : 2 did not confer the physiological Ca^{2+}-sensitivity on the ATPase activity of reconstituted actomyosin in the presence of tropomyosin. Therefore, it is reasonable to conclude that troponin is a complex protein of a molecular weight of

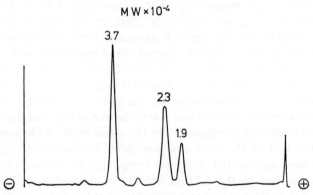

Fig. 3. Densitometer trace of the electrophoretogram of troponin. About 20μg of protein was loaded. 10% acrylamide, 0.27% methylenebisacrylamide, 0.1% SDS, pH 7.3, 7 mA/tube. Stained by 0.2% coomassie brilliant blue for 1 hr at 45°C.

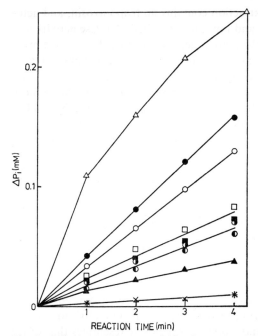

Fig. 4. Effects of different combinations of troponin components on the ATPase activity of actomyosin. 0.04M KCl, 2mM MgCl$_2$, 1mM ATP, pH 7.2, 25°C. 0.16mg/ml myosin, 0.04mg/ml F-actin, 0.02mg/ml tropomyosin. ◐, ◑, control; △, ▲, all the three components added, 0.044mg/ml; +, ×, two components with molecular weights of 23,000 and 37,000 added, 0.034mg/ml; □, ■, two components with molecular weights of 19,000 and 23,000 added, 0.031mg/ml; ○, ●, two components with molecular weights of 19,000 and 37,000 added, 0.04mg/ml. ◐, △, +, □, ○, 0.1mM CaCl$_2$; ◑, ▲, ×, ■, ●, 0.5mM EGTA.

7.9×10^4, and it is composed of a Ca^{2+}-sensitising component (or Ca^{2+}-receptive protein) of molecular weight 1.8–2.2×10^4, an inhibitory component of molecular weight 2.3–2.5×10^4 and a component of molecular weight 3.5–4.0×10^4. Interrelations between the three components of troponin and tropomyosin will be described in detail later in section 6.

4. Properties of the Relaxing Protein

The ATPase activity and superprecipitation by ATP of myosin B (natural actomyosin) are regulated by Ca^{2+} as described above. However, actin obtained by low temperature extraction contains no tropomyosin, and actomyosin reconstituted from it and myosin shows both ATPase activity and superprecipitation, both of which are independent of Ca^{2+}[24]. When myosin B is washed several times at low ionic strength[73], or is lightly treated with trypsin[25], its Ca^{2+}-sensitivity gradually disappears.

Ebashi et al.[25,39,57,58] showed that tropomyosin and troponin are both involved in the inhibition of the superprecipitation of myosin B by EGTA, thus providing a strong indication of the difference between actomyosin which is sensitive to Ca^{2+} and actomyosin which is not. Neither troponin nor tropomyosin alone is the EGTA-sensitising factor, but both compounds are necessary. The amino acid analysis (Table I) shows that the ratio of troponin to tropomyosin in native tropomyosin is 0.6:1. Hartshorne and Mueller[61] reported the binding ratio for maximum EGTA-sensitising activity to be 1.3:1, while Yasui et al.[74] found optimal activity at binding ratios in the range of 1:1 to 3:1. Arai and Watanabe[59] obtained a ratio of 0.4:1 by gel electrophoresis, and between 0.5:1 and 1:1 by ultracentrifugal analysis.

Ebashi and Kodama[57] showed that troponin increases the viscosity of tropomyosin solutions. Staprans and Watanabe[72] recently reported that the circular dichroism spectrum in the aromatic region (300 to 250nm) of 1:1 mixture of troponin and tropomyosin closely resembles that of the relaxing protein, but differs from the calculated spectrum for the mixture. A change in absorption at 278nm can be associated with the binding of troponin to tropomyosin.

The values of 4.4×10^4[59] and 5.0–5.3×10^4[75] have been found for the molecular weight of troponin. However, the value of 7.9×10^4 has been suggested by more recent estimates of molecular weights of the components of troponin (section 3). Mueller[76,77] and Briggs and Fuchs[78] suggested that SH groups are responsible for the relaxing activity. However, Staprans et al.[79] discovered that when the relaxing protein is incubated with NEM, its SH content is reduced from 4.9 to 1.3 moles per 10^5g, while its relaxing activity is affected very little. Ebashi et al.[39] also showed that SH groups are not involved in the function of troponin.

An important property of troponin is its ability to complex strongly with Ca^{2+}[39,59,74,80,81]. Yasui et al.[74] found that 2.4 moles of Ca^{2+} bind to 10^5 g of troponin, and that the association constant is 2.4×10^6 M^{-1}. The equivalent values for tropomyosin were, respectively, 0.35 mole/10^5g and 1.5×10^6 M^{-1}. The amount of Ca^{2+} which binds to 10^5g of troponin has also been variously reported to be 4 moles by Ebashi et al.[39], 3 moles by Arai and

Watanabe[59] and 2.2 moles by Fuchs and Briggs[80]. Fuchs[81a] has recently measured the affinities of various ions for troponin by a Biogel P-10 column chromatography. These affinities of divalent cations depend on their ionic radii, being highest with an ionic radius of 1Å, and less at either smaller or greater radii. The fact that Ca^{2+} has a much greater affinity for troponin than for tropomyosin is thought to be important in the mechanism of regulation of muscle contraction by troponin and tropomyosin. Cardiac troponin has a much higher affinity for Sr^{2+} than has skeletal troponin. Ebashi et al.[39] found that the relative sensitivities towards Ca^{2+} and Sr^{2+} of myosin-actin-tropomyosin-troponin preparations obtained from cardiac and skeletal muscles depend on the origin of the troponin (skeletal or cardiac) but not of the other proteins. Wakabayashi and Ebashi[82] demonstrated that in the presence of a saturating amount of Ca^{2+}, the sedimentation pattern of troponin shows a single sharp peak, which broadens and often splits into two or more peaks when the Ca^{2+} is removed by EGTA. Disc electrophoresis shows similar results.* Thus troponin appears to be a Ca^{2+}-receptor protein.

Ebashi and Kodama[57,58] investigated whether troponin, with its high propensity to complex Ca^{2+}, directly regulates the ATPase activity of actomyosin by interaction with F-actin, or whether the troponin first interacts with the tropomyosin, which would mean that the ATPase activity of actomyosin is regulated through the tropomyosin. They found that troponin has a very strong affinity for tropomyosin and markedly accelerates its aggregation, as mentioned above. If troponin is partially removed from a myofibril by light treatment with trypsin, it can still bind with the thin filaments so long as some tropomyosin remains. However, after strong trypsin treatment which removes even the tropomyosin, troponin can no longer combine with the thin filaments[13]. An immunological electron microscopy study of troponin with ferritin antibody showed that troponin molecules bind to the thin filaments at intervals of about 400Å[84]. Assuming that (1) the molecular weight of troponin is about 53,000, (2) its binding ratio with tropomyosin is about 0.6:1 and (3) it combines with the thin filaments at intervals of 400Å, Ebashi et al.[39] postulated that both troponin and tropomyosin occur on the thin filaments, in such a way that two troponin molecules bind at the crossover points of the actin filaments

---

* Han and Benson[83] found that the fluorescence emission spectrum of troponin changed on the addition of Ca^{2+}. However, a high concentration of Ca^{2+} is necessary (10–100 μM) and the change is slow, which casts doubt upon the physiological significance of this interaction.

and two tropomyosin molecules bind between the cross-over points. This suggests that a complex composed of 6-7 actin monomers, 1 tropomyosin molecule and 1 troponin molecule is the structural unit of the thin filaments[84a].

5. Mechanism of Action of the Relaxing Protein

We shall first discuss the effect of Ca^{2+} and ATP on actomyosin-ATPase in the presence of regulatory proteins[85]. ATP plays a dual role, since in high concentrations it relaxes muscle (chapter 1), but while the close relationship between this phenomenon and the effect of Ca^{2+} has frequently been pointed out, it has not been elucidated completely. Hence we have determined the kinetics of the reaction at low ionic strength between ATP, H-meromyosin and F-actin. When F-actin which has been extracted at room temperature and which contains regulatory proteins is used, and when the concentration of ATP is low, the rate of ATPase reaction of

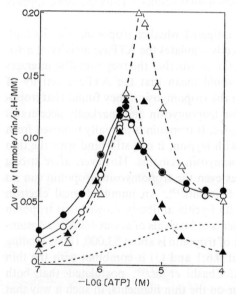

Fig. 5. The effects of Ca^{2+} and EGTA on acto-H-meromyosin-ATPase. Δv shows the difference in ATPase activity between acto-H-meromyosin and H-meromyosin. 0.25 mg/ml H-meromyosin, 0.125mg/ml F-actin, 2mM $MgCl_2$, 50mM KCl, 0.42mM PEP, 66μg/ml PK, 20mM Tris-maleate, pH 7.0, 24°C. ○, ●, pure F-actin; △, ▲, F-actin extracted at room temperature; ○, △, 0.1mM $CaCl_2$; ●, ▲, 0.1mM EGTA; v (······) shows the activity of H-meromyosin.

acto-H-meromyosin increases with increasing ATP concentration and reaches a maximum at about $3\mu M$. Further increase in the ATP concentration causes considerable inhibition; furthermore, the ATPase activity becomes markedly enhanced by traces of Ca^{2+} (Fig. 5). When pure F-actin is used, the substrate inhibition is again observed, but the ATPase activity is not then regulated by Ca^{2+}. The following molecular mechanism explains these results, and is based on the mechanism of reaction of acto-H-meromyosin with ATP, described in chapter 7.

$$\text{F-A-E} + \text{S} \xrightarrow{(1,2)} \text{F-A-E} \genfrac{}{}{0pt}{}{\cdot \text{ADP}}{\cdot \text{P}} \xrightarrow{(4)} \text{E} \genfrac{}{}{0pt}{}{\cdot \text{ADP}}{\cdot \text{P}} + \text{F-A}$$

with recycling arrows labeled (3) and (5).

At ATP concentrations of $3\mu M$ or less, steps (1) and (2) are rate determining, and they are independent of Ca^{2+} and regulatory proteins. However, as the ATP concentration increases, the rates of steps (1) and (2) increase, and the rate of the initial stoichiometric burst is faster than the decrease in light-scattering intensity after addition of $10\mu M$ ATP (Fig. 6A). At such high ATP concentrations, the rate of formation of $E \genfrac{}{}{0pt}{}{\cdot \text{ADP}}{\cdot \text{P}}$, as measured by the initial burst, is so fast that the rate-determining step of the dissociation of acto-H-meromyosin by ATP, *i.e.* for the formation of $E \genfrac{}{}{0pt}{}{\cdot \text{ADP}}{\cdot \text{P}}$, is now step (4). The rate of dissociation by ATP under these conditions is independent of both Ca^{2+} and EGTA (Fig. 6B), so step (4) is also independent of Ca^{2+}.

After the light-scattering intensity had been reduced by the addition of ATP in the presence of Mg^{2+} and a small amount of Ca^{2+} or EGTA, Mg^{2+} was removed by the addition of EDTA and the recovery of light-scattering intensity was observed. This recovery proceeds independently of whether Ca^{2+} or EGTA is present, and its rate is approximately equal to that of the spontaneous recovery accompanying the decomposition of ATP in the presence of Mg^{2+} and a small amount of Ca^{2+}. The recovery rates are all extremely slow, much slower than the binding of actin with H-meromyosin measured under the same conditions, but in the absence of ATP. This indicates that when the F-actin concentration is low, as in these experiments,

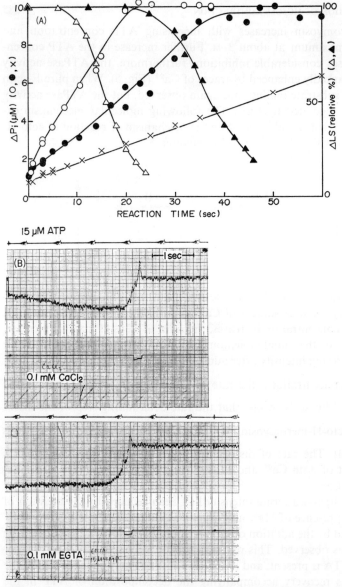

Fig. 6. Rate of change in the light-scattering intensity (ΔLS) and liberation of P_i (ΔP_i) in the reaction of acto-H-meromyosin with ATP. A: 0.25mg/ml H-MM, 0.125mg/ml F-actin extracted at room temperature, 10μM ATP, 50mM KCl, 2mM MgCl$_2$, 20mM Tris-maleate, pH 7.0, 21°C. $\bigcirc, \triangle$, 0.1mM CaCl$_2$; $\bullet, \blacktriangle$, 0.1mM EGTA; $\times$, H-meromyosin alone. B: Time-course of decrease in light-scattering intensity. 0.5mg/ml H-MM, 0.25 mg/ml F-actin extracted at room temperature, 15μM ATP, 2mM MgCl$_2$, 50mM KCl, 20mM Tris-maleate, pH 7.0, 20°C. Lower reflection: decrease in light-scattering intensity. Time: 1 sec/6 div.

step (5) is not the main route for the decomposition of ATP.* Although steps (1), (2), (4) and (5) are independent of Ca^{2+}, the ATPase activity of the system is regulated by Ca^{2+} in the presence of regulatory proteins, and is significantly increased by traces of Ca^{2+} (Figs. 5 and 6A). Hence step (3) is the main route for the decomposition of ATP, and is considerably retarded by the removal of Ca^{2+} when the regulatory proteins are present. However, step (4) is accelerated by ATP (chapter 3, section 7), so that when the ATP concentration is markedly increased, the concentration of F-A-$E \genfrac{}{}{0pt}{}{\cdot \cdot ADP}{\diagdown P}$ decreases and there is substrate inhibition.

Unfortunately, however, the kinetic analyses of the actomyosin-ATP system are much more difficult than those of the acto-H-meromyosin-ATP system, and at the present stage of development, we cannot establish the main reaction step for the 'clearing response' of actomyosin. As discussed in chapter 6, section 6, the following two mechanisms are possible for dissociation of actomyosin. One is the formation of $E_{2S}{}^{1}$: F-A-$E_{2S}{}^{1} \rightarrow$ F-A+$E_{2S}{}^{1}$, and the other is the formation of $E_{2\genfrac{}{}{0pt}{}{\cdot \cdot \cdot ADP}{\cdot \cdot P}}^{1}$, which has been discussed above in detail in the case of acto-H-meromyosin at a low concentration range of ATP. As described in chapter 6, section 6, the concentration of $E_{2S}{}^{1}$ in the myosin-ATP system increases with increase in the concentration of ATP and with lowering temperature. These results seem to support the idea that the main path for the 'clearing response' of actomyosin is the dissociation of actomyosin due to the formation of $E_{2S}{}^{1}$, since it is well known that the 'clearing response' occurs only in the presence of high concentrations of ATP, and the ATP concentration required for the 'clearing response' decreases with lowering temperature (chapter 7, section 7). If we adopt this mechanism, the molecular mechanism of relaxation of muscle (F-A-$E_{2}{}^{1S} \rightarrow$ F-A+$E_{2}{}^{1S}$; inhibition by ATP of the reaction $E_{2S}{}^{1} \rightarrow E_{2\,P}^{1}{}_{ADP}$ by the binding of ATP to a regulatory site, $E_{2S}{}^{1}+S \rightleftharpoons {}^{S}E_{2S}{}^{1}$) becomes to be different from the detachment of the head parts of myosin from F-actin in the contraction cycle (F-A-$E_{2\genfrac{}{}{0pt}{}{\cdot \cdot \cdot ADP}{\cdot \cdot P}}^{1} \rightarrow$ F-A+$E_{2\genfrac{}{}{0pt}{}{\cdot \cdot \cdot ADP}{\cdot \cdot P}}^{1}$) (see chapter 13, section 3). In this monograph, the author is

---

* As described in chapter 7, section 3, step (5) becomes the main path for ATP decomposition either when H-meromyosin is treated with PCMB and β-mercaptoethanol or when the F-actin concentration is very high.

compelled to describe only inconclusively and ambiguously several aspects on the molecular mechanism of contraction, especially those related with relaxation of muscle, because of ambiguity of the mechanism of dissociation of actomyosin under the physiological conditions, as mentioned above.

In any case, the hypothesis that the regulation of actomyosin-ATPase by high concentrations of ATP occurs at different steps from regulation by a trace amount of Ca^{2+} is supported by the effect of Ca^{2+} and ATP on the ATPase of the myosin-actin-regulatory protein system[86]. At low ionic strength, this ATPase is inhibited by high concentrations of ATP, even at high Ca^{2+} concentrations, and substrate inhibition occurs at lower ATP concentrations when the Ca^{2+} concentration is decreased (Fig. 7). This indicates that regulation of this system, by Ca^{2+} and in high concentrations of ATP, occurs in separate reaction steps.

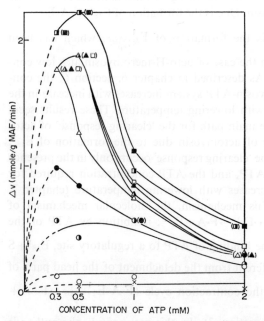

Fig. 7. Effect of Ca^{2+} and ATP on actomyosin-ATPase. Δv shows the difference between the total ATPase activity and that for myosin alone. 0.5mg/ml myosin, 0.01 mg/ml F-A of a complex of relaxing protein (MAF). 0.03M KCl, 2mM free Mg^{2+}, 20 mM Tris-maleate, pH 7.0, 25°C. Concentrations of free Ca^{2+}: ×, 0; ○, 0.31; ◐, 1.19; ●, 5.34; △, 26.55; ▲, 50.6; ▲, 74.51; □, 100; ◨, 125; ■, 150μM.

Secondly, we shall discuss the change in higher order structure of the proteins, and its relationship to the Ca^{2+}-sensitivity of actomyosin in the presence of the relaxing protein. Tonomura et al.[87] investigated the electron paramagnetic resonance (EPR) spectra of troponin, tropomyosin and F-actin, which had been spin-labelled with analogues of NEM, ITC and IAA in order to find the change in conformation of the proteins induced by Ca^{2+}. The SH groups of troponin, tropomyosin and F-actin were spin-labelled with N-2,2,6,6 tetramethyl piperidine nitroxide maleimide (MAL*) and N-2,2,6,6 tetramethyl piperidine nitroxide iodoacetamide (IAA*), and the NH_2 groups of these proteins with N-2,2,6,6 tetramethyl piperidine nitroxide isothiocyanate (ITC*). Modification of the SH groups of the relaxing protein with MAL* does not affect its activity (section 4). The effect of Ca^{2+} on the EPR spectra was then determined (Table II). The EPR spectrum of spin-labelled tropomyosin changes in the presence of both troponin and a trace of Ca^{2+}, and this change is accentuated by the additional presence of F-actin (Fig. 8). The concentration of Ca^{2+} necessary to change the EPR spectrum ($1 \mu M$) is almost equal to that needed for regulating the contraction and relaxation of living muscle. Spin-labelling of F-actin shows that its conformation is changed by Ca^{2+} when the relaxing protein is also present. It can be concluded from these results that (1) when Ca^{2+} binds with troponin a structural change is transmitted to the tropomyosin complexed with the troponin, and (2) that this structural change is further transmitted from tropomyosin to the F-actin with which it is com-

TABLE II. Conformation Changes in the Troponin-Tropomyosin-F-Actin System Induced by Ca^{2+}

Protein modified by spin-labelling probe	Spin-labelling probe used	Accessory protein	Effect of Ca^{2+} on conformation
Troponin	MAL*	None	None
	MAL*	Tropomyosin	None
	ITC*	None	None
	ITC*	Tropomyosin	None
Tropomyosin	MAL*	None	None
	MAL*	Troponin	Slight
	MAL*	Troponin+F-actin	Significant
Relaxing protein	MAL*	None	Slight
	MAL*	Myosin	Slight
	MAL*	F-Actin	Significant
F-Actin	MAL*	None	None
	MAL*	Relaxing protein	Slight

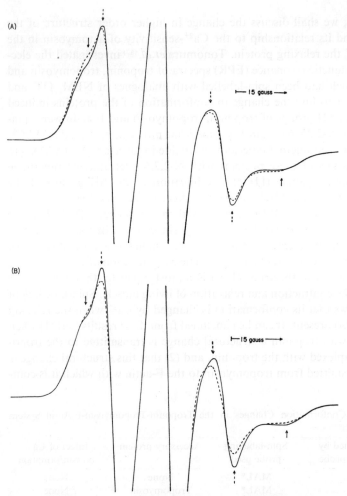

Fig. 8. Effect of Ca^{2+} on the EPR spectrum of MAL*-tropomyosin-troponin in the absence and presence of F-actin. A: 2.34 mg/ml MAL*-tropomyosin, 1.19mg/ml troponin, 0.058M KCl, 1.5mM $MgCl_2$, 20mM Tris-maleate, pH 7.0. ——, 0.2mM EGTA; - - - -, 0.1mM $CaCl_2$. B: 1.87mg/ml MAL*-tropomyosin, 0.6mg/ml troponin, 1.51mg/ml F-actin, 0.064M KCl, 1.5mM $MgCl_2$, 20mM Tris-maleate, pH 7.0.——, 0.2mM EGTA; - - - -, 0.1mM $CaCl_2$. ↓ and ↓ indicate a strongly immobilised signal and a weakly immobilised one, respectively.

bined. The structural change in F-actin is enhanced by tropomyosin while the change in tropomyosin is also intensified by the presence of F-actin.

Recently, Ishiwata and Fujime[88,88a] measured the flexibility of the F-actin-tropomyosin-troponin complex by quasi-elastic light-scattering, and concluded that there is a change in the flexibility of the entire complex at a free Ca^{2+} concentration of about $1\mu M$. However, there remains a possibility that the change is due to a change in the interactions between the actin filaments, but not due to a intramolecular change in the flexibility of the complex. Thus it appears that when Ca^{2+} is released from the sarcoplasmic reticulum it binds with troponin, thereby inducing a conformational change in F-actin *via* tropomyosin. Hence, it is reasonable to assume that the binding between myosin and the thin filament composed of tropomyosin, troponin and actin is strengthened by Ca^{2+} to give rise to the actomyosin-ATPase. H. E. Huxley[89] and Vibert *et al.*[89a] recently found new reflections in the diffraction pattern of active muscle which indicate a structural change in the actin-containing filaments upon stimulation. We can hope that the movements of tropomyosin and troponin upon the actin-filament after excitation will be visualised by X-ray analyses in the near future.

6. *Functions of the Components of the Relaxing Protein*

In this section we will describe the effects of tropomyosin and the three components of troponin on actomyosin-ATPase, in order to gain further insight into the molecular mechanism of regulation of muscle contraction by the relaxing protein.

Hartshorne and Mueller[62,63] observed that the Mg^{2+}-activated ATPase of actomyosin is stimulated by troponin and tropomyosin. Recently, Stewart and Levy[90] reported that 'native' tropomyosin increases the rate of superprecipitation of actomyosin by ATP. Therefore, we[91] investigated which component(s) of 'native' tropomyosin show this 'potentiating action' on the actomyosin-ATP system, and found that tropomyosin alone activates the ATPase activity of actomyosin.

As shown in Fig. 9, in 40mM KCl, tropomyosin activates actomyosin-ATPase considerably when the concentration of Mg-ATP is $30\mu M$, although it does not affect myosin-ATPase. The minimum amount of tropomyosin necessary for the maximum activation of the ATPase is almost equal to the amount of actin. Tropomyosin also activates actomyosin-ATPase which has been made insensitive to the substrate inhibition by PCMB-treatment of its constituent myosin, but has no effect on the ATPase of the treated myosin itself. Furthermore, when the concentration of Mg-

ATP is increased to 1mM, ATPase of the control actomyosin is markedly inhibited by tropomyosin. Thus, in the presence of tropomyosin, the actomyosin-ATPase reaction exhibits a marked 'substrate inhibition'[91a].

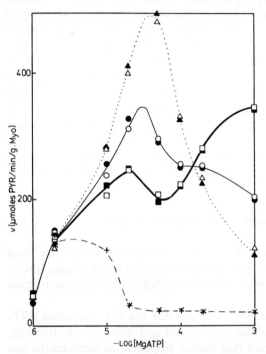

Fig. 9. Dependence of the Mg^{2+}-activated ATPase activity of actomyosin on the concentration of ATP and its alteration by tropomyosin and the inhibitory component of troponin. 0.099mg/ml myosin, 0.025mg/ml F-actin, 0.088mg/ml PK, 0.7mM PEP, 40 mM KCl. The $MgCl_2$ concentration is 1mM higher than that of ATP. pH 7.2, 25°C. ○, ●, control; △, ▲, 0.03mg/ml tropomyosin; □, ■, 0.04mg/ml inhibitory component; ×, +, 0.04mg/ml inhibitory component and 0.03mg/ml tropomyosin. ○, △, □, ×, 0.1mM $CaCl_2$; ●, ▲, ■, +, 0.5mM EGTA.

We have also investigated the properties of the component of the relaxing protein which inhibits the actomyosin-ATPase. In 1966 we[92] observed that the Mg^{2+}-activated ATPase activity of myosin B treated with trypsin is much lower, even in the presence of Ca^{2+}, than that of reconstituted actomyosin. Later, we[71a] observed that the ATPase activity of myosin in the presence of Mg^{2+} is only slightly enhanced by F-actin which has been extracted at room temperature and treated with trypsin (1/200 w/w) for 5 min, and the ATPase activity of actomyosin thus obtained is then independent of Ca^{2+} (Fig. 10). In contrast, the myosin-ATPase is greatly activated by F-actin which has not been treated with trypsin, but is then markedly dependent on Ca^{2+}. Furthermore, F-actin extracted at 0°C considerably activates the myosin-ATPase, both in the presence and absence of Ca^{2+}, and this activation is unaffected by treatment of the F-actin with trypsin.

F-Actin was extracted at room temperature, treated with trypsin, and then dialysed against 2mM Tris-HCl at pH 7.6 for 48 hr. The precipitate from the dialysate after the addition of 0.6mM $MgCl_2$ contained actin, which activated the ATPase in the same way as actin extracted at 0°C, and

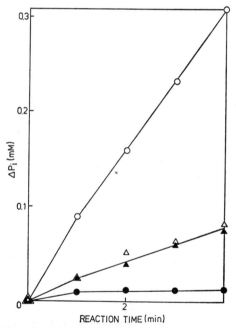

Fig. 10. Activation of the myosin-ATPase by F-actin extracted at room temperature, before and after treatment of the F-actin with trypsin. 0.2mg/ml myosin, 0.05mg/ml F-actin extracted at room temperature, 40mM KCl, 2mM $MgCl_2$, 1mM ATP, pH 7.3, 25°C. ○, ●, before; △, ▲, after treatment of F-actin with trypsin (1/200 w/w) at 25°C for 5 min. ○, △, 0.1mM $CaCl_2$; ●, ▲, 0.5mM EGTA.

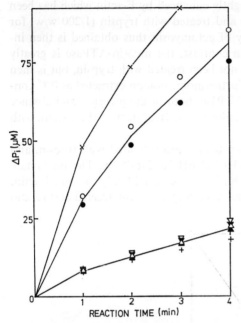

Fig. 11. Inhibition of actomyosin-ATPase by the inhibitory factor. The inhibitory factor was obtained by treatment with TCA of F-actin extracted at room temperature and then digested by trypsin. 0.1mg/ml myosin, 0.026mg/ml F-actin, 1mM ATP, 3mM $MgCl_2$, 40mM KCl, pH 7.3, 25°C. ○, ●, control; △, ▲, 0.05mg/ml inhibitor; ▽, ▼, 0.05mg/ml inhibitor+0.04mg/ml tropomyosin; ×, +, 0.05mg/ml inhibitor+0.04 mg/ml tropomyosin+0.04mg/ml troponin. ○, △, ▽, ×, 0.1mM $CaCl_2$; ●, ▲, ▼, +, 0.5mM EGTA.

the supernatant contained the inhibitory factor. The inhibitory factor was also obtained by reaction of the trypsin-treated F-actin with 2% TCA, since the factor is stable to TCA, while F-actin is unstable. The inhibitory factor composes less than 30% of the original actin preparation. Actomyosin reconstituted from myosin and purified F-actin has a high Mg^{2+}-activated ATPase activity which is independent of Ca^{2+} (Fig. 11); addition of the inhibitory factor markedly decreases its ATPase activity. However, on further addition of a mixture of troponin and tropomyosin, the activity in the presence of 0.5mM EGTA remains inhibited, but the activity in the presence of 0.1mM $CaCl_2$ becomes greater than that of the purified actomyosin. The inhibitory factor is precipitated in 0.4M LiCl at pH 4.6, and we attempted to purify it by chromatography on SE-Sephadex and DEAE-cellulose, but without success.

The ATPase activity of actomyosin is markedly inhibited in the presence of both tropomyosin and the inhibitory component, although the inhibitory component alone inhibits the ATPase slightly in the presence of 50μM Mg-ATP and activates it in the presence of 1mM Mg-ATP (Fig. 9)[91a]. Furthermore, the ATPase activity in the presence of the inhibitory component and/or tropomyosin is independent of Ca^{2+}, as stated above. When the inhibitory component was isolated from troponin and digested with trypsin, the molecular weight decreased from 2.2–2.4 to 1.0×10^4, and the inhibitory effect in the presence of tropomyosin decreased to about 50% of the original (Fig. 12).* Furthermore, actomyosin was rendered

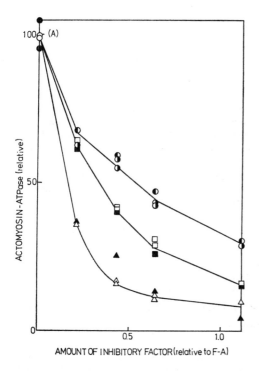

* Recently, Wilkinson et al.[70a] have also suggested that the inhibitory component with a molecular weight of 23,000 can be digested by proteolytic enzymes to a component with a molecular weight of 14,000, which still retains the inhibitory activity.

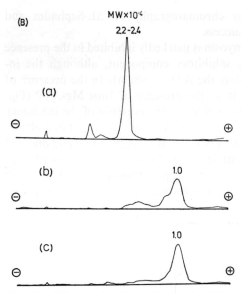

Fig. 12. Effect of digestion with trypsin of the inhibitory component isolated from troponin. A: Inhibition of actomyosin-ATPase by the inhibitory component. The inhibitory component was isolated from purified troponin by chromatography on DEAE-Sephadex A-50 in 6M urea and 1mM DTT at pH 8.0. 127μg/ml myosin, 33μg/ml F-actin, 25 μg/ml tropomyosin, 40 mM KCl, 2mM MgCl$_2$, 1mM ATP, pH 7.2, 25°C. The inhibitory component was digested with 1/300 (w/w) of trypsin at 25°C for 0 (△, ▲), 3 (□, ■) or 5 min (◐, ◑). △, □, ◐, 0.05mM CaCl$_2$; ▲, ■, ◑, 0.25mM EGTA. B: Densitometer traces of the SDS-gel electrophoresis of the inhibitory component. The inhibitory component was digested with 1/300 (w/w) of trypsin at 25°C for 0 (a), 3 (b) or 5 min (c). 10% acrylamide, 0.27% methylenebisacrylamide. 0.1% SDS, pH 7.2. 7 mA/tube. Stained by 0.2% coomassie brilliant blue for 1 hr at 45°C.

sensitive to Ca^{2+} by the combination of troponin components, even when the inhibitory component had been digested with trypsin. The ATPase activity of actomyosin in the presence of other regulatory proteins is activated markedly by minute amounts of Ca^{2+} ions after adding the inhibitory component, which has been isolated from troponin and digested with trypsin (Fig. 13). In contrast, the activity is inhibited markedly by Ca^{2+} ions after adding the inhibitory component isolated from F-actin by extracting at room-temperature and digesting with trypsin. It was also observed that the actomyosin-ATPase in the presence tropomyosin was inhibited markedly by Ca^{2+} ions, when only the inhibitory component and the Ca^{2+}-sensitising component of troponin were added at a weight ratio of 1: 4. These results show clearly that it is the conformation of the inhibitory component which determines the ATPase activity of actomyosin, and that an active subfragment of the inhibitory component is obtained by digestion with trypsin.

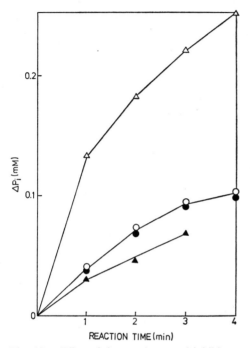

Fig. 13. Effect of the trypsin-treated inhibitory component of troponin on the actomyosin-ATPase activity. 0.16mg/ml myosin, 40μg/ml F-actin, 20μg/ml tropomyosin, 40mM KCl, 2mM $MgCl_2$, 1mM ATP, pH 7.2, 25°C. The trypsin-treated inhibitory component, IN^t, was obtained by digestion of the component with trypsin (300: 1 w/w) for 3 min at 25°C. ○, ●, control; △, ▲, 39.4μg/ml of a mixture of IN^t, the Ca^{2+}-sensitising component and the component of troponin of molecular weight 37,000. ○, △, 0.1mM $CaCl_2$; ●, ▲, 0.5mM EGTA.

The Ca^{2+}-sensitising component does not affect actomyosin-ATPase both in the presence and absence of tropomyosin, and the ATPase is unaffected by minute amounts of Ca^{2+}. As shown in Fig. 14, the Ca^{2+}-sensitising component eliminates the effects of the inhibitory component on the actomyosin-ATPase reaction. Thus, the addition of both the inhibitory and the Ca^{2+}-sensitising components does not affect actomyosin-ATPase in the absence of tropomyosin, but it changes the 'substrate inhibition' of the ATPase in the presence of tropomyosin to be sensitive to minute amounts of Ca^{2+}. However, the Ca^{2+}-sensitivity in the presence of tropomyosin and these two components of troponin is too small to explain the physiological regulation of the ATPase by $Ca^{2+(91a)}$.

The troponin component with a molecular weight of 37,000 inhibits actomyosin-ATPase at low Mg-ATP concentrations and activates the ATPase at high Mg-ATP concentrations, and these effects of this component are eliminated by the further addition of the Ca^{2+}-sensitising factor. These properties of a component with a molecular weight of 37,000 are

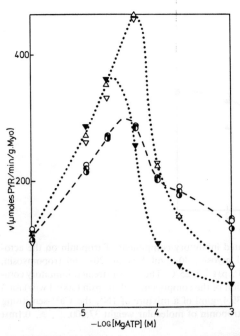

Fig. 14. Dependence of the Mg^{2+}-activated ATPase activity of actomyosin on the concentration of ATP and its alteration by tropomyosin, the inhibitory component and the Ca^{2+}-sensitising component of troponin. Experimental conditions as for Fig. 9. ○, control; △, 0.026mg/ml tropomyosin; ◐, ◑, 0.026mg/ml inhibitory component and Ca^{2+}-sensitising component (weight ratio 1 : 1); ▽, ▼, 0.026mg/ml tropomyosin, and 0.026mg/ml inhibitory component and Ca^{2+}-sensitising component (weight ratio 1 : 1). ○, △, ◐, ▽, 0.1mM $CaCl_2$; ◑, ▼, 0.5mM EGTA.

similar to those of the inhibitory component. However, the effects of a component with a molecular weight of 37,000 are almost independent of tropomyosin. They are also independent of the inhibitory component. The function of a component with a molecular weight of 37,000 is the enhancement of Ca^{2+}-sensitivity of actomyosin-ATPase in the presence of tropomyosin and the other two components of troponin, as shown in Fig. 15[(91a)].

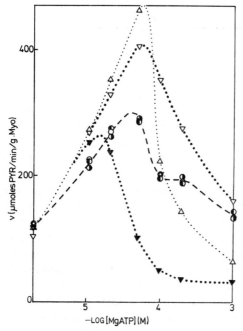

Fig. 15. Dependence of the Mg^{2+}-activated ATPase activity of actomyosin on the concentration of ATP and its alteration by tropomyosin and the three components of troponin. Experimental conditions as for Fig. 9. ○, control; △, 0.026mg/ml tropomyosin; ◐, ◑, 0.033mg/ml three components of troponin (weight ratio 1 : 1 : 1); ▽, ▼, 0.026 mg/ml tropomyosin and 0.033mg/ml three components of troponin (weight ratio 1 : 1 : 1). ○, △, ◐, ▽, 0.1mM $CaCl_2$; ◑, ▼, 0.5mM EGTA.

Thus, the remarkable Ca^{2+}-sensitivity of actomyosin-ATPase appears only in the presence of tropomyosin and all the three components of troponin, as stated above, and the interactions between the three components of troponin and tropomyosin, as revealed by kinetic studies on the ATPase reaction, are summarised as follows:

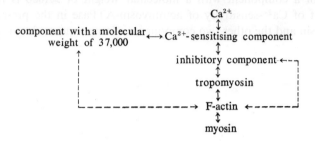

These results strongly support the idea that, when Ca^{2+} is removed from the Ca^{2+}-sensitising component of troponin, a structural change is transmitted to the inhibitory component and it is further transmitted to F-actin *via* tropomyosin. However, the effect of the inhibitory component upon tropomyosin is rather weak, since the interaction between the Ca^{2+}-sensitising component and the inhibitory component is very strong, and the substrate inhibition in the presence of tropomyosin and in the absence of Ca^{2+} is only slightly intensified by adding these two components. When the component with a molecular weight of 37,000 is further added, it interacts with the Ca^{2+}-sensitising component and weakens the interaction between the Ca^{2+}-sensitising component and the inhibitory component. Now the interaction between the inhibitory component and tropomyosin becomes to be appropriately strong, and the inhibitory component can enhance markedly the substrate inhibition of actomyosin-ATPase by tropomyosin after removal of Ca^{2+}. On the other hand, when Ca^{2+} ions bind to the Ca^{2+}-sensitising component, the augmentation by the inhibitory component of the substrate inhibition due to tropomyosin is retracted, and at high ATP concentrations the ATPase activity in the presence of both troponin and tropomyosin becomes higher than that in the presence of tropomyosin alone (Fig. 15), because of the promoting effect of the component with a molecular weight of 37,000 and/or the inhibitory component, as stated above (see Fig. 9). The above conclusions have been deduced from kinetic studies on the ATPase, and they must be re-examined in the light of structural studies on the binding of the components in the near future.

7. Myosin and the Regulation of Muscle Contraction

Finally, we shall briefly discuss the role of myosin in regulation of muscle contraction. It has been established[13,80] that myosin does not bind Ca^{2+} reversively in the range of Ca^{2+} concentrations which cause muscle contraction. Nevertheless, several workers have presented results which strongly suggest that the myosin molecule is also involved in the regulation process.

Weber[93] showed that the Ca^{2+} concentration required for contraction depends on the nature of the nucleoside triphosphate base. It is lower for ITP than for ATP, and decreases with decreasing nucleotide concentration. However, ATP is not bound to tropomyosin or troponin, and the binding of Ca^{2+} to troponin is not altered by ATP[39]. Therefore, one may speculate that the degree of saturation of myosin by nucleotide determines the extent to which troponin must be saturated by Ca^{2+} for contraction to be possible. Kendrick-Jones et al.[94] have shown that in molluscan muscle the factors which regulate contraction by interacting with Ca^{2+} are associated with myosin. However, tropomyosin does not appear to be necessary for the regulation of molluscan actomyosin by Ca^{2+}. As described in chapter 7, section 7, when vertebrate myosin is treated with PCMB and then with β-mercaptoethanol, the intrinsic Ca^{2+} and the light component, g_2, are partially removed and the actomyosin-ATPase becomes insensitive to Ca^{2+} and to substrate inhibition. Therefore, it seems to be very interesting to know how the interaction between the light and heavy chains in molluscan myosin is reversibly regulated by Ca^{2+}. And it is reasonable to suppose that the regulatory system in the myosin molecule was first appeared evolutionally, and then it was replaced by the tropomyosin-troponin system in vertebrates.

All these results show that the submolecular structure of myosin is involved in the regulation of the myosin-actin-ATP system, and indicate that the submolecular or secondary structure of myosin changes upon stimulation and that this change plays an important role in the conversion of the ATPase activity from the myosin- to the actomyosin-type.

REFERENCES

1 J. del Castillo and B. Katz, *Progr. Biophys. Biophys. Chem.*, **6**, 121 (1956).
2 B. Katz, *Johns Hopkins Hosp. Bull.*, **102**, 275 (1958).

3   A. L. Hodgkin, "The Conduction of the Nervous Impulse," Liverpool Univ. Press, Liverpool (1964).
4   B. Katz, "Nerve, Muscle, and Synapse," McGraw-Hill Book Co., New York (1966).
5   I. Tasaki, "Nerve Excitation: A Macromolecular Approach," Charles C. Thomas, Springfield (1968).
5a K. S. Cole, "Membranes, Ions and Impulses," University of California Press, Berkeley (1968).
6   P. Horowicz, in "Biophysics of Physiological and Pharmacological Actions," ed. by A. M. Shanes, American Association for the Advancement of Science, Washington, D. C., p. 217 (1961).
7   S. Ringer, *J. Physiol.*, **4**, 29 (1883).
8   G. R. Mines, *J. Physiol.*, **46**, 188 (1913).
9   F. Ware, A. L. Bennett and A. R. McIntyre, *Federation Proc.*, **14**, 158 (1955).
10 L. V. Heilbrunn and F. J. Wiercinski, *J. Cell. Comp. Physiol.*, **29**, 15 (1947).
11 W. Hasselbach, *Progr. Biophys. Mol. Biol.*, **14**, 167 (1964).
12 A. Weber, *Curr. Topics Bioenergetics*, **1**, 203 (1966).
13 S. Ebashi and M. Endo, *Progr. Biophys. Mol. Biol.*, **18**, 123 (1968).
14 C. C. Ashley, in "Membranes and Ion Transport," ed. by E. E. Bittar, Wiley-Interscience, London, Vol. 2, p. 1 (1970).
15 B. B. Marsh, *Nature*, **167**, 1065 (1951).
16 S. Ebashi and F. Lipmann, *J. Cell Biol.*, **14**, 389 (1962).
17 H. Portzehl, *Biochim. Biophys. Acta*, **26**, 373 (1957).
18 W. Hasselbach and M. Makinose, *Biochem. Z.*, **333**, 518 (1961).
19 S. Watanabe, Y. Tonomura and H. Shiokawa, *J. Biochem.*, **40**, 387 (1953).
20 E. Bozler, *J. Gen. Physiol.*, **38**, 149 (1954).
21 S. Watanabe, *Arch. Biochem. Biophys.*, **54**, 559 (1955).
22 A. Weber, *J. Biol. Chem.*, **234**, 2764 (1959).
23 S. Ebashi, *J. Biochem.*, **48**, 150 (1960).
24 S. V. Perry and T. C. Grey, *Biochem. J.*, **64**, 5P (1956).
25 S. Ebashi and F. Ebashi, *J. Biochem.*, **55**, 604 (1964).
26 T. Ooi, *Seikagaku*, **43**, 387 (1971) (in Japanese).
27 K. Bailey, *Biochem. J.*, **43**, 271 (1948).
28 K. Bailey, *Nature*, **157**, 368 (1946).
29 T.-C. Tsao, P.-H. Tan and C.-M. Peng, *Sci. Sinica*, **5**, 91 (1956).
30 M.-H. Jen and T.-C. Tsao, *Sci. Sinica*, **6**, 317 (1957).
31 D. R. Kominz, F. Saad and K. Laki, in "Conf. Chem. Muscular Contraction," Igaku Shoin, Tokyo, p. 66 (1958).
32 K. Bailey and J. C. Rüegg, *Biochim. Biophys. Acta*, **38**, 239 (1960).
33 E. F. Woods and M. J. Pont, *Biochemistry*, **10**, 270 (1971).
34 T.-C. Tsao, K. Bailey and G. S. Adair, *Biochem. J.*, **49**, 27 (1951).
35 A. Holtzer, R. Clark and S. Lowey, *Biochemistry*, **4**, 2401 (1965).

36 E. F. Woods, *J. Biol. Chem.*, **242**, 2859 (1967).
37 K. Weber and M. Osborn, *J. Biol. Chem.*, **244**, 4406 (1969).
38 D. R. Kominz, F. Saad, J. A. Gladner and K. Laki, *Arch. Biochem. Biophys.*, **70**, 16 (1957).
39 S. Ebashi, A. Kodama and F. Ebashi, *J. Biochem.*, **64**, 465 (1968).
40 R. H. Locker, *Biochim. Biophys. Acta*, **14**, 533 (1954).
41 R. S. Hodges and L. B. Smillie, *Biochem. Biophys. Res. Commun.*, **41**, 987 (1970).
42 W. T. Astbury, R. Reed and L. C. Spark, *Biochem. J.*, **43**, 282 (1948).
43 F. H. C. Crick, *Acta Cryst.*, **6**, 689 (1953).
44 C. Cohen and A. G. Szent-Györgyi, *J. Amer. Chem. Soc.*, **79**, 248 (1957).
45 C. Cohen and W. Longley, *Science*, **152**, 794 (1966).
46 D. L. D. Caspar, C. Cohen and W. Longley, *J. Mol. Biol.*, **41**, 87 (1969).
47 W. Drabikowski and J. Gergely, *J. Biol. Chem.*, **237**, 3412 (1962).
48 A. Martonosi, *J. Biol. Chem.*, **237**, 2795 (1962).
49 W. Drabikowski and E. Nowak, *Eur. J. Biochem.*, **5**, 376 (1968).
50 K. Laki, K. Maruyama and D. R. Kominz, *Arch. Biochem. Biophys.*, **98**, 323 (1962).
51 K. Maruyama, *Arch. Biochem. Biophys.*, **105**, 142 (1964).
52 A. Szent-Györgyi and B. Kaminer, *Proc. Natl. Acad. Sci. U. S.*, **50**, 1033 (1963).
53 N. Azuma and S. Watanabe, *Physiologist*, **7**, 83 (1964).
54 N. Azuma and S. Watanabe, *J. Biol. Chem.*, **240**, 3852 (1965).
55 S. V. Perry, V. Davies and D. Hayter, *Biochem. J.*, **99**, 1c (1966).
56 A. M. Katz, *J. Biol. Chem.*, **241**, 1522 (1966).
57 S. Ebashi and A. Kodama, *J. Biochem.*, **58**, 107 (1965).
58 S. Ebashi and A. Kodama, *J. Biochem.*, **60**, 733 (1966).
58a S. Ebashi and A. Kodama, *J. Biochem.*, **59**, 425 (1966).
59 K. Arai and S. Watanabe, *J. Biol. Chem.*, **243**, 5670 (1968).
60 K. Arai and S. Watanabe, *J. Biochem.*, **64**, 69 (1968).
61 D. J. Hartshorne and H. Mueller, *Biochim. Biophys. Acta*, **175**, 301 (1969).
62 D. J. Hartshorne and H. Mueller, *J. Biol. Chem.*, **242**, 3089 (1967).
63 D. J. Hartshorne and H. Mueller, *Biochem. Biophys. Res. Commun.*, **31**, 647 (1968).
64 D. J. Hartshorne, M. Theiner and H. Mueller, *Biochim. Biophys. Acta*, **175**, 320 (1969).
65 D. J. Hartshorne, *J. Gen. Physiol.*, **55**, 585 (1970).
66 M. C. Schaub and S. V. Perry, *Biochem. J.*, **115**, 993 (1969).
67 M. C. Schaub and S. V. Perry, *Biochem. J.*, **123**, 367 (1971).
68 D. J. Hartshorne and H. Y. Pyun, *Biochim. Biophys. Acta*, **229**, 698 (1971).
69 M. L. Greaser and J. Gergely, *J. Biol. Chem.*, **246**, 4226 (1971).
69a M. L. Greaser and J. Gergely, *Federation Proc.*, **29**, 463 (1970).

69b M. C. Schaub, S. V. Perry and W. Häcker, *Biochem. J.*, **126**, 237 (1972).
69c I. Staprans, H. Takahashi, M. P. Russell and S. Watanabe, *J. Biochem.*, **72**, 723 (1972).
70   A. C. Murray and C. M. Kay, *Biochem. Biophys. Res. Commun.*, **44**, 237 (1971).
70a J. M. Wilkinson, S. V. Perry, H. A. Cole and I. P. Trayer, *Biochem. J.*, **127**, 215 (1972).
71   S. Ebashi, T. Wakabayashi and F. Ebashi, *J. Biochem.*, **69**, 441 (1971).
71a M. Shigekawa and Y. Tonomura, *J. Biochem.*, **72**, 957 (1972).
72   I. Staprans and S. Watanabe, *J. Biol. Chem.*, **245**, 5962 (1970).
73   M. C. Schaub, D. J. Hartshorne and S. V. Perry, *Biochem. J.*, **104**, 263 (1967).
74   B. Yasui, F. Fuchs and F. N. Briggs, *J. Biol. Chem.*, **243**, 735 (1968).
75   T. Wakabayashi, cited in Ref. *39*.
76   H. Mueller, *Nature*, **209**, 1128 (1966).
77   H. Mueller, *Biochem. Z.*, **345**, 300 (1966).
78   F. N. Briggs and F. Fuchs, Information Exchange Group #4, No. 122 (1966).
79   I. Staprans, K. Arai and S. Watanabe, *J. Biochem.*, **64**, 65 (1968).
80   F. Fuchs and F. N. Briggs, *J. Gen. Physiol.*, **51**, 655 (1968).
81   W. Drabikowski, B. Barylko, R. Dabrowska and E. Nowak, *Bull. Acad. Polon. Sci. Cl. II*, **16**, 397 (1968).
81a F. Fuchs, *Biochim. Biophys. Acta*, **245**, 221 (1971).
82   T. Wakabayashi and S. Ebashi, *J. Biochem.*, **64**, 731 (1968).
83   M. H. Han and E. S. Benson, *Biochem. Biophys. Res. Commun.*, **38**, 378 (1970).
84   I. Ohtsuki, T. Masaki, Y. Nonomura and S. Ebashi, *J. Biochem.*, **61**, 817 (1967).
84a R. D. Bremel and A. Weber, *Nature New Biology*, **238**, 97 (1972).
85   K. Sekiya and Y. Tonomura, *J. Biochem.*, **69**, 935 (1971).
86   Y. Kozuki and Y. Tonomura, unpublished.
87   Y. Tonomura, S. Watanabe and M. Morales, *Biochemistry*, **8**, 2171 (1969).
88   S. Ishiwata and S. Fujime, *J. Phys. Soc. Japan*, **30**, 303 (1971).
88a S. Ishiwata and S. Fujime, *J. Mol. Biol.*, **68**, 511 (1972).
89   H. E. Huxley, *Biophys. Soc. Abstr.*, **11**, 235a (1971).
89a P. J. Vibert, J. C. Haselgrove, J. Lowy and F. R. Poulsen, *Nature New Biology*, **236**, 182 (1972).
90   J. M. Stewart and H. M. Levy, *J. Biol. Chem.*, **245**, 5764 (1970).
91   M. Shigekawa and Y. Tonomura, *J. Biochem.*, **71**, 147 (1972).
91a M. Shigekawa and Y. Tonomura, *J. Biochem.*, in press.
92   Y. Kozuki and Y. Tonomura, unpublished.
93   A. Weber, *J. Gen. Physiol.*, **53**, 781 (1969).
94   J. Kendrick-Jones, W. Lehman and A. G. Szent-Györgyi, *J. Mol. Biol.*, **54**, 313 (1970).

11

THE Ca^{2+}-Mg^{2+}-DEPENDENT ATPase AND THE UPTAKE OF Ca^{2+} BY THE FRAGMENTED SARCOPLASMIC RETICULUM*

The main topic of this chapter is the molecular mechanism of uptake of Ca^{2+} by the sarcoplasmic reticulum (SR), which is thought to play an important role in the relaxation of muscle. It is only fairly recently that isolated fragments of SR have been shown to bind strongly with Ca^{2+} in the presence of Mg^{2+} and ATP. Despite a great deal of subsequent work on the SR, there are still many unsolved problems, one of which is the molecular mechanism of the accumulation of Ca^{2+} by the SR.

In order to attack this problem, we have studied the reaction mechanism of the SR-ATPase in the presence of Ca^{2+}. Our goals in this work were to clarify further the molecular mechanisms of muscle relaxation and of active transport of cations in general, and to determine the relationship between the molecular mechanisms of cation transport and muscle contraction. Biological transport of solutes usually involves three major steps:

* Contributor: Taibo Yamamoto

recognition, translocation and release. Of necessity, recognition must occur at the system boundary, most frequently at the membrane boundary of cells and organelles. Translocation through the boundary, release at the other side, and, ultimately, movement across the entire cell or system complete the transport process. Thus the mechanism of transport appears to be very similar to the sliding mechanism of muscle contraction (chapter 1), which involves the specific binding of F-actin by myosin, translocation or sliding of F-actin coupled with splitting of ATP by myosin, and then release or dissociation of F-actin from myosin.

First of all it will be helpful to review the pump mechanism for cation transport. The essential features of this mechanism are contained in Shaw's model[1], proposed in 1956, although more detailed schemes have recently been suggested by Skou[2], Albers[3] and Post[4]. Shaw's model[1] postulates interconvertible carriers for the Na^+/K^+ pump, and also requires the presence of ATP as an energy supply coupled to the conversion of Na^+ carrier into K^+ carrier on the outside of the membrane. This has been generally accepted as the basic model for active transport of cations, but there are still problems concerning the relationship between the cation carrier and the ATPase reaction. With regard to biological motility in general, it is most important to demonstrate whether translocation of the cation site through the membrane really is coupled to the ATPase reaction, and if so, at which step of the ATPase reaction it occurs. Hence, we must uncover the elementary steps of the transport ATPase reaction. If it can be shown that a cation on the inside of the membrane is involved in an elementary reaction step in one direction while the same cation on the outside of the membrane is involved in the reverse of this step, the law of microscopic reversibility would then establish that translocation of the cation site through the membrane is coupled to this elementary step. The Ca^{2+}-Mg^{2+}-dependent ATPase of the fragmented SR is one of the best systems for investigation of this problem, and we shall discuss our studies of the formation and decomposition of the various reaction intermediates, and the coupling of the steps of the ATPase reaction to the uptake of Ca^{2+}. The properties and physiological functions of the enzyme have already been excellently reviewed[5,6,6a], so we shall describe only the general features of the SR.

1. Morphology and Physiological Functions of the Sarcoplasmic Reticulum

A relaxing factor was discovered in muscle homogenates by Marsh[7,8] and later studied by Bendall[9,10]. Its isolation was attempted by Kumagai et al.[11], Portzehl[12,13] and Nagai et al.[14] Kumagai et al.[11] found that most of the relaxing activity is contained in the fraction of the muscle homogenate precipitated by 20% ammonium sulphate. This fraction is similar to the granular ATPase of Kielley and Meyerhof[15,16], which precipitates on centrifugation at $18,000 \times g$ for 1–2 hr. These preparative methods indicate that the relaxing factor is derived from muscle microsomal fractions[14,17]. At about the same time, Bozler[18,19], Watanabe et al.[20,21], A. Weber et al.[22–25], Ebashi[26] and Seidel and Gergely[27] reported that the contraction and relaxation of muscle is regulated by traces of Ca^{2+}. The discovery by Hasselbach and Makinose[28] and Ebashi and Lipmann[26,29–31] that the isolated SR absorbs Ca^{2+} in the presence of Mg^{2+} and ATP revealed the important role of the SR in muscle relaxation.

Many workers have studied the morphology of the SR by electron microscopy. Porter et al.[32–36] and Peachey[37] made the important discovery that the membrane in the cell is composed of vesicles surrounding the myofibril and a tubular structure (the T-system), which runs perpendicularly to the long axis of the muscle and is in contact with the vesicular structure. The region where the vesicular and tubular structures touch is called the triad (chapter 1, section 2; Fig. 2), and this triad junction has the unique property that the gap between the vesicular and tubular structures is nearly or completely bridged by evaginations from the terminal wall of the sarcoplasmic reticulum cisternae[38,39]. It was shown by Armstrong and Porter[35] and Endo[40] that the sarcolemma continues into the T-system. Capacitance measurements of frog sartorius muscle fibres agree rather well with estimates of the surface area of the T-system in the same muscle[37,41].

A. F. Huxley and Taylor[42] observed local contraction when a microelectrode was in contact with a particular region of the surface of a muscle cell and a weak current applied through it. Since the position of the site responding to this stimulus corresponds well with the position of the opening of the T-system, excitation of the plasma membrane (the sarcolemma) is thought to be transmitted to the inside of the muscle cell through the SR via the T-system, which opens to the surface of the membrane. Caffeine causes a contractile response in skeletal muscle cells without any major change in the resting potential[43–46] or in the actomyosin system[47,48]. Caldwell and Walster[49] showed that this contraction occurs even when caffeine is added to completely depolarised muscle or even when caffeine is introduced directly inside a living fibre by micro-injection. These results have led to the suggestion that caffeine acts by releasing 'bound calcium' from sites within the muscle[50-52]. Moreover, Weber[53,54] has observed caffeine-induced Ca^{2+}-release from muscle reticulum. Fujino et al.[55,56] found that the replacement of a muscle in an isotonic Ringer solution following its treatment with solutions made hypertonic with glycerol induces an irreversible loss of the ability to twitch, unaccompanied by any effect on either the resting or action potential. Howell[57] has shown by electron microscopy that this treatment damages selectively the transverse tubules of frog skeletal muscle and destroys the continuity of the tubules. The treated fibres cannot respond mechanically either to electrical stimulation or to increased K^+ concentrations, but they do contract in the presence of caffeine and relax when the caffeine is removed.

More recently, optical methods have been developed which permit the rapid changes in intracellular calcium concentration to be followed over the duration of a single contraction. Jöbsis and O'Connor[58,59] used murexide as a calcium indicator, while Ashley and Ridgway[60-62] used the bioluminescent protein aequorin[63], which, under normal physiological conditions, only emits light in the presence of Ca^{2+}. According to Ashley and Ridgway[62], when the fibre responds to a single depolarising pulse after the injection of aequorin, the peak of the transient calcium concentration occurs when the rate of increase in tension is at its maximum, and returns almost to its resting value when maximum tension is reached. Relaxation then takes place at resting calcium concentrations.* However, the reaction sequence of aequorin bioluminescence is rather complicated, and the bioluminescence cannot respond rapid changes in Ca^{2+} concentration[64a]. The intensity of light emitted is not a linear function of free Ca^{2+} ions, and dependent on Mg^{2+} concentration. Therefore, the relationship between the transient Ca^{2+} concentration and tension must be re-examined.

On the basis of these results, A. Weber[6], A. F. Huxley[65], Hasselbach[5], Ebashi[29,66], Sandow[52] and others proposed the following scheme for the coupling of excitation to contraction in muscle. In resting muscle, Ca^{2+} is concentrated in the SR so that the Ca^{2+} concentration around the myofibrils is less than that needed for contraction to occur. Excitation of the muscle cell membrane is transmitted to the SR through the T-system, releasing Ca^{2+} from the SR, and this then diffuses into the myofibrils and induces contraction. On the termination of excitation, Ca^{2+} is reabsorbed by the SR and the muscle returns to its original relaxed state.

* The interesting review by Ashley[64] describes the general features of the movement of Ca^{2+} in muscle cells during contraction and relaxation.

2. Ca^{2+}-Uptake and the ATPase of the Sarcoplasmic Reticulum

Hasselbach and Makinose[28,67-69] and Ebashi and Lipmann[31] found that, in the presence of Mg^{2+} and ATP, Ca^{2+} continued to be taken up by fragmented SR isolated from skeletal muscle, even when the concentration of Ca^{2+} in solution had been reduced to as little as 10^{-7} M by the addition of EGTA. Hasselbach and Makinose showed that the Ca^{2+}-uptake is markedly increased when oxalate is present, as well as Mg^{2+} and ATP. Since equimolar amounts of oxalate and Ca^{2+} are accumulated, these authors suggested that calcium oxalate precipitates in the fragmented SR. Electron microscopy shows that in the presence of oxalate the SR vesicles which have accumulated Ca^{2+} show an image of high electron density, which is thought to be due to the calcium oxalate precipitate. Hence, Hasselbach and Makinose suggested that Ca^{2+}-accumulation in the fragmented SR occurs by transport of Ca^{2+} through the membrane into the vesicles[5]. From the relative amounts of the products of the concentrations of Ca^{2+} and oxalate ions inside and outside the vesicles, they concluded that there is active transport of Ca^{2+}. Under conditions in which the calcium oxalate precipitates, the product of the concentrations of Ca^{2+} and oxalate ions within the vesicles is equal to the solubility product of calcium oxalate—2×10^{-7} M^2. In order to stop the transport of Ca^{2+} by the sarcoplasmic reticulum, the Ca^{2+} concentration in the external solution must be reduced by EGTA to 10^{-8}–10^{-9} M in the presence of 5mM oxalate. Hence, the product of the concentrations of Ca^{2+} and oxalate ions in the external solution becomes 1/500–1/5,000 of that in the vesicles, from which it can be inferred that there is active transport of Ca^{2+} by the sarcoplasmic reticulum. In order to maintain such a high concentration gradient, it is necessary to supply 3–5 kcal of energy to accumulate 1 mole of Ca^{2+}. From electron micrographs of muscle microsomal membrane, particularly from vesicles treated with Hg-phenyl azoferritin, Hasselbach and Elfvin[70] proposed that there is a structural difference between the outside and the inside of the membrane. From these results, it is now generally accepted that there is active transport of Ca^{2+} through the SR membrane.*

Inesi and Asai[71], Ikemoto et al.[72], Martonosi[73] and Deamer and Baskin[74] showed that only the outer surface of the microsomal membrane

* In order to demonstrate conclusively active transport of cations, it is necessary to determine the membrane potential, but so far this has been too difficult to measure for the SR.

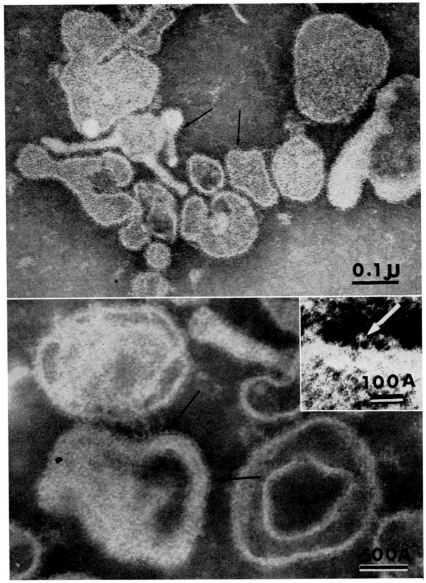

Fig. 1. Negatively stained SR vesicles (potassium phosphotungstate). These photographs clearly show the tripartite structure of SR(↓). (Courtesy of Dr. J. Asai, Nagoya University School of Medicine.)

is covered by spherical particles with diameters of 30–40Å. More recently, Asai* and Ikemoto et al.[74a] showed clearly the existence of head pieces with diameters of 35–55Å and stalks on the base membrane (Fig. 1), and Ikemoto et al.[74a] indicated that all important activities related to the Ca^{2+}-transport are located in the base membrane. Furthermore, Deamer and Baskin[74] revealed by freeze-etch microscopy the existence of 80–90Å particles on the fractured surface. Judging from the purity (section 4) and the molecular weight (section 8), these particles seem to be the ATPase proteins.

Hydrolysis of ATP supplies the energy for the SR to transport Ca^{2+} from the external solution against a concentration gradient. When Ca^{2+} is absent, the SR shows a low, Mg^{2+}-dependent ATPase activity. Hasselbach and Makinose[28,67] found that the addition of Ca^{2+} to the external medium considerably increases its activity. They called this enhanced splitting of ATP in the presence of Ca^{2+} the 'extra ATP-splitting.' The ATPase activity is high only during the transport of Ca^{2+}, and reverts to the low initial value found in the absence of Ca^{2+} when Ca^{2+}-transport stops. Hasselbach and Makinose showed that in the presence of oxalate, 1 mole of ATP is decomposed when 2 moles of Ca^{2+} are removed from the external solution by the SR. The Ca^{2+}-transport system of the SR shows no pronounced nucleoside triphosphate specificity, and there are only relatively minor differences in rates and affinities. Thus ITP, CTP, GTP and deoxy-ATP can also serve as energy donors, while UTP is somewhat less effective[75–77]. Acetyl-phosphate [78,79] and p-nitrophenyl phosphate[80] can also be used as substrates for Ca^{2+}-uptake.

Ohnishi and Ebashi[81,82] measured the rate of binding of Ca^{2+} to the SR, in the absence of oxalate, by following the change in absorption due to the complexing of Ca^{2+} with murexide in the solution. They found that the reaction is much faster than could be ascribed to transport alone, and that the amount of Ca^{2+} taken up and the amount of ATP split in low concentrations of ATP are not necessarily in a fixed ratio[83]. They concluded that the binding of Ca^{2+} to the surface of the SR depends on Mg^{2+} and ATP. In contrast, A. Weber et al.[84] recently measured Ca^{2+}-uptake and splitting of ATP over a broad range of ATP concentrations in the presence of oxalate, and found that, over the entire concentration range, the hydrolysis of 1 mole of ATP resulted in the uptake of 2 moles of Ca^{2+}. As described later, we[85] have also shown that 2 moles of Ca^{2+} are taken up when 1 mole of ATP is hydrolysed in the absence of oxalate, even during the ini-

* J. Asai, unpublished observations.

tial period of the reaction. Thus the uptake of Ca^{2+} by the SR almost certainly involves active transport, so the important problem is now the molecular mechanism of the coupling between the transport of Ca^{2+} and the splitting of ATP.

3. The Reaction Mechanism and the Phosphorylated Intermediate of the Ca^{2+}-Mg^{2+}-dependent ATPase[86,87]

We have used muscle microsomal fractions obtained by the procedure of Imai and Sato[88] to study the Ca^{2+}-Mg^{2+}-dependent ATPase of the SR, particularly the steady-state rate and the amounts of the phosphorylated intermediate. These microsomes have a Ca^{2+}-uptake capacity which is

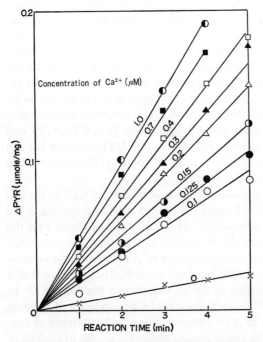

Fig. 2. Kinetics of pyruvate-liberation after addition of ATP to SR at various concentrations of Ca^{2+}. 25μM ATP, 3mM $MgCl_2$, 0.1M KCl, 50mM Tris-maleate (pH 6.5), 0.16mM PEP, 30μg/ml pyruvate kinase at 25°C. The concentrations of free Ca^{2+} were calculated taking 0.625μM as the dissociation constant of the Ca-EGTA complex. ×, 0; ○, 0.1; ●, 0.125; ◐, 0.15; △, 0.2; ▲, 0.3; □, 0.4; ■, 0.7; ◑, 1.0μM free Ca^{2+}.

rather low when oxalate is absent from the reaction medium. The addition of a small amount of Triton X—which mildly cleaves the membrane system—almost completely suppresses the uptake of Ca^{2+}. However, even under these conditions, external Ca^{2+} activates the ATPase and the reaction rate is almost equal to that of the 'extra ATP-splitting.'*

We have measured the kinetics of activation of the SR-ATPase by Ca^{2+} at steady-state when oxalate is absent, but when Mg^{2+} is present. The hydrolysis of ATP by the SR-ATPase was coupled to a pyruvate kinase system, and Fig. 2 shows the rate of formation of pyruvate as the concentration of free Ca^{2+} was varied. According to Weber et al.[84], the vesicles are impermeable to EGTA and EDTA, neither of which has a specific effect on transport, so that these compounds can be used to regulate Ca^{2+} and Mg^{2+} in the medium. The ATPase activity increases tremendously with increase in the external Ca^{2+} concentration. However, there is a low, Mg^{2+}-dependent activity, even in the presence of large amounts of EGTA, which we have taken to be the basic ATPase activity. The difference between this and the total SR-ATPase activity is then the Ca^{2+}-Mg^{2+}-dependent ATPase activity of the SR.

We shall now briefly discuss the role of Mg^{2+} and Ca^{2+} in the ATPase of the SR. In the presence of high concentrations of Mg^{2+} relative to ATP, the Ca^{2+}-dependence of the ATPase rate is independent of Mg^{2+} over a broad range. Hence, Ca^{2+} does not act by first binding with the ATP substrate but instead activates the ATPase by binding with the enzyme, since, if the active substrate were Ca-ATP, its concentration would change when the Mg^{2+} concentration is changed. Thus Mg-ATP is probably the substrate; this point will be discussed further on p. 319.

The Lineweaver-Burk plot of the substrate concentration dependence of the Ca^{2+}-Mg^{2+}-dependent ATPase is composed of two lines, one at high ATP concentrations (100–1,000μM) and the other at low concentrations (1–25μM). Therefore, kinetic studies of the SR-ATPase were carried out at both low and high concentration ranges of ATP. The mechanism of regulation of the ATPase activity by ATP will be described in section 7. We shall now discuss the reaction at steady-state with low concentrations of substrate.

Figure 3 shows the effect of the Ca^{2+} concentration in the external solution upon the Ca^{2+}-Mg^{2+}-dependent ATPase at various low concentrations of ATP. The reciprocal of the ATPase activity (v_o) depends linearly on the

---

* The usefulness of Triton X to solubilise the SR has recently been confirmed by several researchers[88a,b].

reciprocal of the external Ca²⁺ concentration. The slope of the line varies with the ATP concentration, but V_{max}, the extrapolated velocity at an infinite Ca²⁺ concentration, is independent of the ATP concentration. The relationship between the SR-ATPase activity (v_o) and both the external Ca²⁺ and ATP concentrations can be expressed as

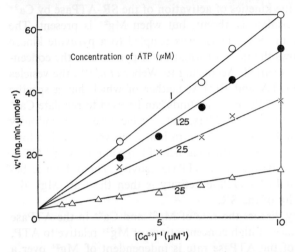

Fig. 3. Dependence of the steady-state rate (v_o) of the Ca²⁺-Mg²⁺-dependent ATPase on Ca²⁺ concentration at low concentrations of ATP. Experimental conditions are as described for Fig. 2. Concentrations of ATP: ○, 1; ●, 1.25; ×, 2.5; △, 25 μM.

$$v_o = \frac{V_{max}}{1 + \frac{\phi_2}{[Ca^{2+}]} + \frac{\phi_2 \phi_1}{[Ca^{2+}][ATP]}} \quad ([Ca^{2+}] \leq 10 \mu M).$$

At low Ca²⁺ concentrations, ATP and Ca²⁺ are thought to be complementary substrates for the SR-ATPase, and to bind to the enzyme in an ordered sequence. ϕ_1 and ϕ_2 are respectively those concentrations of ATP and Ca²⁺—5μM and 0.4μM—at which the reaction rate is half the maximum rate (V_{max}). When the Ca²⁺ concentration in the external medium is greater than 10μM, it inhibits the ATPase. Under these conditions the dependence of the reaction rate (v_o) upon the concentrations of external Ca²⁺ and ATP becomes

$$v_o = \frac{V_{max}}{1 + \phi_3[Ca^{2+}] + \phi_4 \frac{[Ca^{2+}]}{[ATP]}} \quad (10 \mu M \leq [Ca^{2+}]).$$

Thus the inhibition of the ATPase by high concentrations of Ca^{2+} can be nullified to some extent by raising the ATP concentration.

Hasselbach and Makinose[67,89] found that the 'extra-splitting' shown by the fragmented SR during uptake of Ca^{2+} is accompanied by considerable ATP-ADP exchange, and they postulated that during the active transport of Ca^{2+} a high energy phosphorylated intermediate is formed by the enzyme. Ebashi and Lipmann[31] also reported this exchange activity of the SR. Makinose[90] showed that the SR transfers the terminal phosphate of ITP or GTP to ADP, and the terminal phosphate of ATP not only to ADP but also to IDP. We[86,87] have incubated SR in the presence of Ca^{2+} with γ-^{32}P-ATP, and then quenched the whole with TCA, and found significant incorporation of ^{32}P into the SR protein. Similar results were also obtained independently by Makinose[91], and later confirmed by Martonosi[92] and by Inesi et al.[93] Quite recently, Pucell and Martonosi[94] reported that a phosphorylated intermediate is formed by adding acetylphosphate to the SR, and that this intermediate is identical to that formed by ATP.

The incorporation of TCA-stable ^{32}P from γ-^{32}P-ATP into the SR at 25°C and pH 7.0 reaches a maximum within several seconds of the beginning of the reaction. The ATPase activity is almost proportional to the amount of incorporation, which shows a dependence on external Ca^{2+} concentration similar to that of the Ca^{2+}-Mg^{2+}-dependent ATPase activity when there is no inhibition by excessive concentrations of Ca^{2+}. However, in the presence of a large amount of EGTA, the amount of P incorporated into the SR protein is 5% or less of that in the presence of Ca^{2+}. These results suggest that the incorporation in the presence of Ca^{2+} reflects the formation of an intermediate in the Ca^{2+}-Mg^{2+}-dependent ATPase reaction of the SR.

The following mechanism was proposed for the ATPase under conditions where there is no inhibition by high concentrations of Ca^{2+}, and is based on the assumption that the enzyme (E), ATP (S) and Ca^{2+} bind in an ordered sequence, and that the phosphorylated protein (EP) is a reaction intermediate.

$$E + S \rightleftharpoons ES$$
$$ES + Ca^{2+} \rightleftharpoons ESCa$$
$$ESCa \rightleftharpoons EPCa + ADP$$
$$EPCa \longrightarrow E + P_i + Ca^{2+}$$

4. The Formation and Decomposition of the Phosphorylated Intermediate[95]

We[95] have used a simple rapid mixing apparatus of our own design[96] to determine separately the rates of formation and decomposition of EP, in order to verify that it is an intermediate in the ATPase reaction. Since both the level of EP and the ability to take up Ca^{2+} in microsomal fractions obtained by the procedure of Imai and Sato[88] are low, we have used fragmented SR prepared by the method of A. Weber et al.[84] and Martonosi [97], in which Ca^{2+} exerts a marked influence on both the formation of EP and the ATPase activity, except in the presence of EGTA when this effect is almost eliminated.

At 0°C, the amount of EP formed increases linearly with time for about 1 sec after the addition of a relatively low concentration of ATP, and so the initial rate, v_f, of EP-formation is easily measured. Figure 4 shows that, at low concentrations of ATP and over a variety of Ca^{2+} concentrations, there is a satisfactory linear relationship between the reciprocal of the rate of EP-formation, v_f^{-1}, and the reciprocal of the ATP concentration,

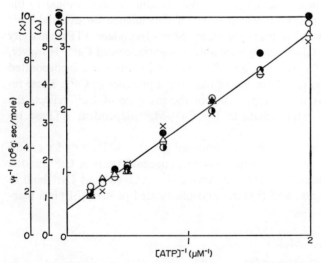

Fig. 4. The relationship between the initial rate, v_f, of formation of EP, and the ATP concentration, [ATP], at various concentrations of Ca^{2+} in the low concentration range of ATP. 0.03mg/ml SR, 5mM $MgCl_2$, 0.1M KCl, 0.5mM $CaCl_2$, 0.1M Tris-maleate, pH 7.0, 0°C. Concentrations of EGTA: ◐, 0.40; ○, 0.45; ●, 0.51; △, 0.54; ×, 0.57mM.

$[ATP]^{-1}$. When the Ca^{2+} concentration is reduced by increasing the EGTA concentration, the maximum value, V_f, of v_f decreases, while the Michaelis constant, K_f, remains constant. This indicates that EP is formed *via* a complex (ES) between the enzyme (E) and substrate (S), and that, at least initially, the bindings of E, S and Ca^{2+} necessary for the formation of EP occur at random. Figure 5 shows that there is also a satisfactory linear relationship between the reciprocal of V_f and the reciprocal of the square of the free Ca^{2+} concentration, indicating that E binds with two Ca^{2+} ions. However, as already described, kinetic analysis of the steady-state reaction suggests that E, S and Ca^{2+} bind in an ordered sequence (Fig. 3). This disagreement may originate in the fact that Fig. 4 refers to the initial rate of formation of EP, while Fig. 3 refers to the overall reaction at steady-state. In the former case, S and Ca^{2+} bind with the free enzyme in a random sequence, while in the latter case, the enzyme is bound to ions (probably Mg^{2+}, as described in section 5), transported to the outside from the inside of the SR as counter ions of Ca^{2+}, so that Ca^{2+} can then bind only after S had bound.

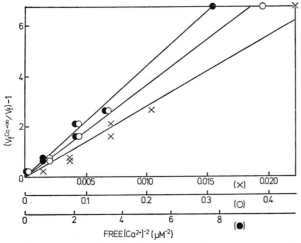

Fig. 5. Double reciprocal plot of the maximum initial rate of formation of EP, (V_f), *versus* the square of the free Ca^{2+} concentration. Experimental conditions are identical to those of Fig. 4. $\frac{V_f^{Ca \to \infty}}{V_f} -1$, where $V_f^{Ca \to \infty}$ is V_f at infinite Ca^{2+} concentration, is plotted against the reciprocal of the square of the concentration of free Ca^{2+}, which is calculated by assuming the apparent dissociation constant of the Ca-EGTA complex to be 0.1 (●), 0.5 (○) or 2.0 (×) μM.

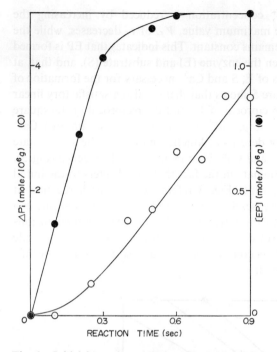

Fig. 6. Initial formation of EP and liberation of P_i in the presence of Triton X. 0.1 mg/ml SR protein, 12.2 μM AT^{32}P, 10 μM CaCl$_2$, 15mM MgCl$_2$, 0.1M KCl, 0.1M Tris-maleate, pH 7.5, 16°C. 20 μl Triton X/mg SR. ○, P_i-liberation; ●, EP-formation.

Figure 6 shows the initial rate of formation of EP and liberation of P_i, using a SR specimen obtained by treating 1mg of SR with 20 μl of Triton X (see p. 313). After this treatment, SR does not take up Ca^{2+}, and an electron micrograph using negative staining shows complete cleavage of the membrane structure. The Triton-treated SR shows a lag in the rate of liberation of P_i compared to the corresponding formation of EP, and there is good agreement with the value calculated from the amount of EP, taking the EP turnover rate as 4.2 sec^{-1}. However, as shown in Fig. 7, v_o/[EP] for the intact SR changes markedly for several seconds after the addition of ATP—thus at pH 7.0 and 15°C its value is about 4 sec^{-1} 0.1–0.2 sec after the addition of 5 μM ATP, while after 10 sec it has decreased to the steady-state value of 0.3–0.5 sec^{-1}. The mechanism of this transition will be discussed later in section 7.

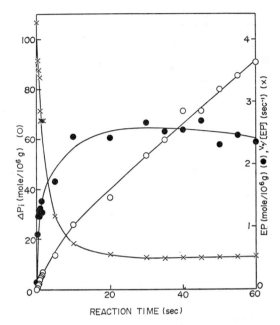

Fig. 7. Transition of the value of $v_o/[EP]$ during the initial phase. 0.01mg /ml SR protein, $5\mu M$ AT^{32}P, 0.47mM CaCl$_2$, 0.5mM EGTA ($10\mu M$ free Ca^{2+}), 1mM MgCl$_2$, 0.1M KCl, 0.1M Tris-maleate, pH 7.0, 15°C. ○, P$_i$-liberation; ●, EP-formation; ×, $v_o/[EP]$.

The rate of decomposition of EP was measured by adding ATP to the fragmented SR in the presence of Ca^{2+} to form EP, and then suppressing this EP-formation by removing Ca^{2+} with EGTA. As shown in Fig. 8, the decrease in EP after the addition of EGTA follows first order kinetics, and the rate constant, k_d, decreases when the time interval between addition of ATP and addition of EGTA is prolonged. This corresponds to the transition of $v_o/[EP]$ mentioned above, and the values of k_d and $v_o/[EP]$ at steady-state are equal. These results show that EP is a true reaction intermediate.

The rate of formation of EP is significantly lower when Mg^{2+} is not added to the system, probably because the substrate for the formation of EP is Mg-ATP. When no external Mg^{2+} is added there is very slow formation of EP derived from the Mg^{2+} contaminating the SR. The formation of EP, at all Mg-ATP concentrations studied, shows no lag phase and is linear with time, at least initially. Hence, the rate constant for the formation of

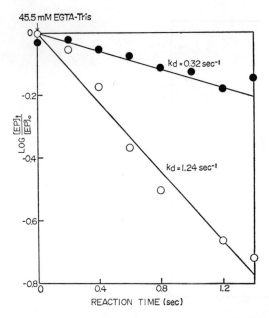

Fig. 8. Rate of decomposition of EP after the addition of EGTA. 0.01mg/ml SR protein, 0.47mM CaCl$_2$, 0.5mM EGTA (10μM free Ca^{2+}), 0.1M KCl, 0.1M Tris-maleate, pH 7.0, 15°C. EP-decomposition was measured by quenching with 45.5mM EGTA-Tris, 0.4 sec (○) and 20 sec (●), after the addition of 5μM AT^{32}P. ○, rate constant for EP-decay, k_d=1.24 sec^{-1}; ●, k_d=0.32 sec^{-1}.

the enzyme-substrate complex from E and Mg-ATP is very large. In addition, there is first-order decay of E^{32}P when its formation is stopped by the addition of EGTA or 'cold' ATP. Even when the SR has completed its uptake of Ca^{2+}, and has a high Ca^{2+} concentration inside the membrane, drastic reduction of the external Ca^{2+} concentration by the addition of EGTA immediately terminates the formation of EP. Thus EP formation requires Ca^{2+} outside the membrane.

We[87] have studied the pH-dependency of the formation and decomposition of EP. Figure 9 shows the dependence upon pH of the EP level and Ca^{2+}-Mg^{2+}-dependent ATPase activity at steady-state. The amount of EP increases considerably with increasing pH, reaching a maximum at pH 8.5 or higher. However, the ATPase activity shows a bell shaped curve with a maximum near pH 7.0. In the presence of 20μM ATP and 10μM

Ca²⁺, inhibition of the ATPase by Ca²⁺ may be disregarded, and the rate is close to V_{max}, so that the following scheme is possible

$$ES \underset{k_{-1}}{\overset{k_{+1}}{\rightleftharpoons}} EP + ADP \overset{k_{+2}}{\longrightarrow} E + P_i + ADP,$$

where E is an active site of the SR, S is Mg-ATP and k_{+1} and k_{+2} are the respective rate constants for the formation and decomposition of EP. The rate at steady-state may be expressed as $v_o = k_{+2}[EP]$. The pH-dependency of log k_{+2}—calculated from $v_o/[EP]$ at various pH's—is shown by the symbol × in Fig. 9, and reaches a maximum ($k°_{+2}$) at pH 7.2 or less. It decreases proportionately to the decrease in [H⁺] above pH 7.5, and can be expressed by

$$k_{+2} = k°_{+2} / \left\{ 1 + \frac{10^{-7.2}}{[H^+]} \right\},$$

from which it can be concluded that a functional group with a pK of 7.2 participates in the decomposition of EP. The rate constant, k_d, for the steady-state decomposition of EP, obtained directly by the method described above at alkaline pH, agrees well with the value of $v_o/[EP]$.

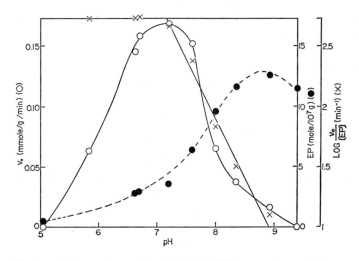

Fig. 9. pH-Dependence of the degree of ³²P-incorporation and the rate of the Ca²⁺-Mg²⁺-dependent ATPase. 20μM AT³²P, 10μM CaCl₂, 5mM MgCl₂, 0.1M KCl, 0.2mg/ml pyruvate kinase, 0.2mM PEP, 0.13mg/ml SR protein, 20mM pH buffer. ●, ³²P-incorporation; ○, ATPase rate, v_o; ×, log $v_o/[EP]$.

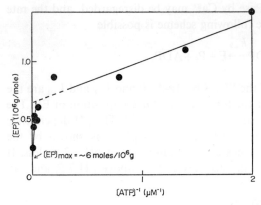

Fig. 10. Dependence of the concentration of EP on the concentration of ATP. 0.47mM CaCl$_2$, 0.5mM EGTA (10 μM free Ca^{2+}), 0.1M KCl, 1mM MgCl$_2$, pH 7.0, 15°C. The value of [EP]$_{max}$ obtained from these data is ~6 moles/10^{6}g.

The identification of EP as the reaction intermediate has prompted us to study its properties. The amount of EP formed in the presence of large amounts of Ca^{2+} and ATP is 5–7 moles/10^6 g protein (Fig. 10; the reason for the two straight lines in the relationship between [EP]$^{-1}$ and [S]$^{-1}$ is discussed on p. 337). However, Vegh et al.[98] found that the enzyme is inactivated by ionising radiation, and obtained a molecular weight of 1.9×10^5 for the Ca^{2+}-Mg^{2+}-dependent ATPase of the SR, assuming that the generation of one ion cluster inactivates a whole enzyme molecule (see section 8 for the molecular weight of the purified ATPase). The agreement of this value with our estimate of the amount of protein per phosphorylation site may be fortuitous, since the molecular weight measurements are rather crude, but the fragmented SR prepared by A. Weber and Martonosi and colleagues nevertheless contains a highly pure Ca^{2+}-Mg^{2+}-dependent ATPase.

We have also investigated whether the EP is a high energy phosphate compound. The rate of decay of EP measured by quenching with EGTA after the formation of EP is considerably increased by adding ADP to the system (Fig. 11). When no ADP is added, the decrease in EP approximately equals the amount of P$_i$ liberated. However, when ADP is added there is almost no P$_i$ liberated from EP, and thin layer chromatography of the products shows that ATP is formed. At pH 8.8 and 15°C, when EP decomposition is particularly slow, the formation of EP by addition of ATP in the presence of Ca^{2+}, followed by quenching with EGTA and the simultaneous addition of 0.2mM ADP, resulted in the almost completely stoichiometric reaction EP+ADP→E+ATP (Fig. 12). Figure 13 shows that when ADP only is added, and not EGTA, a substantial amount of EP is converted to ATP. These results clearly demonstrate that EP is a high energy phosphate compound capable of reacting with ADP to form ATP, but that EP does not bind to ADP under ordinary reaction conditions. In order to study the binding of the phosphoryl group in EP, we have measured the dependence of the stability of EP denatured by TCA upon pH and hydroxylamine concentration[87]. EP is relatively stable at pH 4.0 or below, but at higher pH it becomes increasingly less stable. At pH 5.2 and 25°C it is completely dissociated to P_i by 1M hydroxylamine in only a few minutes. We[99] have demonstrated the formation of a hydroxamate by the following experiment. EP was formed and quenched by the addition of TCA, digested with pepsin at pH 3.0 and chromatographed

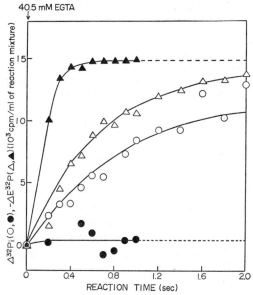

Fig. 11. The effect of ADP on the rate of decay of $E^{32}P$. 0.2mg/ml SR protein, 20μM $CaCl_2$, 1mM $MgCl_2$, 0.1M KCl, 0.1M Tris-maleate, pH 7.0, 15°C. 10 sec after addition of 10μM ATP, 5μM $AT^{32}P$ was added, followed after 10 sec by 40.5mM EGTA to stop $E^{32}P$-formation. The rates of decrease in $E^{32}P$ (△, ▲) and generation of $^{32}P_i$ (○, ●) in the presence and absence of 1mM ADP, respectively, were compared. ○, △, no ADP added ●, ▲, 1mM ADP added.

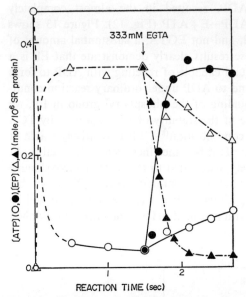

Fig. 12. The formation of ATP from phosphorylated intermediates and ADP (1). 10 mg/ml SR protein, 1mM $MgCl_2$, 50μM $CaCl_2$, 0.15M KCl, 0.1M Tris-HCl, pH 8.8 and 15°C. 1.5 sec after the addition of 4.55μM $AT^{32}P$, 33.3mM EGTA was added to stop EP-formation together with 0.2mM of ADP. The resulting reaction EP+ADP→E+ATP decreases the amount of EP and forms an equivalent amount of ATP. ○, ●, amount of ATP; △, ▲, amount of EP. ○, △, no ADP added; ●, ▲, 0.2mM ADP added.

on Biogel P-2 in order to isolate the P-peptide. Treatment of this peptide with 2-hydroxy-5-nitrobenzyl hydroxylamine liberated roughly equal amounts of P_i and a new peptide bound to 2-hydroxy-5-nitrobenzyl hydroxylamine. This would suggest that the P from ATP is incorporated into the SR protein as an acyl-phosphate. However, hydroxylamine shows no inhibitory effect towards the formation of EP or the ATPase activity of native SR, so the presence of an acyl-phosphate type of EP remains to be established.

5. Elementary Steps in Cation Transport

As described above, Makinose et al.[67,77,89] and A. Weber et al.[84] have shown that the hydrolysis of 1 mole of ATP by SR results in the uptake of 2 moles of Ca^{2+}. Recently we[85] have shown that, at low ATP concentra-

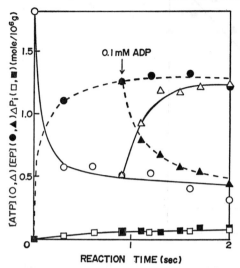

Fig. 13. The formation of ATP from phosphorylated intermediates and ADP (2). 9.9 mg/ml SR, 1mM $MgCl_2$, 50μM $CaCl_2$, 0.1M KCl, 0.1M Tris-HCl, pH 8.8 and 15°C. 0.1 mM ADP was added 0.9 sec after the addition of 18μM $AT^{32}P$. The decrease in EP and the increase in ATP according to the reaction EP+ADP→E+ATP were measured. ○, △, amount of ATP; ●, ▲, amount of EP; □, ■, amount of P_i liberated. ○, ●, □, no ADP added; △, ▲, ■, 0.1mM ADP added.

tions, the Ca^{2+} which is initially incorporated into the fragmented SR is not removed from the SR, even after the addition of a high concentration of EGTA. There are almost 2 moles of such Ca^{2+} initially taken up per mole of ATP hydrolysed in the Ca^{2+}-Mg^{2+}-dependent reaction (Fig. 14). No instantaneous binding of a measurable amount of Ca^{2+} to the SR was also observed on adding ATP by a stopped flow method, using murexide as an indicator for Ca^{2+} ions[99a]. We have also found that the protons and P_i generated by the ATP-hydrolysis are discharged to the outside of the membrane.

The important question here is which step of the Ca^{2+}-Mg^{2+}-dependent ATPase reaction is involved in the transport of Ca^{2+} from the outside to the inside of the membrane, so that it is no longer removable by EGTA added to the outside of the membrane. The formation of $E^{32}P$ from E+ $AT^{32}P$ stops immediately when EGTA or 'cold' ATP is added to the outside of the membrane, so that ATP and Ca^{2+} must be located outside the membrane in the complex $ESCa_2$ (see Fig. 19). In contrast, when the forma-

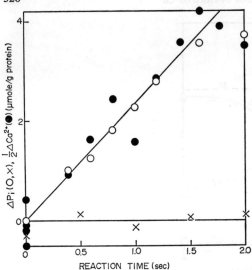

Fig. 14. Initial rate of uptake of Ca^{2+} and splitting of ATP. 0.3mg/ml SR protein, 5μM AT^{32}P, 17μM Ca^{2+}, 5mM MgCl$_2$, 0.1M KCl, 40mM Tris-maleate, pH 6.5, 15°C. ●, 1/2 × Ca^{2+}-uptake; ○, ATP-splitting in the presence of Ca^{2+}; ×, ATP-splitting in the presence of EGTA. The initial concentration of external Ca^{2+} was measured by the Calcein method.

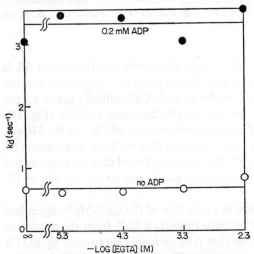

Fig. 15. Effect of external EGTA on the formation of ATP from phosphorylated intermediates and ADP. 5mM MgCl$_2$, 1μM CaCl$_2$, 0.1M KCl, 0.1M Tris-maleate, pH 7.0 and 5°C. 0.1μM AT^{32}P was added to 0.1mg/ml SR to form E^{32}P, followed 3 sec later by 20 μM non-radioactive ATP and 0.2mM ADP. The rate of decrease of E^{32}P according to the reaction E^{32}P+ADP→E+AT^{32}P was measured. The difference between the two gives the rate of this reaction. ●, 0.2mM ADP; ○, no ADP added.

tion of EP is stopped by applying EGTA to the outside of the membrane, and ADP is then added, ATP is produced. Figure 15 shows that the formation of ATP from EP and ADP, which is the reverse of the formation of EP from E and ATP, does not require external Ca^{2+}, and is completely independent of the concentration of Ca^{2+} outside the membrane[95].

It appears that the formation of EP+ADP from E+ATP involves Ca^{2+} outside the membrane, while the reverse process requires the presence of Ca^{2+} inside the membrane, suggesting that Ca^{2+} is transported from the outside of the membrane to the inside when EP is formed. We have confirmed this by the following two experiments. First we studied the formation of ATP from EP and ADP after the membrane structure had been destroyed with Triton X[100]. When $20\mu l$ of Triton X are added per mg of SR, the steady-state decomposition of EP is substantially accelerated

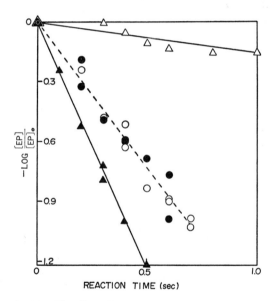

Fig. 16. The effect of Triton X on the decay of EP in the presence of ADP. $2\mu M$ $CaCl_2$, 5mM $MgCl_2$, 0.1M KCl, 0.1M Tris-maleate, pH 7.0 and 10°C. $2.3\mu M$ $AT^{32}P$ was added to 0.1mg/ml SR, followed 3 sec later by 2.5mM EGTA to stop phosphorylation, and the decay of $E^{32}P$ was measured. The difference between the two gives the rate of the reaction $E^{32}P+ADP \rightarrow E+AT^{32}P$. △, ▲, no Triton X added; ○, ●, $20\mu l$ Triton X/mg SR. △, ○, $E^{32}P$-decay after addition of 2.5mM EGTA. ▲, ●, $E^{32}P$-decay after addition of 2.5mM EGTA+0.5mM ADP.

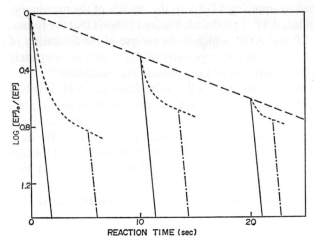

Fig. 17. Schematic representation of the formation of ATP from EP and ADP at pH 9.2 in the presence of EGTA and Ca^{2+}. SR was phosphorylated with $AT^{32}P$ at pH 9.2 and 15°C. — — —, $E^{32}P$-decay after addition of EGTA or non-radioactive ATP. ———, $E^{32}P$-decay when a large amount of ADP was added at various times after the termination of $E^{32}P$-formation with non-radioactive ATP. ------, $E^{32}P$-decay when its formation was quenched with a large amount of EGTA, and a large amount of ADP was added after various time intervals. — · —, $E^{32}P$-decay when $E^{32}P$-formation was stopped with EGTA (— — —) followed by the addition of ADP (------) and then by the addition of a large amount of $CaCl_2$ (— · —).

(as will be described later), but there is no longer a reaction $E^{32}P + ADP \rightarrow E + AT^{32}P$ when the formation of EP is terminated by removing the Ca^{2+} with EGTA and a large amount of ADP is then added (Fig. 16)[95]. However, when non-radioactive ATP is added in the presence of Ca^{2+} in order to terminate the formation of $E^{32}P$, and ADP is then added, this reaction does take place. This indicates that the Ca^{2+} inside the membrane, which is required for this reaction to occur, is only removed by EGTA when the membrane structure has been destroyed by Triton X.

Next we followed the production of ATP from the added ADP and the remaining EP, at various time intervals after the termination of the formation of EP, with EGTA at pH 9.2, when the decomposition of EP is very slow[95]. When formation of EP is stopped with EGTA, the proportion of EP which can react to produce ATP on addition of ADP decreases with time (Fig. 17). A long time after the addition of EGTA most of the EP is no longer able to form ATP, but the addition of a large amount of CaCl$_2$ together with ADP restores this ability. However, if the formation of E^{32}P is suppressed with non-radioactive ATP instead of EGTA, this result is not observed, and E^{32}P and ADP react almost completely to form AT^{32}P. Recently, Duggan and Martonosi[101] showed that the SR membrane is not permeable to Ca^{2+} at neutral pH, but is only so in alkaline conditions. Hence these results can be easily explained by assuming that the formation of ATP from EP and ADP always requires Ca^{2+} inside the membrane, and that at alkaline pH the Ca^{2+} is slowly removed by EGTA outside the membrane. Thus it appears that the formation of EP is accompanied by translocation of the Ca^{2+} site from the outside to the inside of the SR membrane.

It is now necessary to identify the ion which is transported across the membrane in conjunction with the transport of Ca^{2+}. We[85] have found that the acidity change in the external medium during uptake of Ca^{2+} by the fragmented SR is due to the hydrolysis of ATP itself, and that the P$_i$ released from the ATP is not transported to the inside of the SR. Carvalho and Leo[102] investigated the transport of cations coupled with the uptake of Ca^{2+}, and found that the sum of the equivalents of Mg^{2+}, K$^+$ and Ca^{2+} within the SR vesicles is constant. This suggests that there is an efflux of Mg^{2+} and K$^+$ coupled with the influx of Ca^{2+}. We[85] have also shown that the Ca^{2+}-transport rate increases in the presence of K$^+$ and Mg^{2+} when SR prepared in a buffer containing no K$^+$ has been incubated previously with KCl.

The following two experiments indicate that Mg^{2+} is transported out of the membrane as Ca^{2+} is transported in [95]. The values of v_o/[EP], after sufficient preincubation of SR in various concentrations of Mg^{2+}, follows the simple Michaelis equation $v_o/[\text{EP}] = A / \left(1 + \dfrac{K_{\text{Mg}}}{[\text{Mg}^{2+}]}\right)$. At pH 8.5 and 10°C, $K_{\text{Mg}} = 51 \mu\text{M}$. Similar results have also been obtained by Martonosi[103]. Inesi et al.[93] have recently reported that the decomposition of the ^{32}P-membrane complex in the absence of Mg^{2+} follows first-order kinetics and is five times slower than when Mg^{2+} is present, showing that Mg^{2+} is necessary for the decay of EP. This is confirmed by suppressing the formation of EP by adding a large amount of EDTA to chelate the Ca^{2+} and Mg^{2+} (Fig. 18). The decay of EP gradually ceases, but the addition of a large amount of MgCl$_2$ restores it. This gradual decrease in the rate of decay of EP probably indicates that it requires the presence of Mg^{2+} inside the membrane which is slowly removed by EDTA, since the time required to

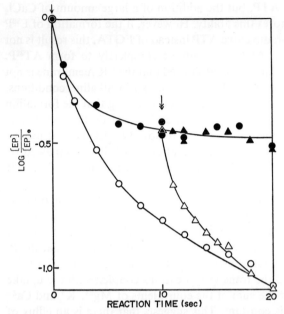

Fig. 18. Inhibition of EP-decay by EDTA. 0.1mg/ml SR was phosphorylated for 2 sec with 12.7μM AT^{32}P in 0.2mM MgCl$_2$, 0.1M KCl, 0.1M Tris -HCl, pH 8.5 and 15.5°C. The amount of Ca^{2+} contaminating the reaction mixture was usually sufficient for the phosphorylation reaction to occur. ○, Ca^{2+} was removed from the system with 0.1mM EGTA at ↓. ●, 0.4mM EDTA was added at ↓ to remove Ca^{2+} and Mg^{2+}; △, 0.4mM EDTA was added at ↓ and 1mM MgCl$_2$ and 0.1mM EGTA were added at ↓; ▲, 0.4 mM EDTA was added at ↓, and 0.1mM EGTA was added at ↓.

stop the decay completely after the addition of EDTA can be considerably reduced by destroying the membrane structure with Triton X.

From the above experiments we can conclude that the formation of 1 mole of EP causes 2 moles of Ca^{2+} to be transported from the outside to the inside of the membrane, where they replace $(1+n)$ moles of Mg^{2+} and $2(1-n)$ moles of K$^+$, of which 1 mole of Mg^{2+} is required for the decay of EP. In this section we have made the operational definition that a divalent cation is outside the membrane if it is immediately removed by a chelating agent outside the membrane, whereas it is inside the membrane when it is not easily removed by a chelating agent outside the membrane, except when the membrane system has been destroyed with Triton X. However, it remains to be determined whether the divalent cation within the mem-

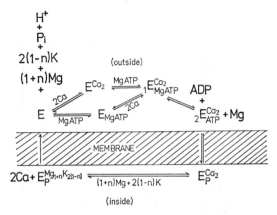

Fig. 19. The mechanism of coupling of ATP-hydrolysis with cation transport across the SR membrane. As to the presence of $_1E^{Ca_2}_{MgATP}$ and $_2E^{Ca_2}_{ATP}$ see p. 339.

brane thus defined is the bulk divalent cation present within the membrane. Weber et al.'s[84] evidence indicates that most of the Ca^{2+} incorporated in the membrane, but not forming a precipitate with oxalate, is bound to the inner membrane structure.

As described above, ATP reacts on the outside of the membrane, and ADP is released from the outside of the membrane when EP is formed. The decay of EP also releases P_i and protons outside the membrane. Figure 19 shows the mechanism of coupling of hydrolysis of ATP by the SR with transport of Ca^{2+} and Mg^{2+}. The translocation of the cation site upon the formation and decomposition of EP is strongly indicated kinetically, but its molecular mechanism remains obscure. With respect to this, it should be added that inducement of conformational changes in the SR by ATP was reported by several researchers[104–106]. In particular, Nakamura et al.[106] spin-labelled SR with N-2,2,6,6-tetramethyl piperidine nitroxide maleimide at pH 8.5 in the presence of ATP, and found that the EPR spectrum of the SR thus obtained is altered by the addition of ATP. This alteration of the EPR spectrum occurs only in the presence of both Mg^{2+} and Ca^{2+}, and is reversibly abolished by removal of the external Ca^{2+}. These results strongly suggest a conformational change in the SR accompanied by the formation and/or decomposition of phosphoryl intermediate.

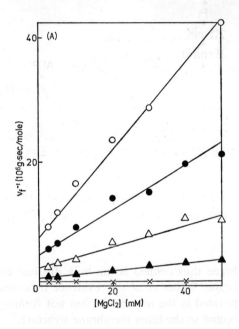

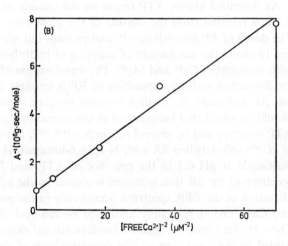

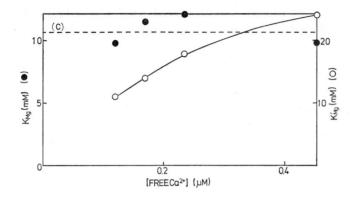

Fig. 20. Competitive inhibition by Mg^{2+} of the Ca^{2+}-dependent formation of EP. 0.07 mg/ml SR, $2\mu M$ $AT^{32}P$, 50mM Tris-maleate, pH 7.0, 0°C. Ionic strength 0.16. A: Plot of reciprocal of v_f versus $[MgCl_2]$. Concentration of free Ca^{2+} was adjusted by EGTA buffer. Assuming the dissociation constant of the Ca-EGTA complex to be $0.14\mu M$, the concentrations of free Ca^{2+} are: ○, 0.12; ●, 0.17; △, 0.23; ▲, 0.45; ×, $1.51\mu M$. $v_f^{-1} = A^{-1}\{1+[Mg^{2+}]/K'_{Mg}\}$, where A and K'_{Mg} are constants which depend only on $[Ca^{2+}]$. B: Dependence of A, obtained from (A), on the concentration of free Ca^{2+}. $A^{-1} = V_f^{-1}\{1+\left(\frac{K_{Ca}}{[Ca^{2+}]}\right)^2\}$, where V_f and K_{Ca} are constants. C: Dependence of K'_{Mg}, obtained from (A), on the concentration of free Ca^{2+}. $K'_{Mg} = K_{Mg}\{1+\left(\frac{[Ca^{2+}]}{K_{Ca}}\right)^2\}$, where K_{Mg} is a constant. ○, K'_{Mg}; ●, K_{Mg}.

6. The Change in Affinity for Calcium and Magnesium Ions

In the preceeding section we described the relationship between the translocation of the cation site and elementary steps of the Ca^{2+}-Mg^{2+}-dependent ATPase of the SR. In this section we will discuss our kinetic studies[107] of recognition of Ca^{2+} and Mg^{2+} ions by the SR and the release of these ions from the SR. Since it is difficult to control the concentrations of internal cations in intact SR vesicles as usually prepared, we have treated the SR with deoxycholate (DOC) to make it soluble, and then purified the ATPase by salt fractionation[108].

First we measured the rate of formation of EP (v_f) over broad concentration ranges of Ca^{2+} (0.12–1.5μM) and Mg^{2+} (2.5–50mM), in order to investigate the competitive inhibition of the Ca^{2+}-dependent formation of EP by Mg^{2+}. As shown in Fig. 20A, v_f is given by

$$v_f^{-1} = A^{-1}\{1+[Mg^{2+}]/K'_{Mg}\},$$

where A and K'_{Mg} are constants which depend on [Ca^{2+}]. As shown in Fig. 20B, A, *i.e.* the value of v_f obtained by extrapolation to [Mg^{2+}]=0, is given by

$$A^{-1}=V_f^{-1}\left\{1+\left(\frac{K_{Ca}}{[Ca^{2+}]}\right)^2\right\},$$

where V_f and K_{Ca} are constants. Hence, the formation of EP requires the binding of 2 moles of Ca^{2+} to 1 mole of the ATPase, as already mentioned on p. 317. Figure 20C shows the dependence of K'_{Mg} on the concentration of free Ca^{2+}. Thus $K'_{Mg}=K_{Mg}\left\{1+\left(\frac{[Ca^{2+}]}{K_{Ca}}\right)^2\right\}$, where K_{Mg} is a constant. These relationships can easily be interpreted by the following reaction scheme

$$ES+Mg^{2+}\underset{K_{Mg}}{\rightleftharpoons}ESMg$$

$$ES+2Ca^{2+}\underset{(K_{Ca})^2}{\rightleftharpoons}ESCa_2\overset{k}{\longrightarrow}EPCa_2+Mg\text{-}ADP$$

$$v_f=\frac{k\varepsilon}{1+\left(\frac{K_{Ca}}{[Ca^{2+}]}\right)^2+\left(\frac{K_{Ca}}{[Ca^{2+}]}\right)^2\frac{[Mg^{2+}]}{K_{Mg}}}=\frac{V_f/\left\{1+\left(\frac{K_{Ca}}{[Ca^{2+}]}\right)^2\right\}}{1+\frac{[Mg^{2+}]}{K_{Mg}\left\{1+\left(\frac{[Ca^{2+}]}{K_{Ca}}\right)^2\right\}}},$$

where S is Mg-ATP and ε is the total concentration of active sites of the ATPase. At an ionic strength of 0.16, pH 7.0 and 0°C, V_f, K_{Ca} and K_{Mg} were 1.33 moles/10^6 g·sec, 0.35μM and 10.6mM respectively. It is interesting to note that Chevallier and Butow[108a] have recently measured the binding of Ca^{2+} to SR in the absence of ATP at pH 7.4 and 4°C, using the equilibrium dialysis method. They found that 10–20 moles of Ca^{2+} bind to 10^{6}g SR with a dissociation constant of 0.4μM and the dissociation constant increases three-fold on adding 1mM MgCl$_2$ and 0.6M KCl. On the other hand, the dissociation constant of binding of Ca^{2+} for EP-formation at pH 7.0 and 0°C is 0.35μM and is independent of ATP (*cf.* p. 317), and its magnitude is two-fold of $\varepsilon=10$–14 moles/10^{6}g (*cf.* p. 322). It is well known that not only Ca^{2+} but also Sr^{2+} is transported into the SR[84,109]. Under the same conditions, the values of V_f, K_{Sr} and K_{Mg} for Sr^{2+} are 2.73 moles/10^6 g·sec, 27.5μM and 7mM respectively. Therefore, 2 moles of both Ca^{2+} and Sr^{2+} bind to 1 mole of ES, the ratio of the binding constants

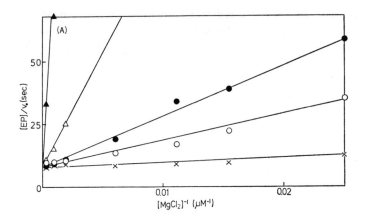

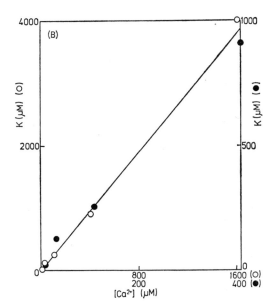

Fig. 21. Competitive inhibition by Ca^{2+} of the Mg^{2+}-dependent decomposition of EP. 0.06mg/ml SR, $2\mu M$ $AT^{32}P$, 0.11M KCl, 50mM Tris-maleate, pH 7.0, 0°C. A: Double reciprocal plot of $v_0/[EP]$ *versus* $[MgCl_2]$. Concentrations of free Ca^{2+}: ×, 10; ○, 30; ●, 110; △, 410; ▲, 1,600μM. $[EP]/v_0 = D^{-1}\{1+K[Mg^{2+}]^{-1}\}$, where D is a constant independent of $[Ca^{2+}]$. B: Dependence of K, obtained from (A), on the concentration of Ca^{2+}. $K = \dfrac{K_{Mg}}{K_{Ca}}[Ca^{2+}]$, where K_{Mg} and K_{Ca} are constants.

being 79:1. These results are consistent with those on the transport of Sr^{2+} into the SR. Thus, according to A. Weber[84], the rate of transport of Sr^{2+} is similar to that of Ca^{2+}, while the affinity of Sr^{2+} is much smaller than that of Ca^{2+}. On the other hand, 1 mole of Mg^{2+} binds to 1 mole of ES competitively with Ca^{2+}, the ratio of the binding constants of Ca^{2+} and Mg^{2+} being $2-3 \times 10^4$:1.

Next we measured $v_o/[EP]$ over broad ranges of Mg^{2+} (40μM–3mM) and Ca^{2+} concentration (10μM–1.6mM), in order to study the competitive inhibition by Ca^{2+} of the Mg^{2+}-dependent decomposition of EP. As shown in Figure 21A, $v_o/[EP]$ is given by

$$[EP]/v_o = D^{-1}\{1 + K[Mg^{2+}]^{-1}\},$$

where D is a constant independent of $[Ca^{2+}]$. The slope, K, increases with increase in the concentration of Ca^{2+}, and is given by $K = \frac{K_{Mg}}{K_{Ca}}[Ca^{2+}]$, where K_{Mg} and K_{Ca} are constants, as shown in Fig. 21B. The value of K_{Mg}/K_{Ca} is 2.5. These results can be explained by the reaction scheme

$$EPMg + Ca^{2+} \underset{K_{Ca}/K_{Mg}}{\rightleftharpoons} EPCa + Mg^{2+}$$

$$EPMg \xrightarrow{k} E + P_i + Mg^{2+}$$

$$v_o = \frac{k\varepsilon}{1 + \frac{K_{Mg}[Ca^{2+}]}{K_{Ca}[Mg^{2+}]}}.$$

Thus the formation of EP, i.e. the translocation of the cation site, accompanies very large changes in the affinities for Ca^{2+} and Mg^{2+}. After the translocation, the ratio of the concentrations of Mg^{2+} and Ca^{2+} necessary for half saturation changes from $2-3 \times 10^4$:1 to 2.5:1. Furthermore, the ratio 2.5:1 of K_{Mg}/K_{Ca} is consistent with the Ca^{2+}-transport inhibition by accumulated Ca^{2+} in the SR (cf. the next section).

7. Regulation of the Ca^{2+}-Mg^{2+}-dependent ATPase

The Ca^{2+}-Mg^{2+}-dependent ATPase of the SR might be regulated by any of several processes and we shall describe two of these, emphasising their relationship to the overall reaction scheme.

First is the acceleration of the ATPase reaction by a high concentration of ATP. This phenomenon is also observed with myosin-ATPase (chapter 3), and with the Na^+-K^+-dependent membrane ATPase (chapter

12). As already mentioned in section 3, the dependence of the steady-state rate (v_o) of the Ca^{2+}-Mg^{2+}- dependent ATPase reaction upon the concentrations of external Ca^{2+} and ATP is given by

$$v_o = \frac{V_{max}}{1+\dfrac{\phi_2}{[Ca^{2+}]}+\dfrac{\phi_2\phi_1}{[Ca^{2+}][ATP]}} \quad ([Ca^{2+}] \leq 10\mu M)$$

in both high and low ATP concentration ranges. However, the constant ϕ_1 for ATP in the high substrate concentration region is much higher than that in the low concentration region. The effects of temperature, pH and treatment with N-ethyl maleimide on the ATPase rate of a given specimen are identical in both low and high substrate concentration ranges, indicating that the same enzyme performs two different Ca^{2+}-Mg^{2+}-dependent ATPase reactions with different K_m values[86]. This is supported by measurements of the dependence upon the ATP concentration of the amount of phosphorylated intermediate and the reaction rate at steady-state.

Figure 22 shows measurements of v_o and [EP] at steady-state over a broad range of ATP concentrations. $v_o/[EP]$ is a constant, independent of

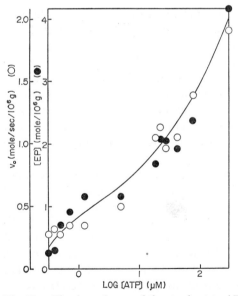

Fig. 22. The dependences of the steady-state ATPase rate and the EP-level on ATP concentration. 0.47mM $CaCl_2$, 0.5mM EGTA (10μM free Ca^{2+}), 0.1M KCl, 1mM $MgCl_2$, pH 7.0, 15°C. ○, steady-state ATPase rate, v_o; ●, concentration of EP, [EP].

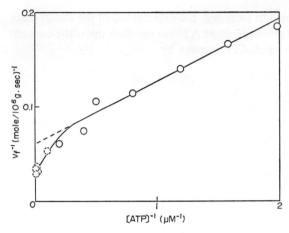

Fig. 23. Double reciprocal plot of the rate constant, v_f, for EP-formation *versus* ATP concentration, [ATP]. 0.02 or 0.1mg/ml SR, 1mM $MgCl_2$, 20μM $CaCl_2$, 0.1M KCl, 0.1M Tris-HCl, pH 8.8, 15°C. ○ indicates a lower limit for v_f.

the ATP concentration, so that there must be a single Ca^{2+}-Mg^{2+}-dependent ATPase for which the rate of decomposition of EP is not affected by a high concentration of ATP, and the formation of EP must be assumed to be accelerated. This is confirmed by the relationship between the rate constant, v_f, for the initial phase of formation of EP and the ATP concentration, [ATP]. Figure 23 shows that at pH 8.8 and 15°C there is a linear relationship between v_f^{-1} and $[ATP]^{-1}$ at values of $[ATP]^{-1}$ equal to or greater than 0.5μM^{-1}, while at lower $[ATP]^{-1}$ values v_f^{-1} falls below this line. Thus the rate of formation of EP is accelerated by a high concentration of ATP.

There is no lag in the formation of EP (Figs. 6 and 7), and the decomposition of $E^{32}P$ after addition of EGTA or 'cold' ATP follows first order kinetics, also with no lag (Fig. 8). Hence, all the reaction steps from the initial reactants to the formation of $ESCa_2$ (see Fig. 19) are in quasi-equilibrium. We have found that the rate of formation of ATP from EP and 13.6μM ADP at pH 7.0 and 15°C is not affected by the addition of 1 mM ATP, and the reaction step EP+ADP→ES is not accelerated by ATP. Therefore, the reverse reaction from ES to EP cannot be accelerated by ATP. These results cannot be explained by the mechanism involving only ES and $ESCa_2$ as bound products from the enzyme and ATP, since the species $E+S+2Ca^{2+}$ are in quasi-equilibrium with $ESCa_2$, and the step $ESCa_2 \to EPCa_2 + ADP$ must be accelerated by a high concentration of ATP. Hence, we propose the following mechanism for the Ca^{2+}-Mg^{2+}-dependent ATPase of the SR, in which EP is formed from the enzyme-substrate-Ca^{2+} complex by two sequential steps, (3) and (4), and ATP at high concentrations shifts the equilibrium (3) towards $_2E_{ATP^o}^{Ca_2^o}$. The scheme is similar to the mechanisms by which EP intermediates are formed by myosin-ATPase (chapter 3), and by the Na^+-K^+-dependent membrane ATPase (chapter 12) (a simplified form is given below, more details being given in Fig. 19).

$$E + Mg^o ATP^o + 2Ca^o \rightleftharpoons {_1E_{Mg^o ATP^o}^{Ca_2^o}} \tag{1,2}$$

$$_1E_{Mg^o ATP^o}^{Ca_2^o} \rightleftharpoons {_2E_{ATP^o}^{Ca_2^o}} + Mg^{o*} \tag{3}$$

$$_2E_{ATP^o}^{Ca_2^o} \rightleftharpoons E_P^{Ca_2^i} + ADP^o \tag{4}$$

$$E_P^{Ca_2^i} + Mg^i \rightleftharpoons E_P^{Mg^i} + 2Ca^i \tag{5}$$

$$E_P^{Mg^i} \rightleftharpoons E + Mg^o + P_i^o + H^{+,o} \tag{6}$$

The superscripts i and o indicate respectively the inside and outside of the membrane.

A second possible regulatory process involves the large transition of $v_o/[EP]$ in the initial phase of the reaction (Fig. 7). This transition also occurs when there is a large amount (5mM) of oxalate present which forms a precipitate within the SR after binding with Ca^{2+}. There is a significant transition even at pH 9.2 (Fig. 24) and the amount of P_i discharged before its completion, about 1 mole/10^6 g, is much smaller than the concentration of phosphorylation sites, suggesting that the transition is not derived from a change in the distribution of Ca^{2+} and Mg^{2+} inside and outside the membrane, caused by Ca^{2+}-transport or the accompanying Mg^{2+}-transport. The following experiment confirms this.

$AT^{32}P$ was added to SR in the presence of Mg^{2+} and Ca^{2+} at pH 7.5 and 15°C to effect Ca^{2+}-uptake by the SR and the formation of $E^{32}P$. The reaction was then quenched with EGTA, and after 10 sec $CaCl_2$ was added to restore the formation of $E^{32}P$ (Fig. 25). The transition of $v_o/[EP]$

---

* Evidence that Mg^{2+} is not bound to $_2E_{ATP^o}^{Ca_2^o}$ is presented in Ref. 95.

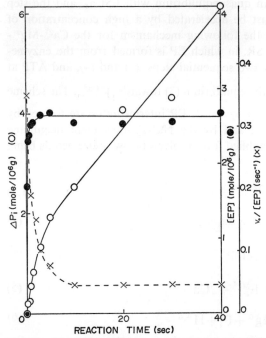

Fig. 24. The initial $v_0/[EP]$ transition at pH 9.2. 0.02mg/ml SR, $5\mu M$ $AT^{32}P$, $20\mu M$ $CaCl_2$, 0.1M KCl, 1mM $MgCl_2$, 0.1M Tris-HCl, pH 9.2, 15°C. ○, P_i-liberation; ●, EP-formation; ×, $v_0/[EP]$.

after the addition of $CaCl_2$ was much smaller than that observed when the reaction was started by the addition of $AT^{32}P$ to SR in the presence of Mg^{2+} and Ca^{2+} (Fig. 7). However, the transition was identical to that when the reaction was initiated by the addition of Ca^{2+} to SR in the presence of Mg^{2+}, EGTA and ATP, even though in the latter case Ca^{2+} had not been taken up by the SR when the reaction began. Thus the same transition occurs regardless of whether Ca^{2+} has accumulated in the SR or not. In the presence of Triton, which destroys the membrane structure, the transition almost completely disappears (Fig. 26), so that one can conclude that it is due to a conformational change in the membrane structure induced by ATP.

In the above experiments at $5\mu M$ ATP concentration, the transition was very fast and we could not determine the value of $v_0/[EP]$ at zero time by extrapolation. However, decreasing the ATP concentration decreased the

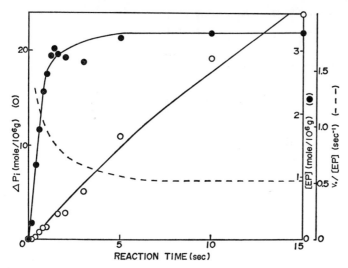

Fig. 25. The $v_0/[EP]$ transition when reaction is initiated by addition of Ca^{2+}. 0.1M KCl, 1mM $MgCl_2$, 0.1M Tris-maleate, pH 7.5, 15°C. 0.1mg/ml SR was incubated with 10μM $CaCl_2$, and 5μM ATP for 10 sec, and 1mM EGTA and 0.5μM $AT^{32}P$ then added. 0.94mM $CaCl_2$ (10 μM free Ca^{2+}) was added after the initial mixture had stood for 10 sec. ○, P_i-liberation; ●, EP-formation; — — —, $v_0/[EP]$.

transition rate so that we could obtain $v_0/[EP]$ at zero time. There is a dramatic decrease in the value of $v_0/[EP]$ to one hundredth of the initial value in only a few seconds after the addition of ATP (Fig. 27). The transition occurs to completion at ATP concentrations much lower than the concentration of phosphorylation sites, ε, on the SR. Hence, we conclude that it is co-operative. If we adopt the mechanism proposed by Changeux et al.[110], we may assume that the ATPase can take only one of two states—an active state, R, and an inactive state, T—corresponding to values for $v_0/[EP]$ of 45 sec^{-1} and 0.4 sec^{-1}, respectively. Figure 27 shows the percentage of the ATPase in the R state at various intervals after the addition of ATP estimated on this basis. When Triton is added, there is no transition and the $v_0/[EP]$ value is much larger than that for the T state but much smaller than that for the R state. The reason for this is unclear, but it may be due to secondary effects such as partial denaturation of the protein portion of the ATPase and extraction of phospholipids by Triton.

However, these results do not exclude the possibility that the decomposition of EP is inhibited by the uptake of Ca^{2+} itself, since the Mg^{2+}-dependent decomposition of EP is competitively inhibited by Ca^{2+}, as

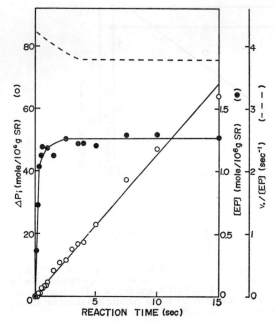

Fig. 26. Removal of the initial $v_0/[EP]$ transition by Triton X. 0.1mg/ml SR, 12.2μM AT^{32}P, 10μM CaCl$_2$, 15mM MgCl$_2$, 0.1M KCl, 0.1M Tris-maleate, pH 7.5, 16°C, 20μl Triton X/mg SR. ○, P$_i$-liberation; ●, EP-formation; — — —, $v_0/[EP]$.

mentioned in the preceeding section. A. Weber[111] has recently suggested that such inhibition of the ATPase could be an important factor regulating the SR. Thus, when there is considerable uptake of Ca^{2+} into the SR, the ATPase activity is decreased during the initial phase by the mechanism described in this section and by the inhibition by Ca^{2+} taken up by the SR.

With respect to the initial transition of $v_0/[EP]$, we must discuss whether the active transport of Ca^{2+} coupled with ATP-splitting described above is rapid enough to account for the physiological functions of the SR. Ebashi et al.[66,81,82] suggested that the Ca^{2+}-transport by the SR cannot occur at a rate comparable to that of muscle relaxation, so that Ca^{2+}-absorption by the SR is rather more important physiologically. This suggestion arose from Ebashi and colleagues' measurement of the Ca^{2+}-uptake in the initial phase and the ATPase activity at steady-state. The transition of $v_0/[EP]$ in the initial phase is large, so that the ATPase rate at steady-state is much

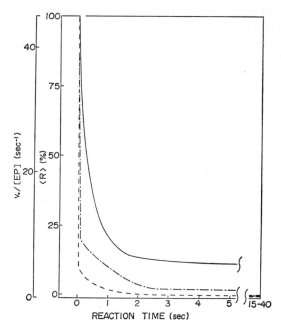

Fig. 27. Transition of $v_0/[EP]$ with time at various ATP concentrations. Initial concentrations of ATP:——, 0.05;—·—, 5.0;————, 13.5μM. $\langle R \rangle$ is the average percentage distribution in state R, assuming that $v_0/[EP]$ for the ATPase can take only the value for one of two states, R and T, corresponding to 45 sec^{-1} and 0.4 sec^{-1}, respectively.

lower than the rate of uptake of Ca^{2+} during the initial phase. This understandably led Ebashi and colleagues to stress that there is no stoichiometric relationship between the two.

The value of $v_0/[EP]$ for the SR during the relaxation of living muscle should lie between that at time zero, given on p. 341, and that at steady-state, but we cannot determine it precisely. We shall assume tentatively a value of 10 sec^{-1}, which lies between 45 sec^{-1} and 0.4 sec^{-1} at 15°C.* Since there is 9mg[112]–13mg[37] of SR protein per gram of muscle, the concentration of ATPase sites is about 6 moles/10^6 g of SR protein, or about 60 μM in muscle. In the presence of sufficient ATP, the rate of degradation of

* Quite recently, the initial phase of Ca^{2+}-uptake by the SR was measured by Inesi and Scarpa[99a]. According to their results, the value of turnover rate of Ca^{2+}-uptake ($=2 \times v_0[EP]$) at 25°C was approximately 10 sec^{-1}.

EP determines the overall reaction rate. The temperature dependence of the Ca^{2+}-Mg^{2+}-dependent ATPase of the SR is very large ($\Delta H^{\ddagger} = 26.2$ kcal/mole[86]), so if the rate constant at 15°C is 10 sec^{-1}, its calculated value at 30°C is about 120 sec^{-1}. Since the decomposition of 1 mole of ATP results in the transport of 2 moles of Ca^{2+}, and a further 2 moles of Ca^{2+} are rapidly transported into the membrane per mole of sites forming EP, within 25 msec the SR removes 360 (overall reaction) + 120 (EP-formation) = 480 μM of Ca^{2+} from the contractile system. The half-life for relaxation of frog sartorius muscle and cat skeletal muscle is 25 msec at room temperature. Maximum tension is generated when the Ca^{2+} concentration is 100 μM, so we can conclude that the Ca^{2+}-transport coupled to the splitting of ATP by the SR occurs at a rate which is capable of inducing the physiological rate of relaxation.

8. Remaining Problems

Three important problems remain. The first is, if the ATPase is a translocase, is it also a carrier protein for Ca^{2+}? We shall discuss this again in chapter 13 in terms of the molecular mechanism of active transport, but it should be noted that a protein which binds Ca^{2+} has been isolated by Wasserman et al.[113] from the intestinal membranes of vertebrates in which Ca^{2+}-transport systems are present, and the presence of a Ca^{2+}-dependent ATPase is also reported by Melancon and DeLuca[114]. A soluble, heat-labile factor with a high affinity for binding Ca^{2+} has been isolated from mitochondria[115], which actively take up Ca^{2+}. Holland and Perry[116] found that during development the ATPase forms first, and reaches almost maximum activity in rabbit muscle after 8–10 days. The activity of the coupling system subsequently rises rapidly, increasing the capacity and efficiency of Ca^{2+}-transport. These studies suggest the possibility that there is a carrier protein other than the translocase in the SR, and this possible involvement of a Ca^{2+}-binding protein in the transport of Ca^{2+} will be a major field of study in the future.

The second problem is the reconstitution of the active SR by self-assembly of the components. The purified SR contains 24μg of lipid P per mg protein[116a]. More than 90% (w/w) of the total lipid are phospholipids, and phosphatidylcholine (73% of total lipid P), phosphatidylethanolamine (14%) and phosphatidylinositol (9%) are the major phospholipids[116a]. Treatment of skeletal muscle microsomes with phospholipase C causes the inhibition of Ca^{2+}-transport and ATPase activity simultaneously with the hydrolysis of membrane lecithin into diglycerides and phosphorylcholine. Addition of synthetic lecithin or lysolecithin to microsomes treated with phospholipase C restores both ATPase activity and Ca^{2+}-transport[103, 117–119]. The SR-ATPase is also destroyed by treatment with phospholipase A[120]. Martonosi et al.[121] have suggested that membrane phospholipids are required for the decomposition of EP, since their hydrolysis by phospholipases inhibits the hydrolysis of ATP without comparable inhibition of the formation of EP. Similar results have also been reported by Meissner and Fleischer[121a]. Contrary to phospholipids, cholesterol and other neutral lipids can be easily extracted with non-polar solvents in the absence of water and do not play any direct role in the active transport of Ca^{2+}[122].

Thus the SR-ATPase is a lipoprotein, and its purification is rather difficult. However, a soluble ATPase was obtained from skeletal muscle microsomes by extraction with DOC[97,123], and was purified by salt fractionation[108]. The ATPase of the SR purified by MacLennan et al. exists as particles of minimum diameter about 90Å when dispersed with DOC[124]. When the dispersing agent is removed, the enzyme spontaneously forms vesicular membranes with globular subunits. The phospholipid and neutral lipid content of the ATPase is identical to that of the SR. The molecular weight of the protein is 102,000, and each molecule appears to contain one phosphorylation site. Besides ATPase, MacLennan and Wong[125] isolated from SR a protein of molecular weight 44,000, which contains many glutamic and aspartic acid residues (36% of the total amino acid residues). This protein seems to be found inside the SR membrane, and binds about 800 moles Ca^{2+} per 10^6 g protein with a dissociation constant of 4×10^{-5} M. Therefore, they suggested that the protein stores Ca^{2+} in the SR. The proteins of the fragmented SR isolated by the method of Martonosi and colleagues[126,127] were resolved by polyacrylamide gel electrophoresis into several fractions ranging in molecular weight from 30,000 to about 300,000. Using as a marker the protein-bound ^{32}P incorporated by previous exposure to $AT^{32}P$, the ATPase enzyme was identified with a major protein fraction of molecular weight 106,000[127]–115,000 [116a], which equals that of the purified ATPase given above. A value of 100,000 has been reported as the molecular weight of the ATPase purified by Ca^{2+}-precipitation and a Sepharose 4B column chromatography of the SR, which was solubilised with Triton X[88b]. On the other hand, Yu and Masoro[128,129] made the SR soluble[130] with SDS and purified it by Sepharose column chromatography. The protein fraction contained more than 90% of the original SR protein, and was composed of large molecular weight aggregates of low molecular weight subunits (6,500–10,000). Gel electrophoresis, Sephadex gel filtration and analysis of the C-terminal and N-terminal amino acid residues suggest that all the subunits are identical or very similar. The reason for the differences between the results of Yu and Masoro and those of other workers is uncertain, but may originate in the different species used to prepare the SR, and more probably in the proteolysis of SR protein during storage.

From these studies, we may anticipate that methods of isolation of the components of the SR and of reconstitution of the active SR from its isolated components will be improved significantly in the near future. However, there has so far been no successful reconstitution of SR which can actively take up Ca^{2+}.

The third problem, which is most important physiologically, concerns the mechanism of release of Ca^{2+} from the SR. The release of Ca^{2+} induced by caffeine was studied by A. Weber[53,54], as mentioned above. Martonosi and Feretos[75], Hasselbach and Seraydarian[131] and A. Weber [132] have demonstrated that when SH groups are blocked there is rapid release of Ca^{2+}. Recently, A. Weber[132] made the interesting observation that the efflux of Ca^{2+} from SR in a Ca^{2+}-free medium becomes rapid when ATP or Mg^{2+} is removed. Lee and his co-workers[133,134] found that the passage of square wave pulses through a suspension of fragmented SR causes a significant release of Ca^{2+} from a fraction which had been previously bound to the SR. A similar result was also reported by Scales and McIntosh[135], but Lee's results were not confirmed by van der Klost[136].

One intriguing possibility for the coupling between the depolarisation of the plasma membrane and the release of Ca^{2+} from the SR is that the release of a membrane bound calcium fraction is the physiological trigger for the release of Ca^{2+} from the main calcium store of the cell—the SR[64,137,138]. Endo et al.[139] and Ford and Podolsky[140,140a] have reported that free Ca^{2+} triggers the release of stored Ca^{2+} from the SR of skinned skeletal muscle fibres immersed in solutions containing low concentrations of Mg^{2+}, and the release process can thus be regenerative. Therefore, it might be possible that excitation causes the release of a small amount of Ca^{2+} from the T-system as the first step, and this process triggers then the release of a large amount of Ca^{2+} from the SR.* Unfortunately, however, the coupling between the T-system and the SR is unknown, and this pro-

---

* See chapter 14, section 3 for an explanation of a molecular mechanism of the trigger-action of a small amount of Ca^{2+} for Ca^{2+}-release from the SR.

blem maybe one of the most important physiological one in the E-C coupling mechanism to be investigated in the near future.

As will be described in chapter 14, section 3, the SR-ATPase reaction is completely reversible, and its equilibrium point shifts very rapidly. Then, it is also possible that the release of Ca^{2+} from the SR occurs by shifting the equilibrium of the ATPase reaction. For example, Nakamaru and Schwartz[141,141a] showed that when the initial pH of 6.5 is decreased by 1/2 to 1 unit Ca^{2+} is released from the SR in an amount sufficient to affect contraction. The use of ionophores might also be beneficial to clarify the mechanism of Ca^{2+}-release from the SR, as in the case of mitochondrial membranes.

9. Other Ca^{2+}-Mg^{2+}-dependent ATPases of Membranes and Regulation of the Mg^{2+}-dependent ATPase by Ca^{2+} Ions and Cyclic AMP

To conclude this chapter, we will briefly describe the properties of the Ca^{2+}-Mg^{2+}-activated ATPase of the plasmic membrane. In 1961, Dunham and Glynn[142] described an ATPase from human erythrocytes which is stimulated by low concentrations of Ca^{2+}. Since then, this enzyme has been the subject of several investigations, and has been related to the active transport of Ca^{2+} across the red blood cell membrane[143–148a]. More recently, Romero and Whittam[149] obtained evidence that the permeability of red blood cell membranes to Na^+ and K^+ is regulated by internal Ca^{2+}, which in turn is controlled by a calcium pump which utilises ATP. A Mg^{2+}-Ca^{2+}-activated ATPase has also been found in brain microsomal fractions[150–153].

Sustained oscillations in enzymic reactions have now been clearly established by work on the photosynthetic[153a], glycolytic[153b,154–161] and peroxidase systems[162,163].* Oscillations of H^+ and K^+ concentrations in heart mitochondria have also been observed[166]. Oscillations in membrane-bound enzymic systems might well be expected on theoretical grounds[167] or on the basis of the oscillatory local response of the calcium-deficient axon[168-170].

The mechanism by which depolarisation alters the permeability to cations is not known, but most probably the current causing excitation displaces Ca^{2+} from strategic sites in the membrane[171-173]. This theory is attractive since, as described above, high concentrations of Ca^{2+} make excitable cells inert, while low concentrations labilise them. On the other

* See Refs. *164, 165, 165a* for oscillations in enzyme reactions.

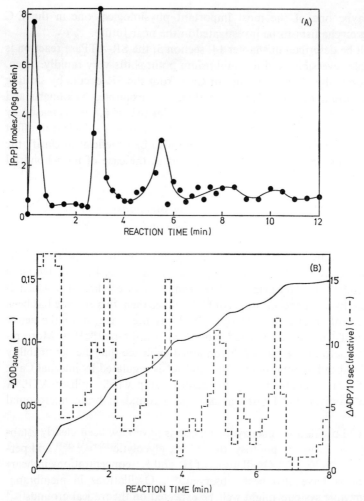

Fig. 28. Oscillations in the amount of PrP and rate of liberation of ADP for microsomal ATPase from the outer medulla of kidney. Microsomal fractions were used after treatment with 0.5% Triton X-100 at pH 7.1, 15°C for 10 min. Ca^{2+} sufficient to allow oscillation was a contaminant in the reaction mixture. A: Amount of EP, 0.11mg/ml protein, 1mM ATP, 1mM $MgCl_2$, pH 7.5, 15°C. B: Rate of liberation of ADP. The ATPase reaction was coupled with pyruvate kinase and lactate dehydrogenase, and the kinetics of ADP-liberation was measured by following the decrease in optical density at 340 nm due to the oxidation of NADH (ATP+H_2O $\xrightarrow{ATPase}$ ADP+P_i; ADP+PEP $\xrightarrow{pyruvate\ kinase}$ ATP+pyruvate; pyruvate+NADH $\xrightleftharpoons{lactate\ dehydrogenase}$ lactate+ NAD). 0.05mg/ml protein, 0.35mM ATP, 3mM PEP, 40units/ml pyruvate kinase, 7.5 units/ml lactate dehydrogenase, 0.3mM NADH, 1mM $MgCl_2$, 2mM KCl, pH 7.5, 15°C.

hand, Singer and Tasaki[174,175] have explained the results of experiments upon internally perfused squid giant axon by assuming that the external layer of the resting membrane is occupied primarily by Ca^{2+} and the internal surface by K^+, and that an outward, stimulating current replaces Ca^{2+} in the membrane by K^+, and this induces conformational changes leading to excitation. Therefore, it is interesting that the ATPase activity and amount of phosphate bound to the microsomal protein (PrP) from the outer medulla of kidney show damped oscillations with an average frequency of 0.5 min^{-1} only in the presence of both Ca^{2+} and Mg^{2+}[176]. Figure 28A shows that the variation with time of the amount of PrP after the addition of 1mM ATP in 1mM $MgCl_2$ at pH 7.5 and 15°C shows a prominent damped oscillation with a frequency of 0.5 min^{-1}. At least five cycles are observed with this preparation, and the effect is seen both in the presence and absence of Triton X. PrP is stable to TCA and unstable to alkaline pH, as is EP of the transport ATPase. However, PrP is stable to hydroxylamine, while EP is unstable to hydroxylamine (see section 4).

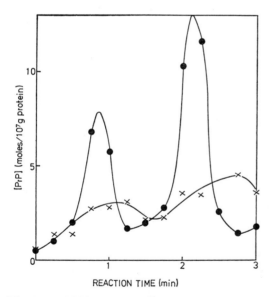

Fig. 29. Inhibition of the oscillation in the amount of PrP by a minute amount of cyclic AMP. Microsomal fractions were prepared from gray matter of bovine brain. 0.16 mg/ml protein, 0.25mM ATP, 1mM $MgCl_2$, pH 7.5, 37°C. ●, control; ×, + 5μM cyclic AMP.

Furthermore, ADP is also liberated in a stepwise fashion with a frequency of 0.5 min^{-1} (Fig. 28B). Both Ca^{2+} and Mg^{2+} are required for oscillation to occur, and the optimal concentration of Ca^{2+} is about 40 μM. Since the rate of Pr^{32}P-decay after the addition of a large amount of 'cold' ATP with Mg^{2+} is much smaller than the value expected if PrP is the true intermediate of the ATPase, and since the ATPase reaction is slightly affected by small amounts of Ca^{2+}, while the oscillation in the amount of PrP is dependent on Ca^{2+}, it is assumed that the rate of Mg^{2+}-ATPase reaction is regulated by phosphorylation of the microsomal protein, the formation of which is dependent on both Mg^{2+} and Ca^{2+}. A similar oscillation in PrP concentration also takes place with the microsomal ATPase from the gray matter of brain, and it is also very interesting that the oscillation in the amount of PrP is markedly inhibited by adding 0.5 μM cyclic AMP[177,177a,b], and almost completely by adding 1 or 5 μM cyclic AMP (Fig. 29). The oscillation is also inhibited by 1mM theophylline.

The detailed molecular mechanism of the oscillation remains to be clarified. But the remarkable influences of Ca^{2+} ions and cyclic AMP suggest the involvement of cyclic AMP-dependent protein kinase[178-187] and protein phosphate phosphatase regulation[187] in this process. At any rate, it is very interesting that the oscillation of phosphorylation of membrane fractions of kidney and brain is regulated by minute amounts of Ca^{2+} ions and cyclic AMP, which are the main regulators of the metabolic processes[188]. This oscillation induces a marked oscillation in the activity of the Mg^{2+}-dependent ATPase. It may be anticipated that future work on the oscillation of the ATPase will throw fresh light on the oscillatory behaviour of membrane excitation.

REFERENCES

1 T. I. Shaw, cited in I. M. Glynn, *J. Physiol.*, **134**, 278 (1956).
2 J. C. Skou, *Physiol. Rev.*, **45**, 596 (1965).
3 R. W. Albers, *Annu. Rev. Biochem.*, **36**, 727 (1967).
4 R. L. Post, *in* "Regulatory Function of Biological Membranes," ed. by J. Järnefet, Elsevier Publishing Co., Amsterdam, p. 163 (1968).
5 W. Hasselbach, *Progr. Biophys. Mol. Biol.*, **14**, 167 (1964).
6 A. Weber, *Curr. Topics Bioenergetics*, **1**, 203 (1966).
6a A. Martonosi, *in* "Biomembranes," ed. by L. A. Manson, Plenum Press, New York, Vol. 1, chapter 3 (1971).
7 B. B. Marsh, *Nature*, **167**, 1065 (1951).

8. B. B. Marsh, *Biochim. Biophys. Acta*, **9**, 247 (1952).
9. J. R. Bendall, *Nature*, **170**, 1058 (1952).
10. J. R. Bendall, *J. Physiol.*, **121**, 232 (1953).
11. H. Kumagai, S. Ebashi and F. Takeda, *Nature*, **176**, 166 (1955).
12. H. Portzehl, *Biochim. Biophys. Acta*, **26**, 373 (1957).
13. H. Portzehl, *Biochim. Biophys. Acta*, **24**, 474 (1957).
14. T. Nagai, M. Makinose and W. Hasselbach, *Biochim. Biophys. Acta*, **43**, 223 (1960).
15. W. W. Kielley and O. Meyerhof, *J. Biol. Chem.*, **176**, 591 (1948).
16. W. W. Kielley and O. Meyerhof, *J. Biol. Chem.*, **183**, 391 (1950).
17. S. Ebashi, *in* "Conf. Chem. Muscular Contraction," Igaku Shoin, Tokyo, p. 89 (1957).
18. E. Bozler, *J. Gen. Physiol.*, **38**, 149 (1954).
19. E. Bozler, *J. Gen. Physiol.*, **38**, 735 (1955).
20. S. Watanabe, Y. Tonomura and H. Shiokawa, *J. Biochem.*, **40**, 387 (1953).
21. S. Watanabe and W. Sleator, *Arch. Biochem. Biophys.*, **68**, 81 (1957).
22. A. Weber, *J. Biol. Chem.*, **234**, 2764 (1959).
23. A. Weber and S. Winicur, *J. Biol. Chem.*, **236**, 3198 (1961).
24. A. Weber and R. Herz, *Biochem. Biophys. Res. Commun.*, **6**, 364 (1962).
25. A. Weber and R. Herz, *J. Biol. Chem.*, **238**, 599 (1963).
26. S. Ebashi, *J. Biochem.*, **50**, 236 (1961).
27. J. C. Seidel and J. Gergely, *J. Biol. Chem.*, **238**, 3648 (1963).
28. W. Hasselbach and M. Makinose, *Biochem. Z.*, **333**, 518 (1961).
29. S. Ebashi, *J. Biochem.*, **48**, 150 (1960).
30. S. Ebashi, *Progr. Theor. Phys. (Kyoto), Suppl.*, **17**, 35 (1961).
31. S. Ebashi and F. Lipmann, *J. Cell Biol.*, **14**, 389 (1962).
32. H. S. Bennett and K. R. Porter, *Amer. J. Anat.*, **93**, 61 (1953).
33. K. R. Porter and G. E. Palade, *J. Biophys. Biochem. Cytol.*, **3**, 269 (1957).
34. K. R. Porter, *J. Biophys. Biochem. Cytol.*, **10**, No. 4, *Suppl.*, 219 (1961).
35. C. F. Armstrong and K. R. Porter, *J. Cell Biol.*, **22**, 675 (1964).
36. K. R. Porter and C. F. Armstrong, *Scientific American*, **212**, 72 (1965).
37. L. D. Peachey, *J. Cell Biol.*, **25**, No. 3, part 2, 209 (1965).
38. C. F. Armstrong, *J. Cell Biol.*, **39**, 6a (1968).
39. D. E. Kelly, *J. Ultrastructure Res.*, **29**, 37 (1969).
40. M. Endo, *Nature*, **202**, 1115 (1964).
41. G. Falk and P. Fatt, *Proc. Roy. Soc.*, **B160**, 69 (1964).
42. A. F. Huxley and R. E. Taylor, *J. Physiol.*, **144**, 426 (1958).
43. R. E. Taylor, *J. Cell. Comp. Physiol.*, **42**, 103 (1953).
44. J. Axelsson and S. Thesleff, *Acta Physiol. Scand.*, **44**, 55 (1958).
45. D. Conway and T. Sakai, *Proc. Natl. Acad. Sci. U.S.*, **46**, 897 (1960).
46. H. C. Lüttgau and H. Oetliker, *J. Physiol.*, **194**, 51 (1968).
47. S. Korey, *Biochim. Biophys. Acta*, **4**, 58 (1950).

48 W. Hasselbach, *Z. Naturforsh.*, **8b**, 212 (1953).
49 P. C. Caldwell and G. E. Walster, *J. Physiol.*, **169**, 353 (1963).
50 G. B. Frank, *J. Physiol.*, **163**, 254 (1962).
51 G. B. Frank, *Proc. Roy. Soc.*, **B160**, 504 (1964).
52 A. Sandow, *Pharmacol. Rev.*, **17**, 265 (1965).
53 A. Weber and R. Herz, *J. Gen. Physiol.*, **52**, 750 (1968).
54 A. Weber, *J. Gen. Physiol.*, **52**, 760 (1968).
55 M. Fujino, T. Yamaguchi and K. Suzuki, *Nature*, **192**, 1159 (1961).
56 T. Yamaguchi, T. Matsushima, M. Fujino and T. Nagai, *Jap. J. Physiol.*, **12**, 129 (1962).
57 J. N. Howell, *J. Physiol.*, **201**, 515 (1969).
58 F. F. Jöbsis and M. J. O'Connor, *Biochem. Biophys. Res. Commun.*, **25**, 246 (1966).
59 F. F. Jöbsis, *in* "Symposium Biologica Hungarica," ed. by E. Ernst and F. B. Straub, Akademiai Kiado, Budapest, Vol. 8, p. 151 (1968).
60 E. B. Ridgway and C. C. Ashley, *Biochem. Biophys. Res. Commun.*, **29**, 229 (1967).
61 C. C. Ashley and E. B. Ridgway, *Nature*, **219**, 1168 (1968).
62 C. C. Ashley and E. B. Ridgway, *J. Physiol.*, **209**, 105 (1970).
63 O. Shimomura and F. H. Johnson, *Biochemistry*, **8**, 3991 (1969).
64 C. C. Ashley, *in* "Membranes and Ion Transport," ed. by E. E. Bittar, Wiley-Interscience, London, Vol. 2, p. 1 (1970).
64a J. W. Hastings, G. Mitchell, P. H. Mattingley, J. R. Blinks and M. van Leeuwen, *Nature*, **222**, 1047 (1969).
65 A. F. Huxley, *Progr. Biophys. Biophys. Chem.*, **7**, 255 (1957).
66 S. Ebashi and M. Endo, *Progr. Biophys. Mol. Biol.*, **18**, 123 (1968).
67 W. Hasselbach and M. Makinose, *Biochem. Biophys. Res. Commun.*, **7**, 132 (1962).
68 W. Hasselbach and M. Makinose, *Biochem. Z.*, **339**, 94 (1963).
69 W. Hasselbach, *Federation Proc.*, **23**, 909 (1964).
70 W. Hasselbach and L. G. Elfvin, *J. Ultrastructure Res.*, **17**, 598 (1967).
71 G. Inesi and H. Asai, *Arch. Biochem. Biophys.*, **126**, 469 (1968).
72 N. Ikemoto, F. A. Sreter, A. Nakamura and J. Gergely, *J. Ultrastructure Res.*, **23**, 216 (1968).
73 A. Martonosi, *Biochim. Biophys. Acta*, **150**, 694 (1968).
74 D. W. Deamer and R. J. Baskin, *J. Cell Biol.*, **42**, 296 (1969).
74a N. Ikemoto, F. A. Sreter and J. Gergely, *Arch. Biochem. Biophys.*, **147**, 571 (1971).
75 A. Martonosi and R. Feretos, *J. Biol. Chem.*, **239**, 648 (1964).
76 M. E. Carsten and W. F. H. M. Mommaerts, *J. Gen. Physiol.*, **48**, 183 (1964).
77 M. Makinose and R. The, *Biochem. Z.*, **343**, 383 (1965).
78 L. de Meis, *Biochim. Biophys. Acta*, **172**, 343 (1969).

79 L. de Meis, *J. Biol. Chem.*, **244**, 3733 (1969).
80 G. Inesi, *Science*, **171**, 901 (1971).
81 T. Ohnishi and S. Ebashi, *J. Biochem.*, **54**, 506 (1963).
82 T. Ohnishi and S. Ebashi, *J. Biochem.*, **55**, 599 (1964).
83 S. Ebashi and F. Ebashi, *J. Biochem.*, **55**, 504 (1964).
84 A. Weber, R. Herz and I. Reiss, *Biochem. Z.*, **345**, 329 (1966).
85 S. Yamada, T. Yamamoto and Y. Tonomura, *J. Biochem.*, **67**, 789 (1970).
86 T. Yamamoto and Y. Tonomura, *J. Biochem.*, **62**, 558 (1967).
87 T. Yamamoto and Y. Tonomura, *J. Biochem.*, **64**, 137 (1968).
88 K. Imai and R. Sato, *J. Biochem.*, **60**, 274 (1966).
88a B. H. McFarland and G. Inesi, *Arch. Biochem. Biophys.*, **145**, 456 (1971).
88b N. Ikemoto, G. M. Bhatnager and J. Gergely, *Biochem. Biophys. Res. Commun.*, **44**, 1510 (1971).
89 M. Makinose and W. Hasselbach, *Biochem. Z.*, **343**, 360 (1965).
90 M. Makinose, *Biochem. Z.*, **345**, 80 (1966).
91 M. Makinose, *Eur. J. Biochem.*, **10**, 74 (1969).
92 A. Martonosi, *Biochem. Biophys. Res. Commun.*, **29**, 753 (1967).
93 G. Inesi, E. Maring, A. J. Murphy and B. H. McFarland, *Arch. Biochem. Biophys.*, **138**, 285 (1970).
94 A. Pucell and A. Martonosi, *J. Biol. Chem.*, **246**, 3389 (1971).
95 T. Kanazawa, S. Yamada, T. Yamamoto and Y. Tonomura, *J. Biochem.*, **70**, 95 (1971).
96 T. Kanazawa, M. Saito and Y. Tonomura, *J. Biochem.*, **67**, 693 (1970).
97 A. Martonosi, *J. Biol. Chem.*, **243**, 71 (1968).
98 K. Vegh, P. Spiegler, C. Chamberlain and W. F. H. M. Mommaerts, *Biochim. Biophys. Acta*, **163**, 266 (1968).
99 T. Yamamoto, A. Yoda and Y. Tonomura, *J. Biochem.*, **69**, 807 (1971).
99a G. Inesi and A. Scarpa, *Biochemistry*, **11**, 356 (1972).
100 S. Yamada, T. Yamamoto, T. Kanazawa and Y. Tonomura, *J. Biochem.*, **70**, 279 (1971).
101 P. F. Duggan and A. Martonosi, *J. Gen. Physiol.*, **56**, 147 (1970).
102 A. P. Carvalho and B. Leo, *J. Gen. Physiol.*, **50**, 1327 (1967).
103 A. Martonosi, *J. Biol. Chem.*, **244**, 613 (1969).
104 G. Inesi and W. C. Landgraf, *Bioenergetics*, **1**, 355 (1970).
105 J. M. Vanderkooi and A. Martonosi, *Arch. Biochem. Biophys.*, **149**, 99 (1971).
106 H. Nakamura, H. Hori and T. Mitsui, *J. Biochem.*, **72**, 635 (1972).
107 S. Yamada and Y. Tonomura, *J. Biochem.*, **72**, 417 (1972).
108 D. H. MacLennan, *J. Biol. Chem.*, **245**, 4508 (1970).
108a J. Chevallier and R. A. Butow, *Biochemistry*, **10**, 2733 (1971).
109 W. G. van der Klost, *Comp. Biochem. Physiol.*, **15**, 547 (1965).
110 J. Changeux, J. Thiery, I. Tung and C. Kittel, *Proc. Natl. Acad. Sci. U. S.*, **57**, 335 (1967).

111 A. Weber, *J. Gen. Physiol.*, **57**, 50 (1971).
112 Y. Ogawa, *J. Biochem.*, **67**, 667 (1970).
113 R. H. Wasserman, R. A. Corradino and A. N. Taylor, *J. Biol. Chem.*, **243**, 3978 (1968).
114 M. J. Melancon and H. F. DeLuca, *Biochemistry*, **9**, 1658 (1970).
115 A. L. Lehninger, *Biochem. Biophys. Res. Commun.*, **42**, 312 (1971).
116 D. L. Holland and S. V. Perry, *Biochem. J.*, **114**, 161 (1969).
116a G. Meissner and S. Fleischer, *Biochim. Biophys. Acta*, **241**, 356 (1971).
117 A. Martonosi, *Federation Proc.*, **23**, 913 (1964).
118 A. Martonosi, *Biochem. Biophys. Res. Commun.*, **13**, 273 (1963).
119 A. Martonosi, J. Donley and P. A. Halpin, *J. Biol. Chem.*, **243**, 61 (1968).
120 W. Fiehn and W. Hasselbach, *Eur. J. Biochem.*, **13**, 510 (1970).
121 A. Martonosi, J. R. Donley, A. G. Pucell and R. A. Halpin, *Arch. Biochem. Biophys.*, **144**, 529 (1971).
121a G. Meissner and S. Fleischer, *Biochim. Biophys. Acta*, **255**, 19 (1972).
122 W. Drabikowski, M. G. Sarzala, A. Wroniszewska, E. Lagwinska and B. Drzewiecka, *Biochim. Biophys. Acta*, **274**, 158 (1972).
123 Z. Selinger, M. Klein and A. Amsterdam, *Biochim. Biophys. Acta*, **183**, 19 (1969).
124 D. H. MacLennan, P. Seeman, G. H. Iles and C. C. Yip, *J. Biol. Chem.*, **246**, 2702 (1971).
125 D. H. MacLennan and P. T. S. Wong, *Proc. Natl. Acad. Sci. U. S.*, **68**, 1231 (1971).
126 A. Martonosi, *Biochem. Biophys. Res. Commun.*, **36**, 1039 (1969).
127 A. Martonosi and R. A. Halpin, *Arch. Biochem. Biophys.*, **144**, 66 (1971).
128 E. J. Masoro and B. P. Yu, *Biochem. Biophys. Res. Commun.*, **34**, 686 (1969).
129 B. P. Yu and E. J. Masoro, *Biochemistry*, **9**, 2909 (1970).
130 B. P. Yu, F. D. DeMartinis and E. J. Masoro, *Anal. Biochem.*, **24**, 523 (1968).
131 W. Hasselbach and K. Seraydarian, *Biochem. Z.*, **345**, 159 (1966).
132 A. Weber, *J. Gen. Physiol.*, **57**, 64 (1971).
133 K. S. Lee, *Nature*, **207**, 85 (1965).
134 K. S. Lee, H. Ladinsky, S. J. Choi and Y. Kasuya, *J. Gen. Physiol.*, **49**, 689 (1966).
135 B. Scales and D. A. D. McIntosh, *J. Pharmacol. Exp. Therap.*, **160**, 249 (1968).
136 W. G. van der Klost, *Comp. Biochem. Physiol.*, **17**, 75 (1966).
137 G. B. Frank, *Proc. Roy. Soc.*, **B160**, 504 (1964).
138 H. Lorkovic, *Comp. Biochem. Physiol.*, **22**, 799 (1967).
139 M. Endo, M. Tanaka and Y. Ogawa, *Nature*, **228**, 34 (1970).
140 L. E. Ford and R. J. Podolsky, *Science*, **167**, 58 (1970).
140a L. E. Ford and R. J. Podolsky, *J. Physiol.*, **223**, 21 (1972).

141 Y. Nakamaru and A. Schwartz, *Biochem. Biophys. Res. Commun.*, **41**, 830 (1970).
141a Y. Nakamaru and A. Schwartz, *J. Gen. Physiol.*, **59**, 22 (1972).
142 E. T. Dunham and I. M. Glynn, *J. Physiol.*, **156**, 274 (1961).
143 H. J. Schatzmann, *Experientia*, **22**, 364 (1966).
144 F. F. Vincenzi and H. J. Schatzmann, *Helv. Physiol. Pharmacol. Acta*, **25**, CR. 233 (1967).
145 H. J. Schatzmann, *Protides Biol. Fluids (Proc. 15th Colloq., Bruges, 1967)*, **15**, 251 (1967).
146 H. J. Schatzmann and F. F. Vincenzi, *J. Physiol.*, **201**, 369 (1969).
147 K. S. Lee and B. C. Shin, *J. Gen. Physiol.*, **54**, 713 (1969).
148 Y. N. Cha, B. C. Shin and K. S. Lee, *J. Gen. Physiol.*, **57**, 202 (1971).
148a H. J. Schatzmann and G. L. Rossi, *Biochim. Biophys. Acta*, **241**, 379 (1971).
149 P. J. Romero and R. Whittam, *J. Physiol.*, **214**, 481 (1971).
150 Y. Nakamaru, *J. Biochem.*, **63**, 626 (1968).
151 Y. Nakamaru, M. Kosakai and K. Konishi, *Arch. Biochem. Biophys.*, **120**, 15 (1967).
152 Y. Nakamaru and K. Konishi, *Biochim. Biophys. Acta*, **159**, 206 (1968).
153 Y. Nakamaru and A. Schwartz, *Arch. Biochem. Biophys.*, **144**, 16 (1971).
153a A. T. Wilson and M. Calvin, *J. Amer. Chem. Soc.*, **77**, 5948 (1955).
153b L. N. M. Duysens and J. Amesz, *Biochim. Biophys. Acta*, **24**, 19 (1957).
154 B. Chance, R. W. Esterbrook and A. Ghosh, *Proc. Natl. Acad. Sci. U. S.*, **51**, 1244 (1964).
155 F. A. Hommes and F. M. A. H. Schuurmans Stekhoven, *Biochim. Biophys. Acta*, **86**, 427 (1964).
156 B. Chance, B. Hess and A. Betz, *Biochim. Biophys. Res. Commun.*, **16**, 182 (1964).
157 B. Chance, B. Schoener and S. Elsaesser, *J. Biol. Chem.*, **240**, 3170 (1965).
158 R. Frenkel, *Arch. Biochem. Biophys.*, **115**, 112 (1966).
159 E. K. Pye and B. Chance, *Proc. Natl. Acad. Sci. U. S.*, **55**, 888 (1966).
160 J. Higgins, *Proc. Natl. Acad. Sci. U. S.*, **51**, 989 (1964).
161 J. Higgins, *in* "Control of Energy Metabolism," ed. by B. Chance, R. W. Esterbrook and J. R. Williamson, Academic Press, New York, p. 13 (1965).
162 I. Yamazaki and K. Yokota, *Biochim. Biophys. Acta*, **132**, 310 (1967).
163 S. Nakamura, K. Yokota and I. Yamazaki, *Nature*, **222**, 794 (1969).
164 C. Walter, "Enzyme Kinetics, Open and Closed Systems," The Ronald Press, New York, chapter 8 (1966).
165 J. Higgins, *J. Ind. Eng. Chem.*, **59**, 19 (1967).
165a B. Hess and A. Boiteux, *Annu. Rev. Biochem.*, **40**, 237 (1971).
166 B. Chance and T. Yoshioka, *Arch. Biochem. Biophys.*, **117**, 451 (1966).
167 R. A. Spangler and F. M. Snell, *Nature*, **191**, 457 (1961).

168 A. Arvanitaki, *Arch. Int. Physiol.*, **49**, 209 (1939).
169 T. H. Bulloch and C. A. Harridge, "Structure and Function of the Nervous System of Invertebrates," W. H. Freeman and Co., New York, Vol. 1, p. 149 (1966).
170 A. M. Shanes, *Pharmacol. Rev.*, **10**, 59 (1958).
171 B. Frankenhaeuser, *J. Physiol.*, **137**, 245 (1957).
172 B. Frankenhaeuser and A. L. Hodgkin, *J. Physiol.*, **137**, 218 (1957).
173 A. L. Hodgkin, "The Conduction of the Nervous Impulse," Liverpool Univ. Press, Liverpool (1964).
174 I. Singer and I. Tasaki, in "Biological Membranes, Physical Fact and Function," ed. by D. Chapman, Academic Press, New York, p. 347 (1968).
175 I. Tasaki, "Nerve Excitation: A Macromolecular Approach," Thomas, Springfield, Illinois (1968).
176 Y. Fukushima and Y. Tonomura, *J. Biochem.*, **72** 623 (1972).
177 G. A. Robinson, R. W. Butcher and E. W. Sutherland, *Annu. Rev. Biochem.*, **37**, 149 (1968).
177a J. G. Hardman, G. A. Robinson and E. W. Sutherland, *Annu. Rev. Physiol.*, **33**, 311 (1971).
177b J. P. Jost and H. V. Rickenberg, *Annu. Rev. Biochem.*, **40**, 741 (1971).
178 D. A. Walsh, J. P. Perkins and E. G. Krebs, *J. Biol. Chem.*, **243**, 3763 (1968).
179 E. Miyamoto, J. F. Kuo and P. Greengard, *J. Biol. Chem.*, **244**, 6395 (1969).
180 J. F. Kuo and P. Greengard, *J. Biol. Chem.*, **245**, 4067 (1970).
181 J. D. Corbin and E. G. Krebs, *Biochem. Biophys. Res. Commun.*, **36**, 328 (1969).
182 J. K. Huttunen, D. Steinberg and S. E. Mayer, *Proc. Natl. Acad. Sci. U.S.*, **67**, 290 (1970).
183 G. N. Gill and L. D. Garren, *Biochem. Biophys. Res. Commun.*, **39**, 335 (1970).
184 H. Maeno, E. M. Johnson and P. Greengard, *J. Biol. Chem.*, **246**, 134 (1971).
185 L. Shlatz and G. V. Marinetti, *Biochem. Biophys. Res. Commun.*, **45**, 51 (1971).
186 M. Weller and R. Rodnight, *Nature*, **225**, 187 (1970).
187 M. Weller and R. Rodnight, *Biochem. J.*, **124**, 393 (1971).
188 H. Rasmussen, *Science*, **170**, 404 (1970).

12

THE Na$^+$-K$^+$-DEPENDENT ATPase OF MEMBRANES

1. Active Transport of Na$^+$ and K$^+$, and the ATPase

As mentioned in chapter 10, Hodgkin and Huxley's ion theory postulates that the prerequisite for excitation of and conduction by muscle and nerve membranes is a non-uniform distribution of Na$^+$ and K$^+$ inside and outside the cell. The control of Na$^+$ and K$^+$ concentration in the cytoplasm also provides the cell with a means of regulation of its volume by adjustment of the internal osmotic pressure, thus freeing the cell from its dependence on a rigid wall structure for the prevention of osmotic lysis. According to Ussing[1], the ratio of influx to efflux of ions for simple diffusion through a membrane is given by

$$\frac{\text{influx}}{\text{efflux}} = \frac{C^o}{C^i} e^{-zF\Delta\phi/RT},$$

where C^i and C^o are the concentrations of ions inside and outside the cell, z is the charge on an ion, and $\Delta\phi$ is the membrane potential. If the transport of ions does not follow this equation, and the metabolism of the cell appears to participate, then there is active transport of that particular ion. Hodgkin and Keynes have obtained evidence that both efflux of Na^+ and influx of K^+ in nerve and muscle involve active transport[2,3], and the studies by Caldwell et al.[4], Gardos[5] and Hoffman et al.[6] indicate that the energy source for Na^+ efflux from both the squid giant axon and erythrocytes is ATP. More recently, Brinley and Mullins[7,8] have presented conclusive evidence that ATP is the energy source for active transport of Na^+. Perfused axons obtained by the methods of Baker et al.[9] and Oikawa et al.[10] transport Na^+ very poorly, so Brinley and Mullins[7,8] inserted a thin-walled porous glass capillary down the axon, and then perfused the capillary with various solutions. The axons transported Na^+ and K^+ normally when they were perfused with solutions resembling the normal axoplasm. By varying the quantities of ATP, arginine phosphate and other energy-rich phosphate compounds in the perfusion solutions, it was demonstrated that ATP is the energy source for the active transport of Na^+[11].*

The biochemical foundation of the active transport of Na^+ and K^+ was discovered by Skou[14] when he found in crab nerve microsome a new type of ATPase, which shows a high activity only when Mg^{2+}, Na^+ and K^+ are present. This ATPase has received much attention, since it is likely to participate in the active transport of Na^+ and K^+, and many excellent reviews of its general properties have appeared[15-20]. The following conclusions may be drawn. (1) The enzyme is present in all cell membranes where active transport of Na^+ and K^+ occurs.** (2) Glynn[22], and Whittam and Ager[23,24] showed that ATP-hydrolysis in erythrocytes is activated by Na^+ inside the cell and K^+ outside the cell, so that the enzyme has vectorial properties in order to perform active transport. Furthermore, Marchesi and Palade[25] found by cytochemical methods that the ATPase activity of red blood cell ghosts causes the released of P_i from sites scattered over the inside of the red blood cell ghost membrane. (3) Sen and Post[25a], Whittam[17] and Glynn[18] demonstrated that the ratio of ions transported per mole of ATP hydrolysed is independent of the concentration gradient over the range studied. K^+/ATP is constant, with a mean value of 2, and is independent of the internal Na^+ concentration, while the mean value

---

* The general features of active transport of cations through membranes are discussed in Refs. *12, 13*.
** See Bonting[21] for the relationship between the Na^+-K^+-dependent ATPase system and the active transport of cations in the various cells and tissues.

of Na⁺/ATP is 3.* (4) The enzyme is inhibited by compounds which inhibit active transport of Na⁺ and K⁺. Ouabain, which is a particularly specific inhibitor of active transport[25b], inhibits both processes in approximately the same concentration range, indicating that the Na⁺-K⁺-dependent ATPase is likely to be associated with active transport.

Samaha and Gergely[26], Rogus et al.[27] and Ash and Schwartz[28] isolated a Na⁺-K⁺-dependent ATPase by high speed centrifugation of deoxycholate-treated homogenates of skeletal muscle. However, attempts to locate the Na⁺-K⁺-dependent ATPase in skeletal muscle have met with limited success, and the evidence that it is positioned at the sarcolemmal membrane is not convincing. Peter[29] recently reported that purified sarcolemma isolated from skeletal muscle contains an ATPase which is stimulated by Na⁺ and K⁺ in the presence of Mg^{2+}. More recently, Sulakhe et al.[30] succeeded in isolating the ATPase of skeletal muscle membrane, which shows much higher activity than that reported by Peter[29].** Since the preparation of this ATPase from skeletal muscle is difficult, brain microsomes treated with NaI or other reagents are commonly used to study the reaction mechanism of the ATPase. Erythrocytes, especially those in which the cell composition has been modified by partial haemolysis[5,6], are generally used to investigate the relationship between hydrolysis of ATP and transport of cations through the membrane, and this will mostly be discussed in chapter 13.

According to Skou[32], the substrate for the Na⁺-K⁺-dependent ATPase is ATP, since with ITP, GTP or UTP there is only slight or no activation by Na⁺ and K⁺ acting together. More recently, Schoner et al.[33] showed that by adding K⁺ in the presence of Na⁺, the rate of hydrolysis of ATP is increased 10 times, that of ITP 4 times and that of GTP 1.2 times. Acetylphosphate is also hydrolysed by the enzyme[33a,b]. Several studies on the partially purified preparations of the Na⁺-K⁺-dependent ATPase indicate that a K⁺-stimulated p-nitrophenyl phosphatase activity is intimately related to the ATPase activity, and may represent the terminal step of the

* The free energy (W) for the transport of 1 g ion equivalent of Na⁺ from the inside to the outside the cell membrane is given by

$$-W = RT \ln \frac{[Na^+]^i}{[Na^+]^o} + F \cdot \Delta\phi.$$

For the transport of Na⁺ in the giant axon of squid in sea water, when $[Na^+]^i = 50mM$, $[Na^+]^o = 460mM$, and $\Delta\phi = 80mV$, W is calculated to be 2,630 cal. Likewise, W for the transport of K⁺ from the outside to the inside of the cell ($[K^+]^i = 400mM$, $[K^+]^o = 10mM$) is 810 cal.

** See Refs. 20, 31 for the activities and motions of Na⁺ and K⁺ in muscle.

sequence of reactions leading to the hydrolysis of ATP[34,35]. However, the exact relationship between the *p*-nitrophenyl phosphatase activity and the ATPase is not known.

The Na+-K+-dependent ATPase activity is inhibited by oligomycin[36,37]. It is inhibited by fusidic acid[37a], which is an inhibitor of G-factor-dependent ribosomal GTPase (*cf.* chapter 14, section 1). It is also lost by modification of amino groups with TBS[38] or acid anhydrides[39], and probably modification of serine residues with diisopropylfluorophosphate[39a], methane sulphonyl chloride or diethyl *p*-nitrophenyl phosphate[40]. Cysteine is also thought to be found near the active site, since the ATPase is inhibited by NEM and the inhibition is retarded by ATP[40a]. Two lines of evidence suggest a histidine at the active site[40b]. (1) The pH-activity curve indicates an ionisable group with a pK of about 7, consistent with the pK of a histidine imidazole. (2) Photo-oxidation of a brain ATPase in the presence of methylene blue rapidly inactivates the enzyme, while ATP markedly slows down the loss of enzymic activity. The detailed roles of these amino acid residues in the ATPase reaction remain to be elucidated, although Robinson[40b] has recently made an interesting suggestion on this point. The role of lysine residue in the transport of cation will be suggested in chapter 13, section 4.

Since its discovery more than a decade ago, the Na+-K+-dependent ATPase has not been isolated in a pure form, although many investigators, including Nakao *et al.*[41], Banerjee *et al.*[42], Dulaney and Touster[43], Duham and Hoffman[44], Towle and Copenhaver[45], and Jørgensen and Skou[46] have attempted to make the enzyme soluble with NaI and nonionic detergents and purify it by gel electrophoresis, gel filtration, density gradient centrifugation and other methods. However, Uesugi *et al.*[47] have recently managed to increase the purity of the ATPase in Lubrol extracts of NaI-treated brain microsomes to 30–50 times over that in the microsomes, using salt and isoelectric precipitation, zonal centrifugation, and ammonium sulphate fractionation. Kline *et al.*[47a] have reported that most of kinetic parameters of the enzyme are not influenced by stages of this purification method. Atkinson *et al.*[47b] also prepared water-soluble ATPase by extraction of NaI-treated brain microsomes with Lubrol, and suggested that 280,000 molecular weight transport ATPase probably comprises 12 subunits of molecular weight 2.5×10^4 arranged as three tetramers of molecular weight 9.8×10^4. Nakao *et al.*[47c] have recently purified the ATPase by an ion exchange cellulose column chromatography from Lubrol extract of NaI-treated brain microsomes. A SDS-gel electrophoresis showed that their preparation contains practically only one peptide with molecular weight of 12×10^4, but the activities scattered over a wide range. Kyte[48] has also described a procedure for purifying the enzyme from renal medulla, which yields a preparation containing two polypeptides, of molecular weights 5.7 and 8.4×10^4, in equimolar amounts. Thus we can expect to obtain a pure preparation of this enzyme in the near future, although there is still disagreement between molecular properties of transport ATPase prepared by various investigators, as mentioned above.

Goldacre[49] has proposed that the active transport of cations occurs as a result of the protein having intrinsically the same function as the contractile protein responsible for muscle contraction. We shall consider this proposal by comparing the reaction mechanism of the contractile ATPase with the results obtained from studies of the membrane ATPase using specimens of brain microsomes treated with a high concentration of NaI by Nakao et al.'s method[41].

2. Phosphorylated Intermediates of the Membrane ATPase

Since Skou[50], Albers et al.[51] and Stahl et al.[52] discovered that the Na^+-K^+-dependent ATPase catalyses ATP-ADP exchange but not ATP-P_i exchange, it has been commonly assumed that a phosphorylated protein is a reaction intermediate. This is supported by the incorporation of ^{32}P from AT^{32}P into the enzyme, as shown by Albers et al.[51], Post et al.[53], Ahmed and Judah[54], Nagano et al.[55] and Hokin et al.[56] This incorporation is significant only in the presence of Na^+, and is suppressed by K^+. Acetyl-phosphate can replace ATP as a phosphorylating agent[33b]. The site of phosphorylation with acetyl-phosphate is shown to be the same as that with ATP. The phosphorylated protein is stable to TCA, unlike the reaction intermediate from myosin-ATPase, and its pH-stability curve, high reactivity with hydroxylamine and degradation by acyl-phosphatase suggest that it has an acyl-phosphate type structure[55,56]. Later, Kahlenberg et al.[57] analysed the hydroxamate obtained by reacting hydroxylamine with this intermediate, and concluded that the carboxyl group of a glutamic acid residue is phosphorylated by ATP. On the other hand, quite recently Post[58] investigated ^{32}P-labelled peptides produced by proteolytic digestion of a phosphorylated intermediate. Pronase digestion of peptides yielded a limit ^{32}P-labelled tripeptide, of which structure appeared to be Ser (or Thr)·Asp(P)·Lys. Recently, Kyte[48] analysed by electrophoresis in SDS the Na^+-dependent phosphorylated intermediate from his purified enzyme, and has estimated for the phosphorylated peak a mobility which corresponds to the larger (molecular weight 8.4×10^4) of the two polypeptides in the enzyme. A similar result has also been reported by Uesugi et al.[47] However, the recovery of phosphorylated intermediate after the SDS-gel electrophoresis was extremely low. Avruch and Fairbanks[58a], using SDS-gel electrophoresis at acidic pH, also reported that even in the case of erythrocyte membranes a protein of apparent molecular weight 105,000 is phosphorylated by ATP. The effects of various cations and ouabain on formation and decomposition of this phosphoprotein suggested that it is an intermediate of the Na^+-K^+-dependent ATPase reaction of erythrocyte membranes.

However, it does not seem possible to decide whether the phosphorylation, found when both Mg^{2+} and Na^+ but not K^+ are present in the medium, is involved in the hydrolysis occurring when all three ions are present, or whether it results from a rather different route for the hydrolysis of ATP being followed under conditions in which the transport of Na^+ is not coupled to the transport of K^+[32,59]. This question must be answered before a scheme showing the intermediate steps in the hydrolysis of ATP by the enzyme system can be constructed. Therefore, we have studied extensively the reaction mechanism using the enzyme obtained by the NaI method[41] from the gray matter of brain, from which it is easy to isolate the membrane ATPase. The formation and decomposition of the phosphorylated protein were measured independently, and compared to the results obtained from the steady-state reaction[60,61]. For this purpose, a simple apparatus was designed to measure changes in the reaction at intervals of about 0.1 sec[61]. All the measurements were performed in the presence of 1mM $MgCl_2$.

3. The Reaction Mechanism of the Membrane ATPase[60,61]

Figure 1 shows $v_o/[EP]$, where v_o is the steady-state rate and [EP] the concentration of phosphorylated protein, plotted against KCl concentration. If the EP is the last reaction intermediate, then $v_o/[EP]$ should equal the decomposition rate constant of EP, k_{+4}. $v_o/[EP]$ increases significantly as the K^+ concentration is increased,* but is independent of the Na^+ concen-

---

* This enzyme preparation contains, in addition to the Na^+-K^+-dependent ATPase, an ATPase which is active only when Mg^{2+} is present. Therefore, the Na^+-K^+-dependent ATPase activity was determined by subtracting the ATPase activity in the absence of both Na^+ and K^+ from the total ATPase activity, and the amount of phosphorylated protein when Na^+ is absent but sufficient K^+ is present was subtracted from the total amount of phosphorylated protein to obtain that portion involved in the transport ATPase of the membrane.

tration. Figure 2 shows that [EP] is proportional to v_o over a broad range of ATP concentrations, and that $v_o/[EP]$ is independent of the ATP concentration.

The decomposition of the phosphorylated intermediate was measured by adding AT^{32}P to the enzyme in the presence of Na$^+$ and Mg^{2+} and then adding EDTA to remove Mg^{2+} and inhibit the formation of EP. The rate of decomposition of the phosphorylated protein can then be determined directly. There is a very rapid K$^+$-dependent drop in the EP concentration immediately after adding EDTA-KCl or ATP-KCl (due to shift of equilibrium: $E_2S \rightleftharpoons EP$, cf. next page), and then the EP-decomposition follows first order kinetics with a rate constant, k_d, which increases with increase in K$^+$ concentration (Fig. 1), but is little affected by Na$^+$. k_d is not generally equal to k_{+4}—thus at 0 and 0.6mM KCl, k_{+4} is, respectively, 1.15 and 1.96 times greater than k_d. When a large amount of non-radioactive ATP is added instead of EDTA, the rate of decomposition of the ^{32}P-protein is unaltered.

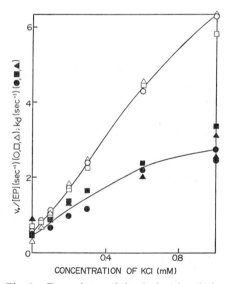

Fig. 1. Dependence of the dephosphorylation reaction on KCl concentration. 1mM MgCl$_2$, 140mM NaCl, 100mM Tris-HCl at pH 8.5, 15°C. ■, ●, rate constant for EP-decomposition, k_d, obtained by the addition of 50mM EDTA. ▲, k_d obtained by the addition of 0.3mM 'cold' ATP instead of 50mM EDTA. ○, □, △, the ratio of the ATPase activity, v_o, (above the base level) to the amount of ^{32}P-incorporation, [EP] (above the base level). Concentrations of AT^{32}P: ○, 1.08; □, 31; △, 301μM.

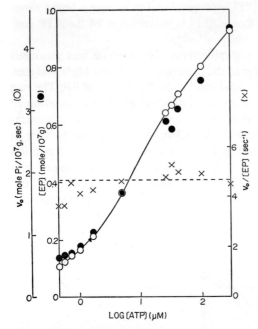

Fig. 2. Dependence of the steady-state rate of ATP-hydrolysis, v_o, and EP concentration, [EP], upon the ATP concentration. 0.6mM KCl, 140mM NaCl, 1mM $MgCl_2$, 100 mM Tris-HCl, pH 8.5, 15°C. ○, v_o; ●, [EP]; ×, v_o/[EP].

The simplest mechanism which explains the results obtained after the addition of EDTA is

$$E_2S \underset{K_3}{\rightleftharpoons} EP \xrightarrow{k_{+4}} E + P_i,*$$

in which the equilibrium between E_2S and EP must be very rapid, since the decay of EP follows first order kinetics. Hence,

$$k_{+4} = \frac{v_o}{[EP]} \quad \text{and} \quad k_d = \frac{k_{+4}}{1+K_3}, \quad \text{where} \quad K_3 = \frac{[E_2S]}{[EP]}.$$

v_o/[EP] and k_d do not depend on ATP, Mg^{2+} or Na^+, and so K_3 and k_{+4} likewise cannot depend on these species, but only on K^+. Values obtained for K_3 and k_{+4} are shown in Table I. The equilibrium between E_2S and EP

* The equilibrium of the step $E_2S \rightleftharpoons EP$ was unaffected by adding ADP[61a]. This indicates that EP contains also ADP and EP is actually E_P^{ADP}.

TABLE I. Rate Constants for the Na$^+$-K$^+$-dependent ATPase 1mM MgCl$_2$, 140mM NaCl, 0.6mM KCl, pH 8.5, 15°C

Constant	Value	Calculated from	Effect of cation
k_{+4}	4.43 sec^{-1} 4.4 sec^{-1}	$v_0/$[EP], steady-state $v_0/$[EP], initial	Increased by K$^+$ Independent of Na$^+$ Independent of Mg^{2+}
K_3	0.96 1.2	k_d, $v_0/$[EP] P$_i$-liberation and EP-decomposition	Increased by K$^+$ Independent of Na$^+$ Independent of Mg^{2+}
k_{+2}[a]	2.1 sec^{-1} 1.73 sec^{-1} 1.62 sec^{-1}	Guggenheim plot of EP-formation, k_d V_f, K_3, ε V_0, K_3, k_{+4}, ε	Increased by Na$^+$ Independent of K$^+$ Dependent on Mg^{2+}
K_1	3.6 μM 1.6 μM	K_f K_0, k_{+2}, K_3, k_{+4}	Independent of Na$^+$ Independent of K$^+$ Dependent on Mg^{2+}
k_{+1}	$\gg 5.5 \times 10^6$ M^{-1}sec^{-1}	τ, K_1	
k_{-1}	$\gg 18$ sec^{-1}	K_1, k_{+1}	
ε	1.57 mole/10^{7}g	[EP]$_{max}$, K_3 $(k_{+2} \gg k_{+4})$[a]	Independent of K$^+$

[a] Step 2, i.e. E$_1$S→E$_2$S, is accelerated by high concentrations of ATP, in that at sufficiently high ATP concentrations, $k_{+2} \gg k_{+4}$.

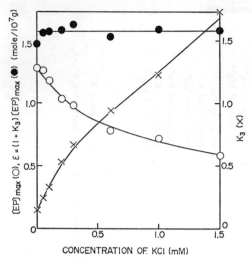

Fig. 3. Determination of the total concentration of phosphorylation sites. $[EP]_{max}$ (○) was measured at 1mM $MgCl_2$, 140mM NaCl, 305μM $AT^{32}P$, pH 8.5, 15°C, but at varying KCl concentrations. The value of K_3 (×) was calculated from the results given in Fig. 1. ε (●) was calculated as $(1+K_3)[EP]_{max}$.

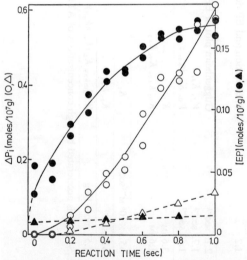

Fig. 4. Initial rates of liberation of P_i and incorporation of ^{32}P. P_i-liberation and ^{32}P-incorporation were measured in 1.0μM $AT^{32}P$, 1.05mg/ml enzyme, 1mM $MgCl_2$, 140 mM NaCl and 0.6mM KCl at pH 8.5 (100mM Tris-HCl) and 15°C. The curve drawn through the open circles is the time-course of P_i-liberation calculated from the rate of ^{32}P-incorporation, assuming that the specific turnover rate of phosphorylated protein is 4.4 sec^{-1}. ●, ^{32}P-incorporation; ○, P_i-liberation; ▲, basal level of ^{32}P-incorporation; △, basal level of P_i-liberation.

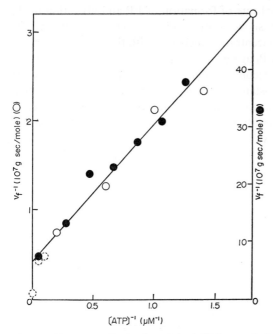

Fig. 5. Lineweaver-Burk plot of the initial rate of phosphorylation. Initial rates of ^{32}P-incorporation, v_f, were measured in 1.05mg/ml enzyme, 1mM MgCl$_2$ and various concentrations of AT^{32}P at pH 8.5 (100mM Tris-HCl) and 15°C in the presence of 140 mM (○, ○) or 0.5mM (●) NaCl. The rate above the base level is plotted. ○, only the lower limits could be obtained because the rates were very high at high ATP concentrations.

shifts to the left upon the addition of K$^+$. Comparison of k_{+4} with k_d shows that K_3 is 0.15 and 0.96 when [K$^+$] is 0 and 0.6mM, respectively. If this mechanism is correct, then the ratio of the rate of liberation of P$_i$ to the rate of decay of EP (after EP-formation has been terminated with EDTA) should also give K_3, and the values of K_3 obtained by these two methods are equal (Table I).

If we assume that the above scheme by itself is sufficient when the reaction occurs in the presence of a large amount of ATP (this assumption is shown to be valid later), then $[EP]_{max} = \varepsilon/(1+K_3)$, where ε is the total concentration of phosphorylation sites. Measurements of $[EP]_{max}$ at different K$^+$ concentrations, and hence different K_3 values, support the validity of this equation (Fig. 3).

Figure 4 shows the initial rate of formation of EP and liberation of P_i. The initial liberation of P_i shows a pre-steady-state closely related to the period in which the EP concentration increases with time.* Moreover, the initial rate of liberation of P_i agrees with that calculated by assuming that the rate constant for the decomposition of EP is equal to the measured value of $v_o/[EP]$ at steady-state. This suggests that the EP is a reaction intermediate. Figure 5 shows the dependence of the initial rate of formation of EP, v_f, upon ATP concentration. The Lineweaver-Burk plot is linear at low concentrations of ATP, so that the following mechanism may be proposed for the overall reaction, including the decomposition of EP.

$$E+S \underset{k_{-1}}{\overset{k_{+1}}{\rightleftharpoons}} E_1S \overset{k_{+2}}{\longrightarrow} E_2S \underset{K_3}{\rightleftharpoons} EP \overset{k_{+4}}{\longrightarrow} E+P$$

The rate of decay of $E^{32}P$ is the same whether its formation is inhibited by EDTA or by a large amount of 'cold' ATP, showing that step 2 is irreversible. Hexum et al.[62] showed by kinetic analysis, following the method of Dixon[63], that Mg-ATP is the real substrate of the reaction, and is therefore S in the above scheme.**

The rate constant for EP-formation is

$$v_f = \frac{k_{+2}\varepsilon}{(1+K_3)\left(1+\dfrac{k_{-1}+k_{+2}}{k_{+1}} \cdot \dfrac{1}{[S]}\right)},$$

the maximum rate of EP-formation is

$$V_f = \frac{k_{+2}\varepsilon}{1+K_3},$$

and the Michaelis constant of EP-formation is given by

$$K_f = \frac{k_{-1}+k_{+2}}{k_{+1}}.$$

---

* Mårdh and Zetterqvist[61b] have later studied the kinetics of EP-formation during the initial phase, using a rapid mixing apparatus. However, they could only indicate that EP is formed rapidly enough to be an intermediate of the ATPase reaction.

** The binding of ATP has recently been investigated by flow dialysis in the presence of EDTA[63a–c]. From these studies, the Mg^{2+}-independent binding of ATP to the active site has been suggested. Mg-ATP might not, therefore, be the real substrate, the formation of complex, E_1MgATP, for phosphorylation possibly occurring instead as,

$$\begin{array}{c} \text{EMg} \\ \overset{Mg}{\nearrow} \quad \overset{ATP}{\searrow} \\ E \qquad\qquad E_1MgATP \xrightarrow{Na} \cdots\cdots \\ \overset{ATP}{\searrow} \quad \overset{Mg}{\nearrow} \\ \text{EATP} \end{array}$$

As mentioned above, V_f depends markedly on Na$^+$, but K_3 does not (Fig. 5), so k_{+2} is also very dependent on Na$^+$. However, the values of k_{+2}, derived from analysis of the steady-state reaction by the method described later, are almost independent of the concentration of K$^+$. While k_{+2} increases with increase in Na$^+$, K_f is independent of Na$^+$ (Fig. 5), and hence the above expression can be simplified to $K_f = k_{-1}/k_{+1} = K_1$. This conclusion is supported by the measurement of the lower limit of k_{-1}, described later. The value of K_1 derived from the formation of EP in the absence of KCl almost equals that obtained from the Michaelis constant, K_o, at steady-state in the presence of KCl (by the method described later), indicating that step 1 is also independent of K$^+$. The mechanism of formation of EP is identical to that for myosin-ATPase and the Ca^{2+}-Mg^{2+}-dependent ATPase of the sarcoplasmic reticulum (SR), and the EP is formed *via* two ES complexes. Since altering the Na$^+$ concentration does not change K_f, but only V_f, it can be concluded that Na$^+$ and ATP bind at random.

$$\begin{array}{ccc}
 & E & \\
S \swarrow & & \searrow Na \\
E_1S & & ENa \\
Na \searrow & & \swarrow S \\
 & E_1SNa & \longrightarrow E_2SNa
\end{array}$$

The value of k_{+2} can also be estimated from measurements of the formation of EP under conditions when $k_{+1}[S] \gg k_{+2}, k_{+4}$. The equation

$$\ln\frac{[EP]_{st}}{[EP]_{st}-[EP]} = \left(k_{+2}+\frac{k_{+4}}{1+K_3}\right)t = (k_{+2}+k_d)t$$

is then valid[64], where $[EP]_{st}$ is the concentration of EP at steady-state. The value of k_{+2} obtained by the Guggenheim method is included in Table I, and agrees well with the value previously derived from V_f, K_3 and ε. If the formation of EP follows the above mechanism, then the induction period is given by

$$\tau = \frac{1}{k_{+1}(K_f+[S])}.$$

Even when the ATP concentration is much lower than K_f, τ is immeasurable, since it is less than 0.05 sec. The lower limits for k_{+1} and k_{-1} derived from this result are shown in Table I, and comparison of k_{-1} with k_{+2} indicates that $k_{-1} \gg k_{+2}$, thereby supporting the contention that $K_f = K_1$.

According to the above scheme, the ATPase rate at steady-state is given by

$$v_o = \frac{k_{+4}\varepsilon}{1+K_3+\dfrac{k_{+4}}{k_{+2}}+\dfrac{(k_{-1}+k_{+2})k_{+4}}{k_{+1}k_{+2}}\cdot\dfrac{1}{[S]}},$$

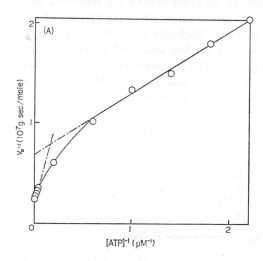

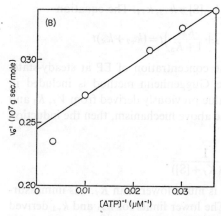

Fig. 6. Lineweaver-Burk plots for the steady-state rate of the ATPase reaction. 1mM MgCl₂, 140mM NaCl, 0.6mM KCl, 100mM Tris-HCl at pH 8.5, 15°C. Points at high ATP concentrations in A are also shown in B.

the maximum rate is

$$V_0 = \frac{k_{+4}\varepsilon}{1+K_3+\dfrac{k_{+4}}{k_{+2}}},$$

and the Michaelis constant is

$$K_0 = \frac{K_f}{1+\dfrac{k_{+2}}{k_{+4}}(1+K_3)}.$$

The values of k_{+2} obtained from V_0, K_3, k_{+4} and ε and of K_f ($=K_1$) obtained from K_0, k_{+2}, K_3 and k_{+4} agree quite well with those measured directly (Table I).

In order for ε to be derived from the above equations, the rate constant for formation of E_2S must be much greater than k_{+4}. However, the values of k_{+4} and k_{+2} estimated from the above reaction mechanism do not satisfy this condition (Table I). Steps 3 and 4 are not dependent on the ATP concentration, and the Lineweaver-Burk plot for the formation of EP shown in Fig. 5 (cf. also Fig. 2) follows a downward curve at high ATP concentrations. The Lineweaver-Burk plot for the steady-state rate also behaves similarly (Fig. 6, cf. also Fig. 2). These results can be explained easily by assuming that step 2 is accelerated by high concentrations of ATP, so that k_{+2} then becomes much larger than k_{+4}. Hence,

$$[EP]_{max} = \frac{\varepsilon}{1+K_3},$$

which agrees with the results shown in Fig. 3. The maximum rate is given by

$$V_0 = \frac{k_{+4}\varepsilon}{1+K_3},$$

and the measured value of V_0 agrees quite well with that calculated from K_3, k_{+4} and ε by this equation. Thus, ATP accelerates the Na$^+$-K$^+$-dependent ATPase by the same molecular mechanism through which it accelerates myosin-ATPase (chapter 3) and the Ca^{2+}-Mg^{2+}-dependent ATPase of the SR (chapter 11). ATP also accelerates the efflux of Na$^+$ from giant axons. Brinley and Mullins[11] found that the rate of efflux of Na$^+$ from giant axons perfused by their method[6,7] is dependent on the internal concentration of ATP, which is very similar to that for the Na$^+$-K$^+$-dependent ATPase shown in Fig. 2. The rate of Na$^+$ efflux rises rapidly with increasing concentration of ATP over the range 0–10μM, and then continues to rise slowly, but steadily, as the ATP concentration is further increased to 10mM.

To conclude this section, we will discuss the mechanism of ATP-ADP exchange by the ATPase. Fahn et al.[65,66] were the first to report that Na$^+$ stimulates the ATP-ADP exchange activity of microsomal fractions from the electric organ of the eel, when a low concentration of MgCl$_2$ is added or when the enzyme is pre-treated with NEM. Na$^+$ also stimulates the ATP-ADP exchange activity of a fraction made soluble from pig kidney membrane preparations[67]. Albers et al.[68,69] suggested that the intermediate EP is actually two species, E_1P and E_2P, formed sequentially. The conversion of E_1P to E_2P would then require Mg^{2+}, and would be effectively irreversible, selectively poisoned by NEM, and partially inhibited by a low concentration of Mg^{2+}. E_1P was postulated to react reversibly with ADP to resynthesise ATP, and E_2P to react irreversibly with water in the presence of K$^+$ to yield P_i. These suggestions were later validated by Post et al.[70] The electrophoretic patterns of the phosphorylated fragments obtained by proteolysis of both reactive forms show that phosphorylation occurs at the same site, and so Post et al.[70] concluded that there is a conformational change at the active centre for phosphorylation during the normal reaction sequence. However, in order to establish the functions of E_1P and E_2P as intermediates of ATPase, detailed kinetic studies must be made at low Mg^{2+} concentrations, in which both E_1P and E_2P are observable, as we did at high Mg^{2+} concentrations in which only the one species, EP, is seen.

Kinetic studies on this line are now in progress in our laboratory, and our preliminary results[61a] suggest the following reaction sequence:

$$E+S \rightleftharpoons E_1S \rightleftharpoons E_2S \rightleftharpoons E\genfrac{}{}{0pt}{}{\cdot ADP}{P} \rightleftharpoons E \sim P + ADP \rightarrow E + P_i + ADP.$$

In the presence of high concentrations of Mg^{2+} ions, the step $E_1S \rightleftharpoons E_2S$ is essentially irreversible and the rate of $E \sim P$ decomposition is much higher than that of ADP-liberation from $E\genfrac{}{}{0pt}{}{\cdot ADP}{P}$, as already shown in p. 368. Furthermore, the step $E_2S \rightleftharpoons E\genfrac{}{}{0pt}{}{\cdot ADP}{P}$ is dependent on K$^+$, and E_2S is formed when K$^+$ ions are added on $E\genfrac{}{}{0pt}{}{\cdot ADP}{P}$. Actually, the formation of ATP is demonstrated when K$^+$ ions are added on the phosphorylated intermediate which is formed in the absence of K$^+$, and the equilibrium constant, K_3, of the step $E_2S \rightleftharpoons E\genfrac{}{}{0pt}{}{\cdot ADP}{P}$ calculated from the

TABLE II. Phospholipid Content of the ATPase Preparation before and after Treatment with Venom

	Total phospholipid	PE	(PS+PI)	(LPS+LPE)	PC	SPh	LPC	Unknown
	(μg phosphorus/mg protein)	(% phosphorus)						
Control	53.2	18.2	15.2	Undetectable	42.5	17.1	0.9	6.0
Treated with venom	49.6	3.0	3.0	29.6	4.8	16.9	39.5	3.0

TABLE III. Effects of Treatment with Phospholipase and Addition of Phospholipid

	(PS+PI) −				(PS+PI) +			
	[EP][a]	$\frac{d[EP]}{dt}$[b]	v_0[a]	$v_0/[EP]$[a]	[EP][a]	$\frac{d[EP]}{dt}$[b]	v_0[a]	$v_0/[EP]$[a]
	(mole/10⁷g)	(mole/10⁷g/sec)	(mole/10⁷g/sec)	(sec⁻¹)	(mole/10⁷g)	(mole/10⁷g/sec)	(mole/10⁷g/sec)	(sec⁻¹)
Control	0.16	4.0	70.5	450	0.32	5.2	115	358
Treated with venom	0.89	2.1	20.3	22.8	0.79	4.5	106.3	135

[a] 100 μM ATP, 6.15 mM MgCl$_2$, 172 mM NaCl, 17.2 mM KCl, 37°C. [b] 2.2 μM ATP, 6.15 mM MgCl$_2$, 172 mM NaCl, 15°C.

amount of ATP formed is almost equal to the value estimated from the kinetic analysis described in p. 364. In the presence of low concentrations of Mg^{2+} ions and especially in the case of the NEM-treated enzyme, a phosphorylated intermediate ($E{\substack{\cdot\text{ADP} \\ \diagdown P}}$) which does not react with ADP appears first, and then an intermediate ($E\sim P$) which decays rapidly by shift of reversible steps on adding ADP appears. Thus, $E\sim P$ and $E{\substack{\cdot\text{ADP} \\ \diagdown P}}$ corresponds respectively to E_1P and E_2P, as defined by Albers, Post and others, and they are both high energy phosphorylated intermediates. Since the step $E_2S \rightleftharpoons E{\substack{\cdot\text{ADP} \\ \diagdown P}}$ is dependent on K^+, both the ADP-ATP exchange activity and the rate of phosphorylation of the ATPase are dependent on K^+ ions, as recently reported by Banerjee and Wong[70a], and Siegel and Goodwin[70b].

4. The Role of Phospholipids in the Transport ATPase

It appears that the reaction mechanism of the Na^+-K^+-dependent ATPase is similar to that for myosin-ATPase, but the Na^+-K^+-dependent ATPase requires the presence of phospholipids, as does the Ca^{2+}-Mg^{2+}-dependent ATPase of the SR (chapter 11), while myosin-ATPase does not. Interesting information has recently been obtained on the physiological role of the large amounts of phospholipid present in the Na^+-K^+-dependent ATPase. Tanaka et al.[71,72] made a preparation lacking the ouabain-sensitive Na^+-K^+-dependent ATPase activity by treatment with deoxycholate (DOC), and the addition of acidic phospholipid or lysolecithin to this restored the Na^+-K^+-dependent ATPase. The restored activity is somewhat lower than that of a control, but is inhibited by ouabain. Use of this system showed that synthetic alkylphosphates or dialkylphosphates can replace phospholipids, that the hydrocarbon chain can also be saturated, and that negatively charged sulphate or carboxyl groups can be substituted, although the recovery of activity may be low. Wheeler and Whittam[73] have also demonstrated that PS* also restores the Na^+-K^+-dependent ATPase activity.

We[74] have treated Na^+-K^+-dependent ATPase specimens with phospholipase C or Naja naja venom (phospholipase A) in order to degrade the phospholipid. Both treatments give similar results, and we shall describe

* See table of abbreviations for the full names of phospholipids referred to in this section.

mainly the venom-treatment in the presence of bovine serum albumin, which absorbs the fatty acids produced. The ATPase preparation used contains a large amount of phospholipid—about $50\mu g$ phosphorus (P) per mg protein (Table II)—and its reaction with phospholipase C indicates that the ATPase activity decreases proportionately to the decrease in the amount of phospholipid. When venom is used, as expected from its substrate specificity, PE, (PS+PI) and PC decrease while (LPS+LPE) and LPC increase considerably. There is no change in the amount of SPh. The treatment with venom decreases the rate of formation of EP in the presence of Na^+ to about half that of the control in the presence of Na^+, but the rate of decomposition of EP in the presence of Na^+ and K^+ decreases to about one twentieth that of the control (Table III). The ATPase activity of the specimen treated with venom is substantially recovered with PI or PS or, better, both, giving a rate of EP-formation equal to the control, but of EP-decomposition rate only half that of the control (Table III).

In the above experiments, it is necessary to ascertain whether there are effects due to the products formed from PL by phospholipase. In the case of the venom-treatment (phospholipase A), we removed the fatty acids formed by adding bovine serum albumin, but other products, lysoderivatives, bind with the membrane (Table II). Hence, the possibility must be considered that the ATPase is inhibited by these lysoderivatives, and is reactivated when they are displaced by added PL. However, the addition of LPC to the preparation treated with venom actually causes some activation, rather than inhibition, of the Na^+-K^+-dependent ATPase, and the addition of PL to the venom-treated specimen does not release the lysoderivatives. Reaction with phospholipase C instead of venom gives similar results. Together with the effect of DOC-treatment, these results show that phospholipid is necessary for the formation of micromedia near the active site of the ATPase, and for the decomposition of EP. More recently, Taniguchi and Iida[75] have shown that the rate of binding between ouabain and the protein moiety decreases markedly with phospholipase-treatment, and that the rate can be restored to the original level by adding PI and PS.

It has not been possible to make simultaneous measurements of Na^+- and K^+-transport and the ATPase reaction using the Na^+-K^+-dependent ATPase preparation, and the strength of binding between each reaction intermediate and Na^+ and K^+ is so far unknown. It will be feasible to propose a molecular mechanism of active transport from the kinetic evidence only when such measurements have been carried out. However, the similarities between the transport ATPase and the contractile ATPase strongly suggest that their molecular mechanisms are similar. In the case of the Ca^{2+}-Mg^{2+}-dependent ATPase of the fragmented SR, there is a relationship between the formation and decomposition of EP and the transport of Ca^{2+} and Mg^{2+} (chapter 11). These results are very useful in considering the mechanism of active transport of Na^{2+} and K^+, and this will be discussed in the following chapter.

5. The Transport ATPase and Myosin-ATPase

Chapter 11 dealt with the Ca^{2+}-Mg^{2+}-dependent ATPase of the SR, while this chapter has been concerned with the Na^+-K^+-dependent ATPase bound to membranes. We shall now compare the advantages and disadvantages of using these two ATPases and the myosin-ATPase for experimental investigations.

It is often necessary to obtain a pure enzyme in order to examine its reaction mechanism in detail. Myosin-ATPase is excellent in this respect, and the structure of its active site is better known than that of any other ATPase. A second point is the stability of the phosphorylated intermediate. The phosphorylated intermediate of the transport ATPase is stable to TCA, and this facilitates the study of its reaction mechanism. This can be seen by comparing the analysis of transport ATPase, discussed in chapters 11 and 12, with that of myosin-ATPase, which has a TCA-labile phosphorylated intermediate, discussed in chapter 3. A third consideration is the stability of the enzyme itself. The Na^+-K^+-dependent ATPase and myosin-ATPase are stable, but the Ca^{2+}-Mg^{2+}-dependent ATPase of the SR is rather unstable, which has made it difficult to analyse its reaction mechanism quantitatively. This partly explains the inadequate quantitative analysis in chapter 11. A fourth question is whether the enzyme preparation is capable of performing a physiological function, *i.e.*, contraction or cation transport. The superprecipitation of actomyosin and the Ca^{2+}-uptake by the Ca^{2+}-Mg^{2+}-dependent ATPase of the fragmented SR greatly help elucidation of the physiological functions of the ATPases.

REFERENCES

1 H. H. Ussing, *Acta Physiol. Scand.*, **19**, 43 (1949).
2 R. D. Keynes, *in* "Membrane Transport and Metabolism," ed. by A. Klein-

zeller and A. Kotyk, Publishing House of the Czechoslovak Academy of Sciences, Praha, p. 131 (1961).
3 B. Katz, "Nerve, Muscle, and Synapse," McGraw-Hill Book Co., New York (1966).
4 P. C. Caldwell, A. L. Hodgkin, R. D. Keynes and T. I. Shaw, *J. Physiol.*, **152**, 561 (1960).
5 G. Gardos. *Acta Physiol. Hung.*, **6**, 191 (1954).
6 J. F. Hoffman, D. C. Tosteson and R. Whittam, *Nature*, **185**, 186 (1960).
7 F. J. Brinley and L. J. Mullins, *J. Gen. Physiol.*, **50**, 2303 (1967).
8 L. J. Mullins and F. J. Brinley, *J. Gen. Physiol.*, **50**, 2333 (1967).
9 P. F. Baker, A. L. Hodgkin and T. I. Shaw, *J. Physiol.*, **164**, 330 (1962).
10 T. Oikawa, C. S. Spyropoulas, I. Tasaki and T. Teorell, *Acta Physiol. Scand.*, **52**, 195 (1961).
11 F. J. Brinley and L. J. Mullins, *J. Gen. Physiol.*, **52**, 181 (1968).
12 H. N. Christensen, "Biological Transport," W. A. Benjamin, New York (1962).
13 W. D. Stein, "The Movement of Molecules Across Cell Membranes," Academic Press, New York (1967).
14 J. C. Skou, *Biochim. Biophys. Acta*, **23**, 394 (1957).
15 J. C. Skou, *in* "Membrane Transport and Metabolism," ed. by A. Kleinzeller and A. Kotyk, Publishing House of the Czechoslovak Academy of Sciences, Praha, p. 228 (1961).
16 J. C. Skou, *Progr. Biophys. Mol. Biol.*, **14**, 131 (1964).
17 R. Whittam, *in* "The Neurosciences, A Study Program," ed. by G. C. Quarton, T. Melnechuk and F. O. Schmitt, The Rockefeller Univ. Press, New York, p. 313 (1967).
18 I. M. Glynn, *British Med. Bull.*, **24**, 165 (1968).
19 P. C. Caldwell, *Curr. Topics Bioenergetics*, **3**, 251 (1969).
20 P. C. Caldwell, *Physiol. Rev.*, **48**, 1 (1968).
21 S. L. Bonting, *in* "Membranes and Ion Transport," ed. by E. E. Bittar, Wiley-Interscience, London, Vol. 1, p. 257 (1970).
22 I. M. Glynn, *J. Physiol.*, **160**, 18P (1962).
23 R. Whittam, *Biochem. J.*, **84**, 110 (1962).
24 R. Whittam and M. E. Ager, *Biochem. J.*, **93**, 337 (1964).
25 V. T. Marchesi and G. E. Palade, *J. Cell Biol.*, **35**, 385 (1967).
25a A. K. Sen and R. L. Post, *J. Biol. Chem.*, **239**, 345 (1964).
25b H. J. Schatzmann, *Helv. Physiol. Pharmacol. Acta*, **11**, 346 (1953).
26 F. J. Samaha and J. Gergely, *Arch. Biochem. Biophys.*, **109**, 76 (1965).
27 E. Rogus, T. Price and K. L. Zierler, *J. Gen. Physiol.*, **54**, 188 (1969).
28 A. S. F. Ash and A. Schwartz, *Biochem. J.*, **118**, 20P (1970).
29 J. B. Peter, *Biochem. Biophys. Res. Commun.*, **40**, 1362 (1970).

30 P. V. Sulakhe, M. Fedelesova, D. B. McNamara and N. S. Dhalla, *Biochem. Biophys. Res. Commun.*, **42**, 793 (1971).
31 C. C. Ashley, in "Membranes and Ion Transport," ed. by E. E. Bittar, Wiley-Interscience, London, Vol. 2, p. 1 (1970).
32 J. C. Skou, *Physiol. Rev.*, **45**, 596 (1965).
33 W. Schoner, R. Beusch and R. Kramer, *Eur. J. Biochem.*, **7**, 102 (1968).
33a Y. Israel and E. Titus, *Biochim. Biophys. Acta*, **139**, 450 (1967).
33b G. H. Bond, H. Bader and R. L. Post, *Biochim. Biophys. Acta*, **241**, 57 (1971).
34 R. Whittam and K. P. Wheeler, *Annu. Rev. Physiol.*, **32**, 21 (1970).
35 R. W. Albers, *Annu. Rev. Biochem.*, **36**, 727 (1967).
36 J. Järnefelt, *Biochim. Biophys. Acta*, **59**, 643 (1962).
37 F. F. Jöbsis and H. J. Vreman, *Biochim. Biophys. Acta*, **73**, 346 (1963).
37a H. Matsui and M. Nakao, presented at a Japan-U.S. Seminar, Tokyo (1972).
38 A. Schwartz, H. S. Bachelard and H. McIlwain, *Biochem. J.*, **84**, 626 (1962).
39 I. Pull, *Biochem. J.*, **119**, 377 (1970).
39a L. E. Hokin and A. Yoda, *Biochim. Biophys. Acta,* **97**, 594 (1965).
40 G. Sachs, E. Z. Finley, T. Tsuji and B. I. Hirschowitz, *Arch. Biochem. Biophys.*, **134**, 497 (1969).
40a J. C. Skou and C. Hilberg, *Biochim. Biophys. Acta*, **110**, 359 (1965).
40b J. D. Robinson, *Nature*, **233**, 419 (1971).
41 T. Nakao, Y. Tashima, K. Nagano and M. Nakao, *Biochem. Biophys. Res. Commun.*, **19**, 755 (1965).
42 S. P. Banerjee, I. L. Dwosh, V. K. Khanna and A. K. Sen, *Biochim. Biophys. Acta*, **211**, 345 (1970).
43 J. T. Dulaney and O. Touster, *Biochim. Biophys. Acta*, **196**, 29 (1970).
44 P. B. Duham and J. F. Hoffman, *Proc. Natl. Acad. Sci. U.S.*, **66**, 936 (1970).
45 D. W. Towle and J. H. Copenhaver, *Biochim. Biophys. Acta*, **203**, 124 (1970).
46 P. L. Jørgensen and J. C. Skou, *Biochem. Biophys. Res. Commun.*, **37**, 39 (1969).
47 S. Uesugi, N. C. Dulak, J. F. Dixon, T. D. Hexum, J. L. Dahl, J. F. Perdue and L. E. Hokin, *J. Biol. Chem.*, **246**, 531 (1971).
47a M. H. Kline, T. D. Hexum, J. L. Dahl and L. E. Hokin, *Arch. Biochem. Biophys.*, **147**, 781 (1971).
47b A. Atkinson, A. D. Gatenby and A. G. Lowe, *Nature New Biology*, **233**, 145 (1971).
47c M. Nakao, T. Nakao, H. Ohta, F. Nagai, K. Kawai, Y. Fujihira and K. Nagano, presented at a Japan-U.S. Seminar, Tokyo (1972).
48 J. Kyte, *J. Biol. Chem.*, **246**, 4157 (1971).
49 R. J. Goldacre, *Int. Rev. Cytol.*, **1**, 135 (1952).

50. J. C. Skou, *Biochim. Biophys. Acta*, **42**, 6 (1960).
51. R. W. Albers, S. Fahn and G. J. Koval, *Proc. Natl. Acad. Sci. U.S.*, **50**, 474 (1963).
52. W. L. Stahl, A. Sattin and H. McIlwain, *Biochem. J.*, **99**, 404 (1966).
53. R. L. Post, A. K. Sen and A. S. Rosenthal, *J. Biol. Chem.*, **240**, 1437 (1965).
54. K. Ahmed and J. D. Judah, *Biochim. Biophys. Acta*, **104**, 112 (1965).
55. K. Nagano, T. Kanazawa, M. Mizuno, Y. Tashima, T. Nakao and M. Nakao, *Biochem. Biophys. Res. Commun*, **19**, 759 (1965).
56. L. E. Hokin, P. S. Sastry, P. R. Galsworthy and A. Yoda, *Proc. Natl. Acad. Sci. U.S.*, **54**, 177 (1965).
57. A. Kahlenberg, P. R. Galsworthy and L. E. Hokin, *Science*, **157**, 434 (1967).
58. R. L. Post, presented at a Japan-U.S. Seminar, Tokyo (1972).
58a. J. Avruch and G. Fairbanks, *Proc. Natl. Acad. Sci. U.S.*, **69**, 1216 (1972).
59. J. C. Skou and C. Hilberg, *Biochim. Biophys. Acta*, **185**, 198 (1969).
60. T. Kanazawa, M. Saito and Y. Tonomura, *J. Biochem.*, **61**, 555 (1967).
61. T. Kanazawa, M. Saito and Y. Tonomura, *J. Biochem.*, **67**, 693 (1970).
61a. Y. Fukushima and Y. Tonomura, unpublished.
61b. S. Mårdh and Ö. Zetterqvist, *Biochim. Biophys. Acta*, **255**, 231 (1972).
62. T. Hexum, F. E. Samson and R. H. Himes, *Biochim. Biophys. Acta*, **212**, 322 (1970).
63. M. Dixon and E. C. Webb, "Enzymes," 2nd ed., Longmans & Green Co., London (1964).
63a. C. Hegyvary and R. L. Post, *J. Biol. Chem.*, **246**, 5234 (1971).
63b. J. Jensen and J. G. Nørby, *Biochim. Biophys. Acta*, **233**, 395 (1971).
63c. O. Hansen, J. Jensen and J. G. Nørby, *Nature New Biology*, **234**, 122 (1971).
64. E. A. Guggenheim, *Phil. Mag.*, **2**, 538 (1926).
65. S. Fahn, G. J. Koval and R. W. Albers, *J. Biol. Chem.*, **241**, 1882 (1966).
66. S. Fahn, M. R. Murley, G. J. Koval and R. W. Albers, *J. Biol. Chem.*, **241**, 1890 (1966).
67. R. Rensi, *Biochim. Biophys. Acta*, **198**, 113 (1970).
68. G. J. Siegel and R. W. Albers, *J. Biol. Chem.*, **242**, 4972 (1967).
69. S. Fahn, G. J. Koval and R. W. Albers, *J. Biol. Chem.*, **243**, 1993 (1968).
70. R. L. Post, S. Kume, T. Tobin, B. Orcutt and A. K. Sen, *J. Cell Physiol.*, **54**, 306s (1969).
70a. S. P. Banerjee and S. M. E. Wong, *J. Biol. Chem.*, **247**, 5409 (1972).
70b. G. J. Siegel and B. Goodwin, *J. Biol. Chem.*, **247**, 3630 (1972).
71. R. Tanaka and K. P. Strickland, *Arch. Biochem. Biophys.*, **111**, 583 (1965).
72. R. Tanaka and T. Sakamoto, *Biochim. Biophys. Acta*, **193**, 384 (1969).
73. K. P. Wheeler and R. Whittam, *J. Physiol.*, **207**, 303 (1970).
74. K. Taniguchi and Y. Tonomura, *J. Biochem.*, **69**, 543 (1971).
75. K. Taniguchi and S. Iida, *Biochim. Biophys. Acta*, **233**, 831 (1971).

13
THE MOLECULAR MECHANISM OF MUSCLE CONTRACTION AND THE ACTIVE TRANSPORT OF CATIONS*

The investigation of the molecular mechanism of muscle contraction originated with Engelhardt and Ljubimova's[1] discovery of myosin-ATPase and Szent-Györgyi and co-workers'[2] studies of the reaction of actomyosin with ATP (chapter 1). Szent-Györgyi[2] and H. H. Weber and Portzehl[3] later showed that the contraction which ATP induces in a glycerol-treated muscle fibre composed mainly of actomyosin is inherently identical to the contraction of living muscle, and it is generally accepted that the contraction of muscle is caused by the reaction of actomyosin with ATP (chapters 1 and 8). In this monograph we have mainly described our own work on the hydrolysis of ATP by myosin, the mechanism of the conformational changes of myosin and actomyosin with ATP, and their relation to muscle contraction. However, there are still unsolved problems, and we shall now discuss some different theories of muscle contraction

* Contributor: Hiroshi Nakamura

and propose a molecular mechanism which at the moment appears to be the most plausible. We shall then extend this to propose a molecular mechanism for the active transport of cations, based on the mechanism of the transport ATPase discussed in chapters 11 and 12.

1. Theories of the Molecular Mechanism of Muscle Contraction

Morales and Botts[4,5] devised an electrostatic-entropic mechanism for muscle contraction, based on the facts that the addition of ATP to a 0.6M KCl solution of myosin B elongates the myosin B molecule without changing its molecular weight (chapter 6), that the charge on ATP is -3 to -4 at neutral pH (chapter 3), and that the binding of myosin with ADP is much weaker than that with ATP (chapter 3). At neutrality in 0.6M KCl, myosin B is negatively charged, and the adsorption of ATP would increase the negative charge, thereby causing elongation on account of the electrostatic repulsion. However, under normal physiological conditions, when there is a low concentration of KCl but a high concentration of Mg^{2+}, the adsorption of Mg^{2+} would confer a positive charge on the myosin B, and this would be decreased by the adsorption of ATP, resulting in entropic contraction. Decomposition of ATP into ADP and P_i would result in removal of ADP from the myosin B, thereby restoring the positive charge on myosin B and leading to elongation due to electrostatic repulsion. Our results[6,7] show that the addition of PP_i causes a change in shape but no change in the high molecular weight of the major constituent of myosin B (chapter 6). The main factors determining the molecular shape of the myosin B-PP_i complex are an entropy effect and electrostatic repulsion, as predicted by Morales et al. Equilibrium dialysis, light-scattering and other methods have shown that the shape of myosin B changes completely on binding 1 mole of its constituent myosin with 1 mole of PP_i or ATP[8]. However, Morales et al. have calculated that the amount of ATP which would need to be bound for contraction to occur according to their mechanism is of an order of magnitude greater than the observed value. Furthermore, the presence of Mg^{2+} gives a net charge of -2 to the ATP because the complex Mg^{2+}-ATP^{4-} is formed, so that the decrease in electrostatic repulsion upon adsorption of ATP is much smaller than that calculated by Morales et al. Also in disagreement with this hypothesis are studies with ATP analogues, which show that their strength of binding with actomyosin often does not agree with their ability to cause contraction (chapter 9). In summary, the hypothesis is correct in that the shape of myosin B is determined by entropy effects and electrostatic repulsion, but it is difficult to accept that the change in charge upon the adsorption of ATP is itself responsible for contraction.

H. E. Huxley and Hanson[9,10] investigated the fine structure of striated muscle, particularly the change accompanying contraction, and proposed the sliding theory of contraction, upon which several molecular mechanisms have been based (chapters 1 and 8). For example, Spencer and Worthington[11] and Ingels and Thompson[12] attributed the development of the contractile force to electrostatic attraction between the actin and myosin filaments. Warner[13] proposed that tension is generated by the rearrangement of water molecules surrounding the contractile protein, and Elliott et al.[14,15] assumed the existence of strong electrostatic fields in the region of overlap between the two filaments which would electrostrict water. Shear[16,17] assumed that the equilibrium separation between the myofilaments is determined by a balance between electrical double layer repulsive forces and London-van der Waals attractive forces, and that the additional charge generated on the myofilaments when ATP is hydrolysed causes shortening. The data currently available cannot exclude these theoretical mechanisms, but in any case, the work on the myosin-actin-ATP system described in chapters 2–10 is only distantly related to them, except in showing that there exist long-range forces between F-actin filaments (chapter 5). We have no positive evidence which supports any of these mechanisms by demonstrating that the forces postulated are responsible for contraction. It is unlikely that such a static mechanism can produce the free energy change large enough to develop the tension and do work in muscle.

However, two other types of sliding mechanism can be postulated from the studies of the actin-myosin-ATP system. One involves conformational change of actin in the presence of myosin and ATP. The other is based on coupling between the splitting of ATP and movements of the cross-bridges from the thick filament. These cross-bridges are thought to be derived from the head portions of the myosin molecules (chapters 1 and 2).

An example of the first type is Podolsky's[18] suggestion that contraction involves the folding of an actin filament. He assumed that during contraction one end of the thin filament is fixed relative to the thick filament, so that the contractile force is derived from the change in elasticity upon folding of the thin filament. This agrees well with H. E. Huxley et al.'s observation that contraction occurs by sliding of the two filaments relative to each other, the I band changing in length but the A band remaining constant. However, a large conformational change in the actin filament during contraction must be assumed. Aubert[19] and Carlson and Siger[20] emphasised the occurrence of a G⇌F transformation of actin during muscle contraction, and this idea of a marked change in shape of the actin filament during contraction has attracted a great deal of attention since A. G. Szent-Györgyi and Prior[21] discovered an increase in the exchangeability of nucleotides bound to actin during superprecipitation of actomyosin. However, Moos and Eisenberg[22] showed that this increased exchangeability is unrelated to the superprecipitation, since it is induced by aggregates of myosin in the presence of ADP as well as in the presence of ATP (chapters 5 and 7). Furthermore, the nucleotides in actin do not participate in superprecipitation or in the actomyosin-ATPase[23,24] (chapter 7), and we[25] and Martonosi et al.[26] have suggested that actin is always present in the F-form in actomyosin during superprecipitation and in living muscle during contraction. It is difficult to conceive that the G⇌F transformation of actin is directly coupled with muscle contraction, and also that other changes in the actin must accompany contraction. Oosawa et al.[27] proposed a mechanism involving transformation of actin from a helical to a linear polymer, but there is no experimental support for this (chapter 5). No evidence has been obtained for the force-generating change in the structure of F-actin, and the change in flexibility of F-actin described in chapters 6 and 10 must be the basis for the regulation of actin-myosin-ATP system by regulatory proteins and Ca^{2+}.

It is also possible to base a contraction mechanism on a large change in shape of the myosin molecule. A. F. Huxley[28] suggested that there is spontaneous association between a contractile site, capable of oscillating for a certain distance along the backbone of the thick myosin filament, and a binding site on the actin filament. The bond between these sites is cleaved by a reaction requiring energy from metabolic sources, and the probability of binding and cleavage is a function of the position of the contraction site on the myosin filament. Huxley showed that, given a simple form of this function, the sliding of both filaments can adequately explain many of the mechanical and thermodynamic properties of contraction discovered by Hill (chapter 1).

Since the physiology of contraction can be accounted for by A. F. Huxley's sliding mechanism, it is this which has become the central concept in modern mechanisms of contraction, and which is being further developed in two directions. One is to perform more detailed analyses of the physiological phenomena, in order to refine the model. Podolsky et al.[29]

analysed the motion following stepwise decreases in load[30] in terms of a sliding mechanism, and estimated the rate constants for the turnover of cross-bridges during contraction. For any given projection from the myosin filament there appears to be a range of about 100Å along the length of the filament, over which the projection can attach to the actin filament and form a cross-bridge. The site of attachment is then displaced by a distance of the same order before the link is broken. More recently, A. F. Huxley and Simmons[30a] analysed the time-course of the tension change after a sudden change in the length of the fibre at isometric tetanus, from a standpoint different from that of Podolsky et al., i.e., from the elastic properties of the cross-bridge itself and its mode of binding with F-actin. A rapid change in length is associated first a rapid tension change with which parallels the length change (an undamped elasticity) and then with a damped recovery, with a time constant of a millisecond or less, to a tension intermediate to the initial tension and that at the end of the length change. They suggested the existence of two elastic components in series, i.e., a stiff, undamped elasticity assigned to the hinged part of the myosin molecule (S-2) and a damped elasticity, with a time constant of a millisecond or less, assigned to the myosin head (S-1)-actin complex. During contraction each myosin head-actin attachment shifts through a series of stable positions, each with a lower potential energy than the last.

The second method of development is to elucidate the biochemical mechanisms of the many reactions assumed in A. F. Huxley's model. A molecular mechanism in which contraction involves conformational change of the myosin molecule was first presented by ourselves[31] in 1961, and a similar mechanism was proposed by Davies[32]. The scheme described later in this chapter is essentially the same as our earlier proposal [31,33,34], except that it incorporates more recent experimental findings. Hence, in the following section we shall summarise those experimental facts which are important in devising a molecular mechanism for muscle contraction, and the conclusions derived therefrom.

2. Experimental Data Relevant to the Molecular Mechanism of Contraction

The distribution of actin and myosin in muscle fibres, and its change upon contraction, have been elucidated by H. E. Huxley and Hanson, mainly by electron microscopy[35,36] (chapter 1). Myofibrils are composed of a regular arrangement of thick filaments of myosin and thin filaments largely composed of actin. Contraction involves a change in the relative positions of these two sets of filaments—the actin filaments slide between the myosin filaments.

In connection with the sliding theory, we shall discuss five biochemical problems, and then describe the results from X-ray diffraction and electron microscopy studies. First is the relationship between the thick and thin filaments in relaxed muscle. On adding ATP at high ionic strength, the main constituent of myosin B does not dissociate into F-actin and myosin but expands instead (chapter 6). Furthermore, according to Pepe[37-39] glycerol-treated myofibrils in a relaxing medium do not expose H-meromyosin antigenic sites in the region of overlap of the thin and thick filaments. These results suggest that at least the ends of the myosin projections are always maintained in some fixed steric relation to the actin filament.

Second is the interrelation between the many cross-bridges on a thick filament. Since the affinity of the binding sites for Ca^{2+} appears to be uniform along the length of the myofilament in the region of overlap[40], at low Ca^{2+} levels the cross-bridges, which bind to actin, must be further apart than at full activation. Hence, Podolsky and Teichholz[41] concluded that there are no co-operative effects between cross-bridges which influence their kinetic properties, since the relationship between relative force and velocity is the same under both circumstances. This indicates that the cross-bridges along the length of the myofilament operate independently. However, our studies on the viscoelasticity of glycerol-treated muscle fibres (chapter 8, section 5) suggest the interdependence of viscoelasticity of cross-bridges[42].

The third problem is the molecular mechanism of movement of cross-bridges upon stimulation. Morales[5] pointed out that if only the change in chemical binding between actin and myosin during sliding of the filaments is taken into account, it is difficult to explain such a rapid change in position of the actin and myosin filaments. There is some evidence that the higher order structure of myosin is altered by ATP and F-actin (chapter 2, section 8). The conformational changes induced in myosin by ATP and its analogues were first suggested by the protection of the ATPase from denaturation[43,44], but more direct evident has now been obtained from changes in the UV absorption[45-48] and the fluorescence of the tryptophane residues[48a,b] of H-meromyosin, and in the EPR spectrum of spin-labelled myosin[49,50] upon the addition of ATP or ADP.

The fourth question is the biochemical mechanism of the coupling of ATP-hydrolysis with muscle contraction. Myosin is phosphorylated by

ATP to form phosphoryl myosin, and this is essential for contraction (chapters 3 and 7). The direct hydrolysis of phosphoryl myosin activates the myosin molecule, and inhibition of this direct hydrolysis inhibits the superprecipitation of actomyosin, which is thought to be an analogue of contraction. The dissociation of actomyosin occurs by conversion of phosphoryl myosin to the myosin-phosphate-ADP complex (chapter 6), and the promotion of the decomposition of the myosin-phosphate-ADP complex by binding with F-actin is due to an allosteric effect of F-actin on the myosin-ATP complex (chapter 7, section 2). The contraction of an isolated sarcomere is proportional to the amount of ATP decomposed[51,52] (chapter 8) and the ATPase activity of a glycerol-treated muscle fibre is proportional to the size of the overlapping area between the F-actin and myosin filaments[53,54]. When the length of the sarcomere is 1.5μ or less, a considerable decrease in ATPase activity occurs, which can be explained by assuming that the ATPase is coupled to the movements of the cross-bridges from the myosin filaments[54]. Gordon et al.[55] obtained a similar relationship between sarcomere length and development of tension in living muscle, so that this phenomenon can be explained by the coupling of the ATPase activity and of the development of tension with the sliding of the actin filaments over the myosin filaments.

Fifth is the regulation of the myosin-actin-ATP system by Ca^{2+} and regulatory proteins. We have described in chapter 6 how the binding between F-actin and myosin becomes easy to cleave upon conversion of phosphoryl myosin to the myosin-phosphate-ADP complex. Ca^{2+} is another important factor regulating the binding of actin with myosin. In living muscle in the presence of ATP, excitation results in release of Ca^{2+} from the sarcoplasmic reticulum to the contractile system, where it complexes troponin bound to the F-actin-tropomyosin complex[56], resulting in binding between actin and myosin. In chapter 10 we described how complexing with Ca^{2+} induces conformational changes in troponin and tropomyosin, and how this causes further conformational change in the F-actin[57,58]. However the relation of the molecular mechanism of relaxation with that of detachment of cross-bridges from F-actin during the contraction cycle remains to be clarified, as described in chapter 6, section 6 and chapter 10, section 5.

Recently Elliott et al.[59] and H. E. Huxley and Brown[60] made a detailed analysis of the X-ray diffraction pattern of activated muscle fibres. In an actively contracting muscle, the overall repeating periodicities along both the myosin and the actin filaments remain virtually constant[60]. However, large changes in intensity take place in certain of the low-angle reflections, and show that there is movement of the cross-bridges during contraction[60]. The alterations in the pattern indicate that a fairly small change in tilt and/or longitudinal position of the cross-bridges is accompanied by larger changes in their azimuthal and perhaps radial positions. The changes in position of individual cross-bridges are not synchronised with each other. H. E. Huxley[61] observed large difference between the low-angle equatorial X-ray diagrams of muscle at rest and in rigor. The difference can be explained if, in relaxed muscle, the projections do not contact the actin filaments, and if, in rigor, they protrude further from the thick filaments and attach themselves to sites on the actin filaments. H. E. Huxley[9,61] suggested that the linear part of H-meromyosin is flexibly attached to the backbone of the thick filaments and to the globular part of H-meromyosin, so that, by tilting outwards further, it can vary the position of the active end of the cross-bridge (chapter 2, Fig. 2).*

From the results of antibody staining studies, Pepe[37,38] suggested that changes in the position of the cross-bridges may not be an active part of the mechanism of translation. The bridges may serve to transmit tension, while the interaction responsible for translation may be limited to the point of contact between the cross-bridge and the actin filament.

Aronson and Morales[62] have examined the polarisation of tryptophane fluorescence from glycerol-treated rabbit fibre bundles. When the light causing excitation is polarised with its electric vector perpendicular to the fibre axis, the polarisation, $P_\perp$, is $\sim$0.08. Relaxation causes a reversible increase in $P_\perp$ to $\sim$0.125, and isometric contraction produces an intermediate value, $\sim$0.10. They suggested that the change in $P_\perp$ reflects changes in the average orientation of an element of muscle containing tryptophane. More recently, dos Remedios et al.[63] have extended these experiments, and have found that the polarisation, $P_\perp$, which changes with the state of the muscle fibre (relaxation > contraction > rigor), is diagnostic of the contractile state of myosin, and is probably sensitive to the orientation of the cross-bridges from the myosin filament. M. Bárány et al.[64] showed that there is decreased incorporation of NEM into the globular head part (S-1) of myosin during contraction, and suggested that the change in reactivity during contraction is localised in the active ends of the cross-bridges, while the backbone of the filament and its linkages with the cross-bridges remain unaffected.

Reedy et al.[65] showed by electron microscopy that in glycerol-treated insect flight muscle there is a change in the orientation of the cross-bridges

* See Refs. 10, 61a for X-ray diffraction studies on muscle.

between the relaxed and rigor states. From electron microscopy using negative staining methods, Takahashi and Yasui[66] postulated that at the time of the clearing response, which occurs before the superprecipitation of myosin B by ATP, the projection from the myosin filament, if bound to the actin filament, is short and perpendicular to the axis, while if it is not bound to actin, it is long and slanted towards the axis (chapter 7, section 6). Our own electron microscopy observations[67] have suggested that acto-H-meromyosin in the absence of ATP exhibits an arrow-head structure (chapter 6), but that a substantial amount of acto-H-meromyosin dissociates upon addition of ATP. However, when PCMB-DTT-treated acto-H-meromyosin was used, ATP did not dissociate acto-H-meromyosin, as already described in chapter 7, section 7, and induced remarkable changes in the ultrastructure of acto-H-meromyosin[68]. ATP changed the arrow-head structure into a new structure, in which H-meromyosin bound to F-actin formed projections oriented more perpendicularly to the F-actin filament axis than in the absence of ATP, and became fairly rounded in shape. These results suggest that the addition of ATP changes the shape of the head part of the myosin molecule, but one cannot exclude the possibility of artifacts due to the negative staining method. Ignoring this possibility for the moment, we shall tentatively call the arrow-head structure of H-meromyosin, after binding with F-actin in the absence of ATP, the β-state, and the spherical shape, binding perpendicularly to the F-actin filament axis in the presence of ATP, the α-state.

3. The Molecular Mechanism of Muscle Contraction

The previous sections describe experimental results thought to be important in the elucidation of a molecular mechanism of muscle contraction, and we shall now make a systematic summary of these results and of the hypotheses derived from them. (1) Muscle contraction is coupled to the phosphorylation and dephosphorylation of myosin. In particular, the direct hydrolysis of phosphoryl myosin activates the myosin molecule. (2) The shape of the head portion of the myosin molecule changes on addition of ATP. (3) The binding between F-actin and myosin is easily cleaved upon formation of the myosin-ATP complex at high concentrations of ATP or the myosin-phosphate-ADP complex at low concentrations of ATP. The strength of this binding greatly depends on the conformation of both myosin and the complex of F-actin with the relaxing protein, and the conformation of the latter is controlled by traces of Ca^{2+}. (4) In order for

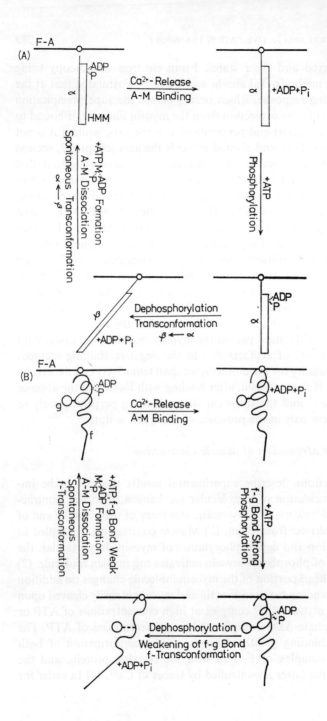

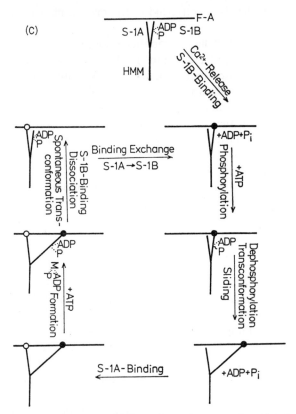

Fig. 1. Molecular mechanism of muscle contraction. F-A (or A) and M indicate F-actin and myosin respectively. A: H-MM represents the head part of myosin. α and β indicate two different conformations of the myosin molecule. B: g and f represent the light (g) and heavy (f) chains of myosin. C: S-1B is the head part of myosin which contains both the active site of the ATPase and a site for binding to F-actin. S-1A contains only a site for binding to F-actin.

a complete cycle of contraction and relaxation to take place, the binding between myosin and F-actin must be cleaved after transconformation of the head part of the myosin molecule coupled with energy release has occurred, and must be restored when the myosin molecule regains its original shape. Based on these ideas, the simplest mechanism for a cyclic process is that the myosin molecule can assume two conformations which determine whether the phosphoryl myosin formed by reaction with ATP is directly hydrolysed, or whether it is converted to the myosin-phosphate-ADP complex.

As mentioned above, we[31,33,34] have proposed a mechanism for the complete cycle of contraction and relaxation involving transconformation of the myosin molecule caused by interaction with ATP and F-actin. We currently favour the scheme shown in Fig. 1A[69], which incorporates the most recent experimental results. This diagram shows only that portion of the myofibril in which the reaction essential to contraction takes place, *i.e.*, the cross-bridge between the F-actin and myosin filaments. Activation occurs by the following process. (1) When the muscle is at rest, myosin is in the α-state and forms a stable myosin-phosphate-ADP complex. (2) Excitation causes release of Ca^{2+} from the sarcoplasmic reticulum, resulting in the complexing of Ca^{2+} with the relaxing protein and changing the conformation of F-actin. Myosin binds with the F-actin, resulting in the rapid liberation of phosphate and ADP from myosin. (3) Myosin in the α-state is phosphorylated by ATP, and the phosphoryl myosin is rapidly dephosphorylated by interaction with F-actin. The energy released causes transconformation of the myosin to the β-form, as a result of which the actin filament is drawn between the myosin filaments. (4) Myosin in the β-state forms a myosin-phosphate-ADP complex after reacting with ATP. The binding between myosin and F-actin is cleaved and the myosin is spontaneously transformed back to the α-form. This restores the system to its original condition. Active transport of Ca^{2+} into the sarcoplasmic reticulum stops the cycle at condition (1), which is the resting state.*

In the molecular model just mentioned, we assumed that both the dissociation of actomyosin at the relaxed state and the detachment of the head part of myosin from F-actin during the contraction cycle occur when $E_2^1{\cdots_{\cdot P}^{ADP}}$ is formed. However, as described in chapter 10, section 5, there remains another possibility that the myosin molecule in the relaxed muscle forms a complex with two ATP molecules ($^S E_2 S^1$), of which conversion to $E_2^1{\cdots_{\cdot P}^{ADP}}$ is inhibited, while the formation of $E_2^1{\cdots_{\cdot P}^{ADP}}$ is the reaction step for the detachment of cross-bridge during the contraction cycle. When this mechanism is adopted, the description for the molecular mecha-

---

* In condition (2) myosin is bound to F-actin in the presence of ATP. Therefore, in rigor in the absence of ATP, it is possible that the shape of the projections from the myosin filament is different from that of condition (2).

nism is modified as follows: (1) When muscle is at rest state, myosin is in the α-form and forms a myosin-ATP complex. (2) Excitation causes release of Ca^{2+} from the sarcoplasmic reticulum, resulting in the complexing of Ca^{2+} with the relaxing protein and changing the conformation of F-actin. Myosin binds with the F-actin. (3) Myosin in the α-state is phosphorylated by ATP, and the phosphoryl myosin is rapidly dephosphorylated by interaction with F-actin. The energy released causes transconformation of the myosin to the β-form, as a result of which the actin filament is drawn between the myosin filaments. (4) Myosin in the β-state forms a myosin-phosphate-ADP complex after reacting with ATP. The binding between myosin and F-actin is cleaved and the myosin is spontaneously transformed back to the α-form. Myosin binds again with the F-actin, resulting in the rapid liberation of phosphate and ADP from myosin, and ATP binds to myosin. This restores the system to its original condition.

At present the molecular mechanism of the conformational change of the head part of myosin, coupled with the phosphorylation and dephosphorylation of myosin, is unclear. However, one interesting possibility is that a change in the strength of binding between the light and heavy chains making up the myosin molecule (chapter 2) results in a conformational change coupled with phosphorylation and dephosphorylation[69]. According to Gershman et al.[70] and Frederiksen and Holtzer[71], there is a reversible conformational change of myosin due to the dissociation and association of light and heavy chains, and the binding with actin requires association between the light and heavy chains. Figure 1B presents a molecular mechanism of this type. The binding between the light and heavy chains weakens when phosphoryl myosin is dephosphorylated, and this induces transconformation of the heavy chains, resulting in sliding of the actin filaments. When the bond between the light and heavy chains is weak, i.e., when the binding between actin and myosin is weak, ATP forms $E_2 {\overset{1}{\underset{P}{\cdots}}} ADP$, and the binding between myosin and actin cleaves. The conformation of the heavy chain reverts spontaneously to its original state, and the heavy chain once again binds strongly with the light chain, allowing the contraction cycle to be repeated.

We will discuss here the functions of the two heads of the myosin molecule in contraction. It is not certain whether the two heads have identical or different structures and functions (chapter 4, section 2). Morales et al.[72,73] consider that they are identical, and that the sliding of the thin filaments relative to the thick filaments coupled with ATP-splitting is due to co-operative action between them. However, many of our results suggest that the two heads are not identical (chapters 2, 3, 4 and 6). Site 1 for simple hydrolysis and site 2 for phosphorylation are present in one of the heads of myosin, S-1B (chapters 2 and 3). The binding between S-1A and F-actin is probably much weaker than that between S-1B and F-actin (chapter 6), and this fact can be incorporated into the molecular mechanism of contraction as follows[74]. S-1B undergoes the conformational change $\alpha \to \beta$ upon direct dephosphorylation of phosphoryl myosin. The binding between S-1B and F-actin is cleaved when the S-1B assumes the β-form, due to the formation of a myosin-phosphate-ADP complex with ATP, resulting in spontaneous reversion of the S-1B to the α-conformation. The relatively weak binding between S-1A joined to the myosin filament and F-actin does not occur when S-1B is in the α-form because of steric hindrance, unlike when the head is isolated from myosin, and weak binding occurs only when the S-1B is in the β-form. Figure 1C shows a molecular mechanism which has been modified to take into account the functions of the two heads of the myosin molecule. The release of Ca^{2+} results in the binding of F-actin with α-S-1B. In this state, reaction with ATP produces phosphoryl S-1B, and its dephosphorylation results in the formation of β-S-1B, leading to sliding of the F-actin. A weak bond between S-1A and F-actin is also formed. Since the binding of β-S-1B with F-actin is now weak, the β-S-1B reacts with ATP to form a phosphate-ADP complex, causing rupture of the bond between S-1B and F-actin. The β-S-1B spontaneously reverts to the α-conformation, steric hindrance then ruptures the weak binding between S-1A and F-actin, and the strong bond between S-1B and F-actin is reformed, returning the system to its original condition.

4. The Molecular Mechanism of Active Transport

Goldacre and Lorch[75] and Goldacre[76] suggested that a rhythmically expanding and contracting surface might, by adsorbing and subsequently desorbing various substances, be a means for concentrating compounds within the cell. This contractile protein hypothesis has been modified and extended by Danielli[77,78], who suggested that protein chains on the outside of the cell membrane which have substances absorbed on them might actually be pulled inside through a micropore by an internal contraction. He also showed that, if ATP provides the energy for contraction, the distribution of phosphatases in the various secretory tissues is appropriate to the direction of secretion. More recently, Opit and Charnock[79] and Jardetzky[80] proposed a mechanism involving protein transconformation accompanied by the formation and decomposition of a phosphorylated intermediate.* Hubbell and McConnell[82] observed that spin-labelled steroids with a paramagnetic five-membered ring attached rigidly to the steroid nucleus are quite soluble in biological membranes, and that their EPR spectra in membranes show a high degree of motion of the nitroxide group, with rotational diffusion frequencies of the order of 10^7 to 10^8 sec^{-1}. More recently, studies on proton nuclear magnetic resonance of the membrane lipids[82a] and on the behavior of spin labels distributed within the lipid component of the membrane indicate the presence of low viscosity regions with the SR membranes. Hence, one can assume that translocation of the carrier (chapter 11) occurs rather easily in the hydrophobic region of the membrane.** However, at the time when the above theories were presented, the reaction mechanism of the transport ATPase had not been sufficiently clarified, and the reaction step responsible for actual transport of ions into the membrane was unknown.

Chapters 11 and 12 discussed the striking detailed resemblance between the reaction mechanisms of the Ca^{2+}-Mg^{2+}-dependent ATPase of the fragmented sarcoplasmic reticulum, the membrane bound Na^+-K^+-dependent ATPase and the contractile ATPase. In all three ATPases, a phosphorylated protein, EP, is a reaction intermediate, and is formed *via* two enzyme-substrate complexes, E_1S and E_2S. The rate constant for the formation of E_1S from the substrate, ATP, and E is immeasurably large, and the formation of E_2S is accelerated by a high concentration of ATP.

The other common feature of muscle contraction and active transport of cations is the fact that they have vectorial properties. Unidirectional change, such as the shortening of muscle cells, which is derived from the polarity of myosin and actin filaments constituting the myofibrils, has already been described in chapter 1. According to our molecular mechanism (Fig. 1C), the directional properties of contraction are derived from the directional properties of the actin filaments and from the positions of the two non-identical head parts of myosin relative to the myosin filament. The vectorial properties of the transport of Na^+, K^+, Ca^{2+} and Mg^{2+} through plasma membranes and the sarcoplasmic reticulum should be

* Ref. *81* gives a critical review of models for Na^+- and K^+-transport.
** See chapters by D. Chapman and G. H. Dodd, by P. Jost *et al.* and by S. J. Singer in Ref. *82b* for the details of the supermolecular organization of membranes. Recent articles by Caspar and Kirschner[82c] and Singer and Nicolson[82d] are also readable.

derived from the vectorial properties of the membrane ATPase. Chapter 11 describes in detail how the effects of Ca^{2+} and Mg^{2+} on each elementary step of the ATPase reaction of the sarcoplasmic reticulum have vectorial properties in good agreement with the idea that the ATPase is the basis of cation transport.

There are two significant differences between the transport ATPase and the actomyosin-ATPase (chapters 11 and 12). Firstly, the transport ATPase requires the presence of phospholipids for the decomposition of EP. Secondly, in the transport ATPase under normal physiological conditions, the splitting of ATP occurs by the direct decomposition of E∼P, while in the contractile ATPase, ATP is hydrolysed *via* both E∼P and E⋯P, and E∼P is very unstable and occurs only transiently. Proteins which bind specifically with sulphate[83], galactose[84], calcium[85,86] and amino acids[87,88] have already been isolated from cell membrane fractions by mild procedures such as osmotic shock.* Therefore, we shall consider the transport ATPase to be a complex of a translocase and a carrier protein for the cations. The decomposition of 1 mole of ATP causes the efflux of 3 moles of Na^+ and the influx of 2 moles of K^+ (chapter 12), and so the carrier protein may be an oligomer. We shall assume that the association of carrier protein with translocase always occurs under normal physiological conditions, since they correspond to actin and myosin in the contractile system, and the intermediate E⋯P which cleaves the binding of actin with myosin does not appear in the transport ATPase under these conditions.

The transport ATPase is activated by K^+ outside and Na^+ inside the membrane (chapter 12). The formation of the phosphorylated intermediate depends on Na^+, while its decomposition depends on K^+. The ATP acts from inside the membrane, and the products ADP and P_i are formed within the membrane. Hence, Post *et al.*[92], Stone[93], and Albers *et al.*[94] proposed that the formation of EP results in the transport of Na^+ from the inside to the outside of the membrane, while the decomposition of EP results in the transport of K^+ from the outside to the inside of the membrane. However, it is exceedingly difficult to measure directly the relationship between the formation of the ATPase intermediates and the translocation of the cation site in the case of the Na^+-K^+-dependent ATPase, so that these ideas have never been verified. On the other hand, in the case of the Ca^{2+}-Mg^{2+}-dependent ATPase of the fragmented SR, one can demonstrate a direct correlation between the elementary steps of the reaction and

* See Refs. *89–91a* concerning the properties of the binding proteins in the transport system.

the transport of Ca^{2+}, since it was shown that the formation of EP results in the transport of Ca^{2+} from the outside to the inside of the membrane, while within the membrane Ca^{2+} complexed to the EP is replaced by Mg^{2+}, resulting in the decomposition of EP (chapter 11). Hence, we shall assume that in the Na^+-K^+-dependent membrane ATPase Na^+ is transported from the inside to the outside of the membrane upon the formation of EP, while the decomposition of EP occurs after exchange of K^+ outside the membrane with Na^+ bound to EP.

An interesting related phenomenon is the exchange diffusion of Na^+ on the plasma membrane. The concept of exchange diffusion was introduced by Ussing[95] to explain the large efflux of labelled Na^+ from frog muscle fibres without a corresponding consumption of energy. Its first direct experimental support was obtained by Keynes and Swan[96], who showed that in freshly dissected frog muscle about half the Na^+ efflux was dependent on the presence of Na^+ in the external medium. According to Caldwell et al.[97], when the ATP/ADP ratio is high, the Na^+ efflux displays its normal sensitivity to removal of external K^+. When there is no ATP, the efflux is reduced to a very low level, but when the ATP/ADP ratio has an intermediate value, the efflux is large and yet quite unaffected by removal of K^+ from the medium. There are indications that this K^+-insensitive Na^+ efflux has a larger exchange diffusion component than usual, because it is appreciably reduced when Li^+ is substituted for external Na^+. The mechanism of exchange diffusion has been studied further by de Weer[98], and by Garrahan and Glynn[99,100], who found that reconstituted mammalian erythrocyte ghosts can also respond similarly. This exchange diffusion supports the above concept that the formation of EP is accompanied by the transport of Na^+ from the inside to the outside of the membrane, and these results can be easily explained by the following mechanism.

$$E+ATP^i+Na^i \quad\quad E+ATP^i+Na^\circ$$
$$\updownarrow \quad\quad\quad\quad\quad \updownarrow$$
$$EATPNa^i \;\rightleftharpoons\; EPNa^\circ+ADP^i$$

A connection between K^+-transport and the dephosphorylation step is suggested by the observation that under normal physiological conditions, although the efflux of K^+ from red blood cells is downhill, it is partly inhibited by cardiac glycosides, and this probably represents a kind of backwash through the Na^+/K^+ pump[101]. If the entry of K^+ into the cell by the pump mechanism involves dephosphorylation and the release of P_i into the cell, the backwash might be expected to reverse these processes and therefore to require the presence of P_i inside the cell. It was found recently that removal of internal P_i abolishes the glycoside-sensitive K^+ efflux [102,103].

For simplification, we shall assume that the $Na^+(Ca^{2+})$ and the $K^+(Mg^{2+})$ sites are the same site in two different forms, and that the $Na^+(Ca^{2+})$ and $K^+(Mg^{2+})$ at the site are not only factors for the activation of the enzyme system but are also the cations which are transported. As described in chapter 11, the formation of EP results in the translocation of the Ca^{2+} site from the outside to the inside of the membrane. Furthermore, comparing the competition between Ca^{2+}, Sr^{2+} and Mg^{2+} ions of the Ca^{2+}-Mg^{2+}-dependent ATPase of the SR with the competition between these cations during the transport through the SR membrane strongly suggests that the cation site for activation of ATPase is the same as the cation site for transport (see chapter 11, section 6). We will assume that there are at least two conformations for the translocase, and that the transitions between them are coupled with chemical reactions identical to those in the myosin-ATPase. These assumptions have received some experimental support. For example, in the presence of Mg^{2+} and ATP, Na^+ increases and K^+ decreases the fluorescence of 1-anilino-8-naphthalene-sulphonic acid bound to the Na^+-K^+-dependent ATPase, and ouabain reverses the K^+-induced changes[104]. Using infrared absorption spectroscopy, Graham and Wallach[105] have suggested that in the presence of Mg-ATP, some peptide linkages of the protein in the erythrocyte membranes undergo a transition from a mixture of α-helical and 'unordered' conformations to an antiparallel β-structure, which is enhanced by the addition of Na^+ and K^+. The transition can be prevented or reversed under conditions which inhibit the ATPase reaction. Recently, Nakamura et al.[106] observed that the EPR spectrum of spin-labelled SR changes upon adding ATP only in the presence of both Mg^{2+} and Ca^{2+}, which are necessary for the formation and decomposition of the phosphorylated intermediate (chapter 11). Following Opit and Charnock[79] and Jardetzky[80], we shall also assume that the affinities of the cation sites for $Na^+(Ca^{2+})$ and $K^+(Mg^{2+})$ are different in the two conformations. In fact, a dramatic change in the affinities of binding of Ca^{2+} and Mg^{2+} to the Ca^{2+}-Mg^{2+}-dependent ATPase occurs when EP is formed (chapter 11, section 6).

Figure 2 shows a molecular mechanism for active transport of Na^+ and K^+ coupled with the ATPase reaction which takes account of the above observations. It is similar to that proposed by Hokin[107], and contains a translocase, TL, capable of assuming two molecular shapes, α and β, a carrier protein, CP, which is strongly bound to TL and has binding sites

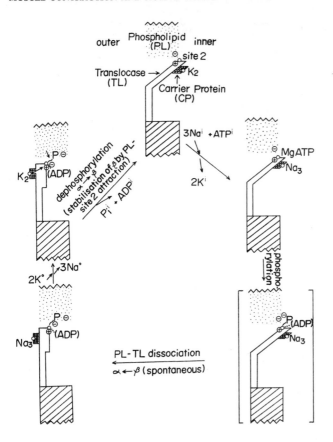

Fig. 2. The molecular mechanism of active transport of ions coupled with the Na^+-K^+-dependent ATPase. Superscripts i and o refer to the inside and the outside of the membrane; CP, ion carrier protein; PL, phospholipid.

for ions, and phospholipid, PL. Since there is a continuous cycle in a transport system, one may select any state as the starting point. We shall begin at the centre top of the Figure. (1) The TL is in the β-shape and CP located within the membrane. By analogy to the contractile system, the β-TL is unstable. According to Schwartz et al.[108], the transport ATPase is inhibited by TBS, which reacts with an amino group of the protein similarly to the case with myosin-ATPase. Therefore, the neighbouring reactive sites of TL are thought to be positively charged, so that the un-

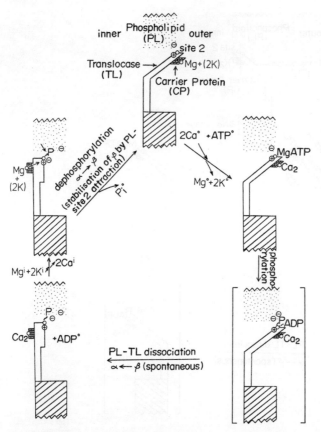

Fig. 3. Molecular mechanism of active transport of Ca^{2+} by the sarcoplasmic reticulum coupled with the Ca^{2+}-Mg^{2+}-dependent ATPase. The caption is the same as for Fig. 2.

stable β-form of TL may be stabilised by electrostatic attraction to the negatively charged phospholipid. (2) When TL is in the β-conformation, the CP prefers to bind with Na^+ rather than with K^+. Replacement of K^+ by Na^+ on the CP results in the binding of Mg-ATP onto site 2. (3) Phosphorylation of TL cleaves the electrostatic linkage between PL and TL because of the negative charge of the phosphate group bound at site 2. This state is transient. (4) As a result, TL spontaneously reverts from the unstable β-form to the stable α-form, and the CP moves to the outside of the membrane. (5) With TL in the α-form, CP now tends to bind with K^+

rather than with Na^+, and Na^+ is replaced by K^+ outside the membrane. Thus the dephosphorylation of phosphoryl TL is effected by CP bound to K^+ and PL. P_i and ADP are released inside the membrane at this stage. The energy released by dephosphorylation is coupled with conversion of TL back to the β-form, thereby restoring the system to its original condition (1).

The mechanism of Ca^{2+}-transport by the Ca^{2+}-Mg^{2+}-dependent ATPase of SR can be easily derived by reversing the outside and inside of the membrane in the above Na^+-K^+-transport mechanism, and by replacing Na^+ by Ca^{2+} and K^+ by Mg^{2+} (Fig. 3).* This is apparent from the identical mechanisms of the Ca^{2+}-Mg^{2+}-dependent ATPase of the SR (chapter 11) and the plasma membrane bound Na^+-K^+-dependent ATPase (chapter 12).

The kinetic methods which have been mainly used in the work described in this monograph cannot, of course, determine the distance of translocation of cation sites in active transport. However, since the membrane is usually 80–100Å thick, translocation would appear to occur over a distance of this order, which is almost the same as the distance moved by the projections from the myosin filament (p. 385). Thus muscle contraction and the active transport of cations, which are very important in biological energy transformation, can now be considered as molecular motions coupled with phosphorylation and dephosphorylation of the key protein by ATP. However, it is regrettable that even for muscle contraction, upon which the most extensive work has been performed, there is no direct experimental verification of the relationship between the movement of the key protein, myosin, and the steps of the reaction of myosin with ATP. The molecular mechanisms for the transport systems are constructed only by analogy with the contractile system. We have already discussed in chapter 12 the advantages of comparing the reaction mechanisms of the various ATPases, and the relationship between the contractile and transport ATPases has been made clearer by elucidating their physiological functions—muscle contraction and active transport of cations. Thus the contractile ATPase causes unidirectional translocation of F-actin coupled with decomposition of ATP, and the phosphoryl intermediate formed is an unstable transient species which changes into the more stable enzyme-

* For simplicity in Fig. 3, we have assumed that two Ca^{2+} ions on the EP are replaced by one Mg^{2+} ion and two K^+ ions. As described in chapter 11, while it is essential for the decomposition of EP that one of the two Ca^{2+} ions is replaced by one Mg^{2+} ion, the remaining Ca^{2+} may be replaced by either Mg^{2+} or K^+ ions.

phosphate-ADP complex, resulting in the cleavage of the binding between the ATPase and the F-actin, which is translocated in coupling with dephosphorylation of the phosphoryl intermediate. The transport ATPase causes cyclic translocation of the cation carrier protein coupled with the decomposition of ATP. In this case, the phosphoryl intermediate is relatively stable and no enzyme-phosphate-ADP complex is formed. However, the conformational change accompanying the translocation of the carrier protein causes a change in cation binding affinity, thereby performing an important physiological function.

5. Evolution of High Energy Phosphate Compounds

To conclude this chapter, we will briefly mention the evolution of high energy phosphate compounds in the primeval period, with emphasis on the evolutionary aspects of contractile and transport ATPases.

The possibility that energy-rich nucleotides were present in the 'primordial soup' has been suggested, and searches for simulations of them have been made in irradiated model systems[109-111] and in chemical systems containing phosphorylating agents[112-116]. Oró and Kimball[117,118] discovered that adenine is produced by heating a concentrated aqueous solution of ammonium cyanate. Gabel and Ponnamperuma[119] obtained 3, 4, 5 and 6 carbon sugars by heating a dilute aqueous solution of formalin with kaolin, and identified ribose among the products. Ponnamperuma et al.[109] found that the UV irradiation of a dilute solution (1mM) of adenine, ribose and a phosphate salt gives adenosine. They[109] also observed the formation of AMP, ADP and ATP upon UV irradiation of a mixture of 'ethyl metaphosphate' with adenosine. 'Ethyl metaphosphate' is a totally artificial reagent, anfi is not thought to have been present on the primitive earth, but it is possible that some other phosphate could show similar activity. Akaboshi et al.[109a,b] have recently reported the conversion of adenosine, AMP and ADP into AMP, ADP and ATP, respectively, with recoiled ^{31}P atoms obtained from the β-decay of $^{31}S_1$. However, ATP can hardly have been the first energy-rich phosphorus compound, or even one of the first of these compounds, although adenine is stable. ATP is rather complicated substance, and many of its analogues are energy-rich in the same way as ATP. Some of them, in which adenine is replaced by another base, occur in nature. Compounds containing another sugar instead of ribose are also worth consideration. Calvin[120], Lipmann[121] and Baltscheffsky[122,123] have suggested ordinary pyrophosphate as a possible starting point; inorganic polyphosphates are still abundant in bacteria[124]. This suggestion is also supported by the discovery of a light-dependent synthesis of PP_i by *Rhodospirillum rubrum chromatophores*[125-127].

Myosin catalyses the hydrolysis of all triphosphate compounds tested so far, even including inorganic triphosphate, although muscle contraction requires the presence of a nucleoside base (chapter 9). Furthermore, the substrate specificities of the transport ATPases are very broad (chapters 11 and 12). In particular, sarcoplasmic reticulum catalyses the hydrolysis not only of nucleoside triphosphates but also of *p*-nitrophenyl phosphate and acetyl-phosphate, which support the Ca^{2+}-uptake. These facts suggest that both the contractile and transport ATPases probably appeared at an early stage of evolution.

REFERENCES

1 W. A. Engelhardt and M. N. Ljubimova, *Nature*, **144**, 668 (1939).
2 A. Szent-Györgyi, "Chemistry of Muscular Contraction," 1st ed. (1947) & 2nd ed. (1951), Academic Press, New York.
3 H. H. Weber and H. Portzehl, *Progr. Biophys. Biophys. Chem.*, **4**, 60 (1954).
4 M. F. Morales and J. Botts, *Arch. Biochem. Biophys.*, **37**, 283 (1952).
5 M. F. Morales, *in* "Enzymes: Units of Biological Structure and Function," ed. by O. H. Gaeber, Academic Press, New York, p. 325 (1956).
6 T. Nihei and Y. Tonomura, *J. Biochem.*, **46**, 1355 (1959).
7 Y. Tonomura and F. Morita, *J. Res. Inst. Catalysis, Hokkaido Univ.*, **7**, 126 (1959).
8 Y. Tonomura and F. Morita, *J. Biochem.*, **46**, 1367 (1959).
9 H. E. Huxley, *Science*, **164**, 1356 (1969).
10 J. Hanson, *Quart. Rev. Biophys.*, **1**, 177 (1968).
11 M. Spencer and C. R. Worthington, *Nature*, **187**, 388 (1960).
12 N. P. Ingels and N. P. Thompson, *Nature*, **211**, 1032 (1966).
13 D. T. Warner, *J. Theor. Biol.*, **26**, 289 (1970).
14 G. F. Elliott, *J. Theor. Biol.*, **21**, 71 (1968).
15 G. F. Elliott, E. M. Rome and M. Spencer, *Nature*, **226**, 417 (1970).
16 D. B. Shear, *Physiol. Chem. Physics*, **1**, 495 (1969).
17 D. B. Shear, *J. Theor. Biol.*, **28**, 531 (1970).
18 R. J. Podolsky, *Annu. N. Y. Acad. Sci.*, **72**, 522 (1959).
19 X. M. Aubert, *Rev. Ferment*, **10**, 51 (1955).
20 F. D. Carlson and A. Siger, *J. Gen. Physiol.*, **44**, 33 (1960).
21 A. G. Szent-Györgyi and G. Prior, *J. Mol. Biol.*, **15**, 515 (1966).
22 C. Moos and E. Eisenberg, *Biochim. Biophys. Acta*, **223**, 221 (1970).
23 T. Tokiwa, T. Shimada and Y. Tonomura, *J. Biochem.*, **61**, 108 (1967).
24 M. Bárány, A. F. Tucci and T. E. Conover, *J. Mol. Biol.*, **19**, 483 (1966).

25　S. Tokura, M. Kasai and Y. Tonomura, unpublished.
26　A. Martonosi, M. A. Gouvea and J. Gergely, *J. Biol. Chem.*, **235**, 1700 (1960).
27　F. Oosawa, S. Asakura and T. Ooi, *Progr. Theor. Phys. (Kyoto), Suppl.*, **17**, 14 (1961).
28　A. F. Huxley, *Progr. Biophys. Biophys. Chem.*, **7**, 255 (1957).
29　R. J. Podolsky, A. C. Nolan and S. A. Zaveler, *Proc. Natl. Acad. Sci. U. S.*, **64**, 504 (1969).
30　M. M. Civan and R. J. Podolsky, *J. Physiol.*, **184**, 511 (1966).
30a　A. F. Huxley and R. M. Simmons, *Nature*, **233**, 533 (1971).
31　Y. Tonomura, K. Yagi, S. Kubo and S. Kitagawa, *J. Res. Inst. Catalysis, Hokkaido Univ.*, **9**, 256 (1961).
32　R. E. Davies, *Nature*, **199**, 1068 (1963).
33　Y. Tonomura, T. Kanazawa and K. Sekiya, *Annu. Rep. Sci. Works, Fac. Sci., Osaka Univ.*, **12**, 1 (1964).
34　Y. Tonomura, *J. Res. Inst. Catalysis, Hokkaido Univ.*, **16**, 323 (1968).
35　H. E. Huxley and J. Hanson, *Symp. Soc. Exp. Biol.*, **9**, 228 (1955).
36　H. E. Huxley, in "The Cell," ed. by J. Brachet and A. E. Mirsky, Academic Press, New York, Vol. 4, p. 365 (1960).
37　F. A. Pepe, *J. Cell Biol.*, **28**, 505 (1966).
38　F. A. Pepe, *J. Mol. Biol.*, **27**, 227 (1967).
39　F. A. Pepe, *Int. Rev. Cytol.*, **24**, 193 (1968).
40　D. C. Hellam and R. J. Podolsky, *J. Physiol.*, **200**, 807 (1969).
41　R. J. Podolsky and L. E. Teichholz, *J. Physiol.*, **211**, 19 (1970).
42　H. Onishi, K. Miki, M. Kaneko and Y. Tonomura, unpublished.
43　L. Ouellet, K. J. Laidler and M. F. Morales, *Arch. Biochem. Biophys.*, **39**, 37 (1952).
44　T. Yasui, Y. Hashimoto and Y. Tonomura, *Arch. Biochem. Biophys.*, **87**, 55 (1960).
45　F. Morita and K. Yagi, *Biochem. Biophys. Res. Commun.*, **22**, 297 (1966).
46　F. Morita, *J. Biol. Chem.*, **242**, 4501 (1967).
47　F. Morita, *Biochim. Biophys. Acta*, **172**, 319 (1969).
48　K. Sekiya and Y. Tonomura, *J. Biochem.*, **61**, 787 (1967).
48a　F. Morita, in "Molecular Mechanisms of Enzyme Action," ed. by Y. Ogura, Y. Tonomura and T. Nakamura, University of Tokyo Press, Tokyo, p. 281 (1972).
48b　M. M. Werber, A. G. Szent-Györgyi and G. D. Fasman, *Biochemistry*, **11**, 2872 (1972).
49　J. C. Seidel, M. Chopek and J. Gergely, *Biochemistry*, **9**, 3265 (1970).
50　D. B. Stone, *Arch. Biochem. Biophys.*, **141**, 378 (1970).
51　K. Takahashi, T. Mori, H. Nakamura and Y. Tonomura, *J. Biochem.*, **57**, 637 (1965).

52 H. Nakamura, T. Mori and Y. Tonomura, *J. Biochem.*, **58**, 582 (1965).
53 P. C. J. Ward, C. Edwards and E. S. Benson, *Proc. Natl. Acad. Sci. U. S.*, **53**, 1377 (1965).
54 Y. Hayashi and Y. Tonomura, *J. Biochem.*, **63**, 101 (1968).
55 A. M. Gordon, A. F. Huxley and F. J. Julian, *J. Physiol.*, **184**, 170 (1966).
56 S. Ebashi, A. Kodama and F. Ebashi, *J. Biochem.*, **64**, 465 (1968).
57 Y. Tonomura, S. Watanabe and M. Morales, *Biochemistry*, **8**, 2171 (1969).
58 S. Ishiwata and S. Fujime, *J. Phys. Soc. Japan*, **30**, 303 (1971).
59 G. F. Elliott, J. Lowey and B. M. Millman, *Nature*, **206**, 1357 (1965).
60 H. E. Huxley and W. Brown, *J. Mol. Biol.*, **30**, 383 (1967).
61 H. E. Huxley, *J. Mol. Biol.*, **37**, 507 (1968).
61a M. Miller and R. T. Tregear, *in* "Contractility of Muscle Cells and Related Processes," ed. by R. J. Podolsky, Prentice-Hall, Engelwood Cliffs, N. J., p. 205 (1971).
62 J. F. Aronson and M. F. Morales, *Biochemistry*, **8**, 4517 (1969).
63 C. G. dos Remedios, R. G. C. Millikan and M. F. Morales, *J. Gen. Physiol.*, **59**, 103 (1972).
64 M. Bárány, K. Bárány and E. Gaetjens, *J. Biol. Chem.*, **246**, 3241 (1971).
65 M. K. Reedy, K. C. Holmes and R. T. Tregear, *Nature*, **207**, 1276 (1966).
66 K. Takahashi and T. Yasui, *J. Biochem.*, **62**, 131 (1967).
67 K. Takeuchi and Y. Tonomura, unpublished.
68 Y. Tonomura, Y. Hayashi and A. Inoue, *Cold Spring Harbor Symp. Quant. Biol.*, in press.
69 Y. Tonomura, H. Nakamura, N. Kinoshita, H. Onishi and M. Shigekawa, *J. Biochem.*, **66**, 599 (1969).
70 L. C. Gershman, P. Dreizen and A. Stracher, *Proc. Natl. Acad. Sci. U. S.*, **56**, 966 (1966).
71 D. W. Frederiksen and A. Holtzer, *Biochemistry*, **7**, 3935 (1968).
72 M. F. Morales, *Proc. Natl. Acad. Sci. U. S.*, **67**, 572 (1970).
73 G. Viniegra-Gonzalez and M. F. Morales, *Bioenergetics*, in press.
74 Y. Hayashi and Y. Tonomura, *J. Biochem.*, **68**, 665 (1970).
75 R. J. Goldacre and I. J. Lorch, *Nature*, **166**, 497 (1950).
76 R. J. Goldacre, *Int. Rev. Cytol.*, **1**, 135 (1952).
77 J. F. Danielli, *Symp. Soc. Exp. Biol.*, **6**, 1 (1952).
78 J. F. Danielli, *Proc. Roy. Soc.*, **B124**, 146 (1954).
79 L. J. Opit and J. S. Charnock, *Nature*, **208**, 471 (1965).
80 O. Jardetzky, *Nature*, **211**, 969 (1966).
81 P. C. Caldwell, *in* "Membranes and Ion Transport," ed. by E. E. Bittar, Wiley-Interscience, London, Vol. 1, p. 433 (1970).
82 W. L. Hubbell and H. M. McConnell, *Proc. Natl. Acad. Sci. U. S.*, **63**, 16 (1969).
82a D. G. Davis and G. Inesi, *Biochim. Biophys. Acta*, **241**, 1 (1971).

82b   L. I. Rothfield, "Structure and Function of Biological Membranes," Academic Press, New York (1971).
82c   D. L. D. Caspar and D. A. Kirschner, *Nature New Biology*, **231**, 46 (1971).
82d   S. J. Singer and G. L. Nicolson, *Science*, **175**, 720 (1971).
83   A. B. Pardee, *J. Biol. Chem.*, **241**, 5886 (1966).
84   C. F. Fox, J. R. Carter and E. P. Kennedy, *Proc. Natl. Acad. Sci. U. S.*, **57**, 698 (1967).
85   R. H. Wasserman, R. A. Corradino and A. N. Taylor, *J. Biol. Chem.*, **243**, 3978 (1968).
86   A. L. Lehninger, *Biochem. Biophys. Res. Commun.*, **42**, 312 (1971).
87   D. L. Oxender, J. R. Piperno and W. R. Penrose, *Federation Proc.*, **26**, 393 (1967).
88   Y. Anraku, *J. Biol. Chem.*, **243**, 3116, 3123, 3128 (1968).
89   A. B. Pardee, *Science*, **162**, 631 (1968).
90   A. Kotyk, *in* "Membranes, Structure and Function," ed. by J. R. Villanueva and F. Ponz, Academic Press, New York, FEBS, Vol. 20, p. 99 (1970).
91   H. R. Kaback, *Annu. Rev. Biochem.*, **39**, 561 (1970).
91a   E. C. C. Lin, *in* "Structure and Function of Biological Membranes," ed. by L. I. Rothfield, Academic Press, New York, p. 285 (1971).
92   R. L. Post, A. K. Sen and A. S. Rosenthal, *J. Biol. Chem.*, **240**, 1437 (1965).
93   A. J. Stone, *Biochim. Biophys. Acta*, **150**, 578 (1968).
94   R. W. Albers, G. J. Koval and C. J. Siegel, *Mol. Pharmacol.*, **4**, 324 (1968).
95   H. H. Ussing, *Physiol. Rev.*, **29**, 127 (1949).
96   R. D. Keynes and R. C. Swan, *J. Physiol.*, **147**, 591 (1959).
97   P. C. Caldwell, A. L. Hodgkin, R. D. Keynes and T. I. Shaw, *J. Physiol.*, **152**, 591 (1960).
98   P. de Weer, *J. Gen. Physiol.*, **56**, 583 (1970).
99   P. J. Garrahan and I. M. Glynn, *Nature*, **207**, 1098 (1965).
100   P. J. Garrahan and I. M. Glynn, *J. Physiol.*, **192**, 159 (1967).
101   I. M. Glynn, *J. Physiol.*, **136**, 148 (1957).
102   I. M. Glynn and U. Lüthi, *J. Gen. Physiol.*, **51**, 385s (1968).
103   I. M. Glynn, V. L. Lew and U. Lüthi, *J. Physiol.*, **207**, 371 (1970).
104   K. Nagai, G. E. Lindenmayer and A. Schwartz, *Arch. Biochem. Biophys.*, **139**, 252 (1970).
105   J. M. Graham and D. F. H. Wallach, *Biochim. Biophys. Acta*, **241**, 180 (1971).
106   H. Nakamura, H. Hori and T. Mitsui, *J. Biochem.*, **72**, 635 (1972).
107   L. E. Hokin, *J. Gen. Physiol.*, **54**, 327s (1969).
108   A. Schwartz, H. S. Bachelard and H. McIlwain, *Biochem. J.*, **84**, 626 (1962).
109   C. Ponnamperuma, C. Sagan and R. Mariner, *Nature*, **199**, 222 (1963).
109a M. Akaboshi, K. Kawai and A. Waki, *Biochim. Biophys. Acta*, **238**, 5 (1971).

109b M. Akaboshi and K. Kawai, *Biochim. Biophys. Acta*, **246**, 194 (1971).
110 G. Steinman, R. M. Lemmon and M. Calvin, *Proc. Natl. Acad. Sci. U.S.*, **52**, 27 (1964).
111 C. Sagan, *in* "The Origins of Prebiological Systems and of Their Molecular Matrices," ed. by S. W. Fox, Academic Press, New York (1965).
112 G. Schramm, H. Grötsch and W. Pollmann, *Angew. Chem.*, **73**, 619 (1961).
113 G. Schramm, *in* "The Origins of Prebiological Systems and of Their Molecular Matrices," ed. by S. W. Fox, Academic Press, New York (1965).
114 J. Rabinowitz, C. Chang and C. Ponnamperuma, *Nature*, **218**, 442 (1968).
115 A. Schwartz and C. Ponnamperuma, *Nature*, **218**, 443 (1968).
116 R. Lohrmann and L. E. Orgel, *Science*, **161**, 64 (1968).
117 J. Oró, *Biochem. Biophys. Res. Commun.*, **2**, 407 (1960).
118 J. Oró and A. P. Kimball, *Arch. Biochem. Biophys.*, **94**, 217 (1961).
119 N. W. Gabel and C. Ponnamperuma, *Nature*, **216**, 453 (1967).
120 M. Calvin, cited by H. Baltscheffsky, *Acta Chem. Scand.*, **21**, 1973 (1967).
121 F. Lipmann, *in* "The Origins of Prebiological Systems and of Their Molecular Matrices," ed. by S. W. Fox, Academic Press, New York (1965).
122 H. Baltscheffsky, *Acta Chem. Scand.*, **21**, 1973 (1967).
123 H. Baltscheffsky, *Federation Eur. Biochem. Soc. Meeting*, Madrid, 1969.
124 F. M. Harold, *Bact. Rev.*, **30**, 772 (1966).
125 H. Baltscheffsky and L. V. von Stedingk, *Biochem. Biophys. Res. Commun.*, **22**, 277 (1966).
126 H. Baltscheffsky, L. V. von Stedingk, H. W. Heldt and M. Klingenberg, *Science*, **153**, 1120 (1966).
127 R. J. Guillory and R. R. Fisher, *Biochem. J.*, **129**, 471 (1972).

14
CONTRACTION, TRANSLOCATION, TRANSPORT AND PHOSPHORYLATION

In the previous chapters we have emphasised the parallels between the molecular mechanisms of muscle contraction and active transport of cations. The key reactions are conformational changes in the ATPase coupled with its phosphorylation of a carboxyl group by ATP and the subsequent dephosphorylation of the phosphorylated intermediate. To conclude this monograph, it is appropriate to review briefly how these conclusions can be applied to other biological energy transduction systems. We will consider two such systems—peptidyl translocation and phosphorylation in membrane systems, particularly in the SR membrane.

1. The G-Factor-linked Ribosomal GTPase

The physiological functions of the nucleoside triphosphatases are closely connected with their catalysis of the hydrolysis of nucleoside triphosphates, and myosin-ATPase, the Na^+-K^+-dependent and Ca^{2+}-Mg^{2+}-dependent

ATPases and the G-factor-linked ribosomal GTPase have all been studied extensively. We have already reviewed the properties and reaction mechanisms of myosin-ATPase and the two transport ATPases. In this section we will discuss the reaction mechanism of the G-factor-linked GTPase, which is highly specific to GTP[1], and is believed to be required for the translocation of peptidyl-tRNA from the aminoacyl (A) to the peptidyl (P) site on the ribosome[2,3].

According to Nishizuka and Lipmann[4,5], peptide synthesis involves four phases. In phase 1, peptidyl-tRNA occupies the P site on the ribosome, while the A site is empty. In phase 2, an aminoacyl-tRNA molecule attaches to the neighbouring free triplet on the messenger RNA and settles on the site A. A new peptide bond is then formed by peptidyl transfer to the free amino group of the amino acid on the incoming aminoacyl-tRNA. The tRNA on the P site, now free of peptidyl, is released in phase 3, leaving the new, longer peptidyl-tRNA on the 'wrong site,' A. The last phase involves translocation of this peptidyl-tRNA from the site A to site P. The peptidyl-tRNA on the P site is released by the addition of puromycin. During this process, the messenger RNA is carried one triplet forward together with the peptidyl-tRNA, and this movement is in some respects similar to the sliding of actin relative to myosin in muscle contraction, which is coupled to the hydrolysis of ATP. Hence, Nishizuka and Lipmann[4,5] have proposed that a G-factor-linked ribosomal GTPase is involved in the translocation.

It has recently been shown by S. S. Thach and R. E. Thach[6] that the messenger RNA is actually moved a distance of approximately three nucleotides in the 5'-direction relative to the ribosome during the translocation of peptidyl-tRNA from the A to P site. A conformational change of the ribosome associated with the GTPase reaction has been proposed as the motive force for translocation[4,7,8]. Quite recently, a conformational change in the ribosome associated with translocation has been indicated by a concomitant increase in hydrogen exchange rate[9] and a change in sedimentation constant[9,10]. Thus there is a remarkable parallel between muscle contraction and peptidyl translocation.

The G-factor[11,12] can be isolated from the supernatant of homogenates of *Escherichia coli* using DEAE-cellulose and gel filtration, and has recently been purified to a crystalline form[13,14]. A ribosome-dependent hydrolysis of GTP which is catalysed by the G-factor occurs in the absence of messenger RNA and aminoacyl-tRNA, *i.e.* in the absence of translocation. This reaction is therefore called an uncoupled GTPase reaction, and its mechanism has been studied by Kaziro et al.[15,16], Parmeggiani and Gottschalk[17], and Brot et al.[18] Kaziro and co-workers[15] showed that the dependence of its rate on GTP concentration follows the Michaelis-Menten equation, and that the GTPase is competitively inhibited by an analogue of GTP, in which the oxygen linking the β- and α-phosphoryls is replaced by -CH_2- (GMP-PCP). All the above researchers found that radioactivity from ^{3}H-GTP becomes bound to ribosomes in the presence of the G-factor, and that the GTP which binds initially is rapidly converted to GDP. The G-factor also associates with ribosomes in the presence of GTP to form a ternary complex. Bodley et al.[19] recently reported that, when the nucleotide level is higher than those of the ribosomes and the G-factor, there is a burst of hydrolysis approximately equivalent to the molar amount of the G-factor, as long as there is an excess of ribosomes over G-factor.

Fusidic acid[20] and related steroidal antibiotics inhibit bacterial protein synthesis by interfering with the ribosome-dependent activity of the G-factor[21]. Bodley et al.[22,23] have observed that fusidic acid facillitates the isolation of a complex between ribosome, G-factor and GDP, while Okura et al.[24] demonstrated by equilibrium dialysis and ultracentrifugal separation that a complex of fusidic acid with ribosome, G-factor and GDP is formed.

The following mechanism for the uncoupled hydrolysis of GTP by ribosome and G-factor is proposed to account for these observations:

$$\text{ribosome} + \text{G-factor} + \text{GTP} \rightleftharpoons \text{ribosome-G-factor-GTP}$$
$$\text{ribosome-G-factor-GTP} \longrightarrow \text{ribosome-G-factor-GDP (P}_i\text{)}$$
$$\text{ribosome-G-factor-GDP (P}_i\text{)} \underset{\text{fusidic acid}}{\overset{\text{inhibition by}}{\rightleftharpoons}} \text{ribosome} + \text{G-factor} + \text{GDP} + \text{P}_i.$$

2. The Intermediates of the Ribosomal GTPase Reaction

The reaction mechanism of the G-factor-linked ribosomal GTPase was clarified considerably by the studies described above. However, some important problems remain to be solved: (1) whether the complex of ribosome, G-factor and GTP is formed in an ordered or random sequence; (2) whether the ribosome-G-factor-GDP complex is an intermediate of the GTPase reaction, and especially (3) whether a phosphorylated intermediate is involved in this enzymatic reaction.

To investigate these questions, we[25] first made kinetic measurements of

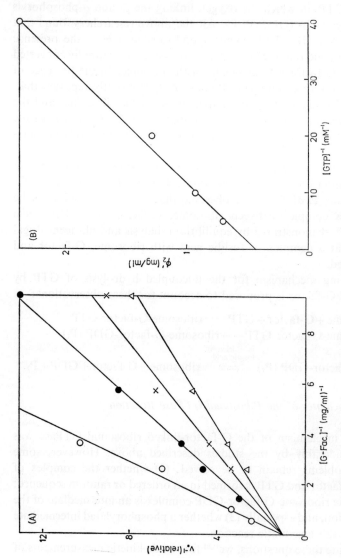

Fig. 1. Dependence of the ribosomal GTPase rate, v_0, on the concentration of GTP and G-factor. 11.25mg/ml ribosome, pH 7.0, 25°C. A: Double reciprocal plot of v_0 against G-factor concentrations from 0.175 to 1.75mg/ml. Concentrations of GTP: ○, 0.025; ●, 0.05; ×, 0.1; △, 0.2mM. $v_0^{-1} = A^{-1}\left\{1 + \frac{\phi'_2}{[\text{G-factor}]}\right\}$, where A^{-1} and ϕ'_2 are the intercept and slope, respectively. B: Dependence of the slope ϕ'_2, calculated from (A) on the GTP concentration. $\phi'_2 = \phi_2\left\{1 + \frac{\phi_1}{[\text{GTP}]}\right\}$.

the uncoupled GTPase reaction at steady-state. Reaction was initiated by adding GTP, was quenched with TCA at appropriate intervals, and the amount of liberated P_i was determined. Figure 1A shows the double reciprocal plot of the GTPase rate, v_o, *versus* the G-factor concentration at various GTP concentrations. At very high concentrations of G-factor, v_o becomes independent of GTP concentration. When the slope, ϕ'_2, is replotted against the reciprocal of the GTP concentration, a straight line is again obtained (Fig. 1B). These results are easily accounted for by the following scheme

$$R + S \rightleftharpoons R \cdot S$$
$$K_s$$
$$R \cdot S + G \rightleftharpoons G \cdot R \cdot S$$
$$K_g$$
$$G \cdot R \cdot S \rightarrow \cdots \rightarrow R + G + P_i + GDP,$$

where R, S and G are ribosome, GTP and G-factor, respectively. Therefore, v_o is given by

$$v_o = \frac{V_o}{1 + \frac{\phi_2}{[G]}\left(1 + \frac{\phi_1}{[S]}\right)}.$$

From the results shown in Fig. 1B, the values of ϕ_2 and ϕ_1 at pH 7.0 and 25°C are estimated to be 0.6mg/ml and 0.1mM, respectively. Our conclusion from kinetic studies, that the GTP and G-factor bind to ribosomes in an ordered sequence, is consistent with the results from binding experiments by other investigators[16-18,26], which indicate that the G-factor can bind to ribosomes only in the presence of GTP.

Figure 2 plots the reciprocal of the rate, v_o^{-1}, against the GDP concentration at various GTP concentrations. Experiments similar to those shown in Fig. 2 at different G-factor concentrations demonstrated that GDP complexes the ribosome competitively with GTP, and that the dissociation constants (K_g) for the binding of the R·GDP and R·GTP complexes with the G-factor are very similar:

$$R + GDP \rightleftharpoons R \cdot GDP$$
$$K_{GDP}$$
$$R \cdot GDP + G \rightleftharpoons G \cdot R \cdot GDP.$$
$$K_g$$

K_{GDP} is 1/10 of ϕ_1, and so GDP binds strongly to ribosomes competitively with GTP.

Secondly, we measured the initial rate of liberation of P_i. As shown in

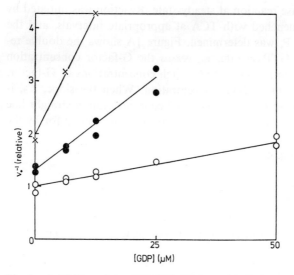

Fig. 2. Inhibition of the GTPase by GDP. 11.25mg/ml ribosome, 0.35mg/ml G-factor, pH 7.0, 25°C. Concentrations of GTP: ×, 0.025; ●, 0.05; ○, 0.2mM.

Fig. 3, there is a short lag period (about 0.15 sec), which is followed by the initial burst. At concentrations of GTP ranging from 1.7 to 68μM and of G-factor ranging from 0.0375 to 0.225mg/ml, the amount of the initial burst of P_i is proportional to the GTPase rate at steady-state, v_o, while the lag period is almost independent of the concentrations of GTP and G-factor. These results suggest that the binding equilibrium $R+G+S \rightleftharpoons$ G·R·S is very rapid, and that the G·R·S thereby formed is converted to an intermediate X, which does not liberate P_i with TCA. X is then transformed into a final intermediate Y, which does liberate P_i, at least with TCA.* The Lineweaver-Burk plot of the GTPase reaction shows a downward deviation at high concentrations of GTP (Fig. 4). The acceleration by high concentrations of GTP is due to the following two effects: (1) the value of ϕ_2 becomes very small, and thus the G-factor binds stoichiometrically to ribosomes in the presence of high concentrations of GTP; (2) the value of V_o in the range of high concentrations of GTP becomes to be about 1.5 fold of that in the range of low concentrations of GTP. These

* See Refs. 27–29 for analysis of the initial lag phase and the initial burst of liberation of products in enzyme reactions.

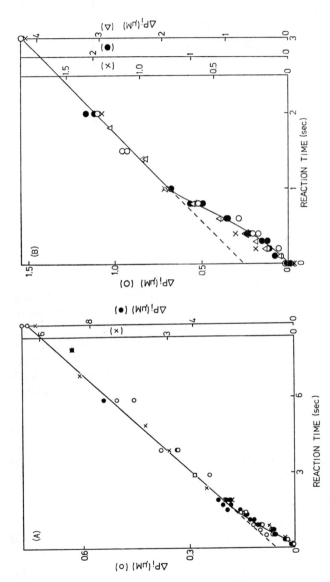

Fig. 3. Initial rate of liberation of P_i by the ribosomal GTPase reaction. The amount of P_i liberated was measured after quenching the reaction with TCA. All measurements were made at pH 7.0. A: Dependence on the concentration of GTP. ○, 1.7 μM GTP, 22.5 mg/ml ribosome, 0.05mg/ml G-factor, 25°C; ×, 34 μM GTP, 11.25mg/ml ribosome, 0.075mg/ml G-factor, 23°C; ●, 68 μM GTP, 11.25mg/ml ribosome, 0.075mg/ml G-factor. B: Dependence on the concentration of G-factor. 9.3 μM GTP, 10mg/ml ribosome, 25°C. ○, 0.056; ×, 0.075; ●, 0.112; △, 0.224mg/ml G-factor.

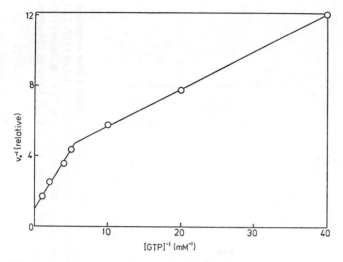

Fig. 4. Acceleration of the GTPase reaction by high concentrations of GTP. 11.25 mg/ml ribosome, 0.282mg/ml G-factor, pH 7.0, 25°C. The reciprocal of the rate at steady-state, v_o, is plotted against the reciprocal of the GTP concentration.

two phenomena can easily be interpreted by assuming that high concentrations of GTP accelerate the transition of G·R·S to X. These results are quite similar to the acceleration by high concentrations of ATP observed with the contractile and transport ATPases (chapters 3, 11 and 12).

The experimental data obtained so far are consistent with the assumption that the reaction mechanism of GTPase parallels those of the ATPases, especially myosin-ATPase. Hence, the mechanism of the GTPase reaction can be represented by

$$R + S \underset{0.1\text{mM}}{\rightleftharpoons} R \cdot S \tag{1}$$

$$R \cdot S + G \underset{1.8\text{mg/ml}}{\rightleftharpoons} G \cdot R_1 \cdot S \tag{2}$$

$$G \cdot R_1 \cdot S \xrightarrow{\overset{\text{acceleration by high conc. of S}}{\downarrow}}_{1.9 \to \gg 1.2 \text{ sec}^{-1}} G \cdot R_2 \cdot S \tag{3}$$

$$G \cdot R_2 \cdot S \xrightarrow{5.1 \text{ sec}^{-1}} G \cdot R_P{}^{GDP} \tag{4}$$

$$G \cdot R_P{}^{GDP} \xrightarrow{1.2 \text{ sec}^{-1}} R + G + P_i + GDP, \tag{5}$$

where $G \cdot R_2 \cdot S$ and $G \cdot R_P{}^{GDP}$ correspond respectively to the intermediates X and Y mentioned above, and the conversion of $G \cdot R_1 \cdot S$ to $G \cdot R_2 \cdot S$ is accelerated by a high concentration of GTP. Furthermore, step (3) is assumed to be irreversible, since the enzyme does not catalyse the GTP-GDP exchange reaction[4]. The dissociation constants and the rate constants of the elementary steps are calculated from the experimental results, and their values at pH 7.0 and 25°C are given in the above reaction scheme. This reaction mechanism has been supported by analog computer simulations of the time-courses of P_i-liberation.

The nature of the binding of GDP and P in $G \cdot R_P{}^{GDP}$ remains to be elucidated, and attempts to isolate a TCA-stable phosphorylated protein, such as that found with transport ATPases (chapter 11 and 12), have been unsuccessful. With some preparations of ribosomes we have found that addition of excess GDP during the initial phase of the reaction causes a transient decrease in the amount of TCA-P_i liberated, suggesting that a TCA-labile phosphorylated intermediate is formed in the GTPase reaction. This intermediate does not contain bound GDP, and is a high energy phosphate compound which converts GDP to GTP. However, some other preparations of ribosomes have exhibited no such decrease in the amount of P_i liberated when a large amount of GDP was added during the initial phase.

Table I summarises the reaction mechanisms of the four nucleoside triphosphatases reviewed in this monograph, and demonstrates the close similarity between the reaction mechanisms of the GTPase and the ATPases, particularly myosin-ATPase. The investigations described above concern the uncoupled GTPase reaction, and they cannot decide which step in the GTPase reaction accompanies a conformational change in the ribosome[9,10]. This problem may be elucidated in the near future by kinetic studies on the effect of messenger RNA and peptidyl-tRNA on the formation and decomposition of intermediates of G-factor-dependent ribosomal GTPase reaction. Another method is measurement of a change in light-scattering intensity due to the conformational change of the ribosome, using a rapid flow method, concomitantly with the kinetics of liberation of P_i by the GTPase reaction. However, at the present stage it is tempting to conclude that the translocation of peptidyl-tRNA is coupled with the decomposition of the $G \cdot R \sim P$ complex, just as the translocation

TABLE I. Functions and Reaction Mechanisms of Nucleoside Triphosphatases

	ES→EP	EP	EP→E+P$_i$	$\frac{[E\sim P]}{[\Sigma EP]}$	Translocation
Ca^{2+}-Mg^{2+}-dependent ATPase of sarcoplasmic reticulum	Ca^{2+}-dependent, accelerated by ATP	TCA-stable, E$\sim$P, E$\sim$P+ADP$\rightleftharpoons$E+ATP	Mg^{2+}-dependent, accelerated by phospholipid	1	Carrier site, 80Å
Na$^+$K$^+$-dependent ATPase of membrane	Na$^+$-dependent, accelerated by ATP	TCA-stable, E$\stackrel{\text{ADP}}{\underset{P}{\sim}}$, E$\sim$P E$\sim$P+ADP$\rightleftharpoons$E$\stackrel{\text{ADP}}{\underset{P}{\sim}}\rightleftharpoons$E+ATP	K$^+$-dependent, accelerated by phospholipid	1	Carrier site, 80Å
Ribosomal GTPase	G-Factor-dependent, accelerated by GTP	TCA-unstable, E$\sim$P (?), E$\stackrel{\text{(GDP)}}{\underset{P}{\cdot\cdot}}$ E$\sim$P+GDP$\rightleftharpoons$E+GTP (?)		(?)	Messenger RNA + peptidyl-tRNA 20Å
Myosin-ATPase	Accelerated by ATP	TCA-unstable, E$\stackrel{\text{ADP}}{\underset{P}{\sim}}$, E$\stackrel{\text{ADP}}{\underset{P}{\cdot\cdot}}$ E$\stackrel{\text{ADP}}{\underset{*P}{\sim}}$+ATP$\rightleftharpoons$E$\stackrel{\text{ADP}}{\underset{P}{\sim}}$+AT*P	Accelerated by F-actin	≤0.1~0.2	F-Actin 100Å

of F-actin is coupled with the decomposition of phosphoryl myosin (chapter 13).* Investigations on the uncoupled reaction also provide neither information about the manner in which the GTPase reaction is initiated when site A is occupied by peptidyl-tRNA and site P is empty, nor how the reaction is terminated when the messenger RNA is carried one triplet forwards (about 20Å) together with peptidyl-tRNA, nor which components of the G-factor-linked ribosomal GTPase correspond to actin and myosin in the contractile ATPase. Nomura[31] recently managed to reconstitute ribosomes by self-assembly of their isolated components. Kischa et al.[32] obtained evidence that the G-factor-dependent GTPase activity can be reconstituted from intact 30S subunits and '50S cores,' prepared according to Meselson et al.[33], by the addition of a 50S acidic ribosomal protein, isolated according to Möller et al.[34] Suzuka[35] recently showed that nitrotroponylation of 50S subunits inhibits the G-factor-dependent GTPase activity and the capacity of binding of GTP to ribosomes. Hamel and Nakamoto[36] isolated a protein factor from 50S ribosomes, which is essential to the G-factor-dependent GTPase activity and the ability to bind GTP, by precipitation with ethanol at 0°C in the presence of 1M NH_4Cl. Consequently, we may hope that the role of each ribosome component in the GTPase reaction coupled with peptidyl translocation, and the relationship between the components of ribosomes and the components of the myosin-F-actin system (heavy and light chains of myosin-actin-tropomyosin-three troponin components), will be established in the near future.

3. Phosphorylation of Transport ATPase Coupled with Cation Translocation

The translocation of carrier protein by transport ATPase is a cyclical process, while the translocation of F-actin by the contractile ATPase occurs unidirectionally (chapter 13). Hence, reversing the cycle of translocation of carrier protein in the transport ATPase should result in formation of

* Kaziro[30] have reported recently that the change of peptidyl-tRNA from the puromycin insensitive state to the sensitive state can be brought about by adding GMP-PCP in the presence of G-factor. This suggests that in the G-factor-dependent ribosomal GTPase reaction, the unstable, energy-rich β form, which is induced by the hydrolysis of GTP, is stabilised by secondary bonds as in the case of transport ATPase (chapter 13, section 4), and that the secondary bonds are broken by the binding of GTP and the spontaneous transition from the β- to the α- form accompanies the translocation. However, it remains possible that the 'translocation' on adding GMP-PCP is an unphysiological process, since it is measured by the reaction with puromycin and its extent and the rate are smaller than those observed on adding GTP.

ATP from ADP and P_i, while there is much less likelihood of reversing the contractile ATPase reaction by stretching. Stretching of a muscle pretreated with fluorodinitrobenzene in a maintained contraction definitely does not resynthesise ATP[37], showing that the reaction cycle in the contractile system cannot be reversed in this way (chapter 1). In contrast, the synthesis of ATP by reversal of the Na^+-pump has been demonstrated by Glynn et al.[38-41] However, the efficiency of ATP-synthesis by reversal of this pump was rather low.

More recently, Makinose and Hasselbach[42] made the important observation that, under conditions in which a fast ADP- and P_i-dependent release of Ca^{2+} from the Ca^{2+}-loaded SR occurs[43], the net outward movement of two moles of Ca^{2+} through the SR membrane is stoichiometrically related to a net formation of one mole of ATP. Similar findings were also reported by Panet and Selinger[43a]. These results suggest that the ATPase reaction of the SR is reversible and the ATPase is phosphorylated by P_i coupled with cation translocation, since ATP is formed from ADP and the phosphorylated intermediate, EP, of the SR-ATPase (chapter 11, section 4).

Makinose and Hasselbach[42] loaded the SR with Ca^{2+} ions by incubating it with Ca^{2+} in the presence of ATP. When the Ca^{2+}-loaded SR prepared by this method was used, the quantitative analysis of the results was difficult because of the presence of ATPase reaction products and of vagueness in the activity of Ca^{2+} ions in the membrane. Therefore, we[44] loaded the SR with Ca^{2+} ions by simple preincubation of a solution containing various concentrations of $CaCl_2$ and 0.1M KCl at pH 7.0 and 0°C for several hours. As shown in Fig. 5, the amount of Ca^{2+} loaded in the SR increased gradually to the saturation value. The half-saturation time was about 10 min, and the saturation values were 34 and 75 moles/10^6g protein when the external Ca^{2+} concentrations were 20 and 80mM, respectively. Thus, the saturation value was not proportional to the external Ca^{2+} concentration. This may be due to the change in the membrane potential with Ca^{2+}-loading or the binding of Ca^{2+} ions to inner structures of the SR. In the following, we will assume that the electrochemical potential of Ca^{2+} ions inside the membrane becomes equal to that outside the membrane after a sufficiently long preincubation time.

The Ca^{2+}-loaded SR was added into 1mM $CaCl_2$, 10mM EGTA, 20mM $MgCl_2$, 5mM $^{32}P_i$ and 90mM KCl at pH 7.0 and 20°C, and the reaction was stopped after appropriate intervals by adding perchloric acid. The amount of P incorporated in SR was measured as described in chapter 11. Com-

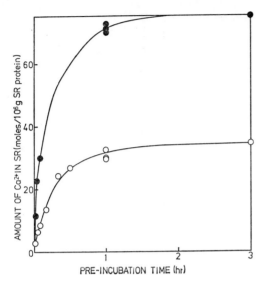

Fig. 5. Time-courses of Ca^{2+}-loading of SR. The SR (20mg/ml) was incubated in solution of total volume 0.025ml, containing 20mM (O) or 80mM (●) $^{45}CaCl_2$ and 0.1M KCl at pH 7.0 and 0°C. At appropriate intervals the SR solution was diluted to 5ml with solution containing 20mM (O) or 80mM (●) $CaCl_2$ and 0.1M KCl, filtrated through a Millipore filter, and the radioactivities of the SR were measured.

mercial preparations of $^{32}P_i$ were contaminated with minute amounts of impurities, which heavily interfered with the phosphorylation experiments since it was difficult to separate them from the SR. Therefore, before using $^{32}P_i$ was boiled in 1N HCl for 6 hr, and purified throughly by column chromatography on Dowex-1. As shown in Fig. 6, incorporation of P into the Ca^{2+}-loaded SR occurred rapidly and amounted to 1.7 moles/10^6g protein when the SR which had been preincubated with 20 mM $CaCl_2$ for 1.5 hr was used. When Ca^{2+}-unloaded SR was used, the amount of P incorporated after adding the SR to 20mM $MgCl_2$, 5mM $^{32}P_i$ and 10mM EGTA was only 0.5 mole/10^6g protein.

Since the phosphorylation of the SR by P_i could be measured easily, the effects of various factors on the phosphorylation reaction were examined[45]. It was found that phosphorylation under the standard conditions occurred rapidly after a very short lag phase, and the half-saturation time was about 0.2 sec. In the following experiments, the amount of

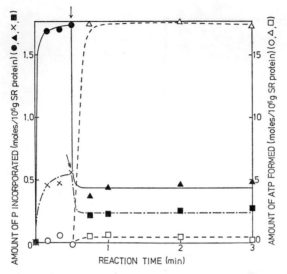

Fig. 6. Incorporation of P into SR-ATPase. ●, ▲, ○, △, to load the SR with Ca^{2+}, the SR (27mg/ml) was incubated with 20mM $CaCl_2$ at pH 7.0 and 0°C for 1.5 hr. The incorporation of P into the SR was started by adding 1.35mg/ml of Ca^{2+}-loaded SR into 1mM $CaCl_2$, 10mM EGTA, 20mM $MgCl_2$, 90mM KCl and 5mM$^{32}P_i$ at pH 7.0 and 20°C. ●, ▲, amounts of P incorporated into SR before and after adding 0.3mM ADP (↓). ○, △, amounts of ATP formed before and after adding 0.3mM ADP (↓). ×, ■, □, SR unloaded with Ca^{2+}. The incorporation of P into the SR was measured in 10mM EGTA, 20mM $MgCl_2$, 90mM KCl and 5mM $^{32}P_i$. ×, ■, amounts of P incorporated to SR before and after adding 0.3mM ADP (↓). □, amount of ATP formed after adding 0.3 mM ADP (↓).

P-incorporation in the steady-state was measured by stopping the reaction 10 sec after starting it. The amount of P incorporated increased with increase in the time of preincubation with Ca^{2+} ions, and reached the saturation value after about 2-hr preincubation. The amount of P incorporated also increased with increase in the concentration of Ca^{2+} ions in the preincubation medium. When the SR was preincubated with 80mM $CaCl_2$ for 2 hr, the amount of P incorporated under the standard conditions reached 4 moles/10^6g protein, which was almost equal to the amount of phosphorylation with ATP, 5–7 moles/10^6g (see chapter 11, section 4).

Figure 7 shows the dependence of the amount of P incorporated into the Ca^{2+}-loaded SR on the concentration of external free Ca^{2+} ions in the

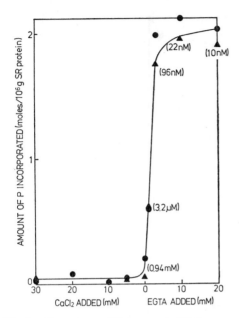

Fig. 7. Dependence of the amount of P incorporated into SR on the concentration of external Ca^{2+} ions. To load the SR with Ca^{2+}, the SR (16.5mg/ml) was incubated with 20mM $CaCl_2$ at pH 7.0 and 0°C for 4.5 (●) or 2.5 (▲) hr. The incorporation of P into the SR was started by adding 0.825mg/ml of Ca^{2+}-loaded SR into 1mM $CaCl_2$, various concentrations of EGTA, 20mM $MgCl_2$, 0.1M KCl and 5mM $^{32}P_i$ and pH 7.0 and 20°C. Values in parentheses indicate the concentrations of external free Ca^{2+} ions, which were calculated by assuming 2.1×10^{-7} M as the dissociation constant of Ca-EGTA complex.

phosphorylation medium in the presence of 5mM $^{32}P_i$ and 20mM $MgCl_2$. This figure clearly indicates that the phosphorylation with P_i occurred only when the concentration of free Ca^{2+} ions outside the membrane was decreased by adding EGTA. Thus, the Ca^{2+} gradient across the membrane is necessary to phosphorylate the SR with P_i. Figure 8 shows the relation between the amount of P incorporated and the Ca^{2+} gradient across the membrane. The electrochemical activity of Ca^{2+} ions inside the membrane ($[Ca^i]$) was assumed to be equal to that of the preincubation medium, as mentioned above, and the activity of Ca^{2+} ions outside the membrane ($[Ca^o]$) was estimated by adopting an appropriate value as the association constant of Ca^{2+} ions with EGTA. As shown in the inserted figure, a straight line was obtained when the reciprocal of the amount of P in-

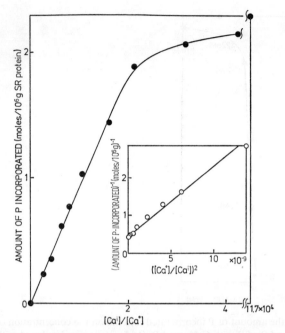

Fig. 8. Dependence of amount of P incorporated on Ca^{2+} gradient. Experimental conditions same as for Fig. 7, except that the preincubated time was 3 hr. The activity of Ca^{2+} ions inside the membrane, $[Ca^i]$, was assumed to be equal to that outside the membrane in the preincubation medium. The concentration of external Ca^{2+} ions $[Ca^o]$) was calculated as described in the legend for Fig. 7.

corporated was replotted against the square of the Ca^{2+} gradient ($[Ca^o]/[Ca^i]$).

When the reciprocal of the amount of P incorporated was plotted against the reciprocal of the concentration of Mg^{2+} ions, a straight line was obtained. The dependence of the amount of P-incorporation on the concentration of Mg^{2+} ions in the phosphorylation medium was unaffected by changing the concentration of Mg^{2+} ions in the preincubation medium over a range of 0 to 20mM. Since we previously showed that the SR membrane is permeable to Mg^{2+} ions (chapter 11, section 5), this result indicates that the concentration of Mg^{2+} ions, not the gradient across the membrane, determines the amount of P incorporated in the steady-state. The dependence of the amount of P-incorporation ($[EP]$) on the reciprocal of the concentration of Ca^{2+} outside the membrane ($[Ca^o]$) was shifted by changing the concentration of Mg^{2+}, and the dependence was given by $\frac{[EP]}{\varepsilon}=1/\left(1+\phi'\frac{[Ca^\circ]^2}{[Mg]}\right)$, where ε is the total concentration of the phosphorylation site and ϕ' is a constant dependent on the concentrations of P_i and Ca^{2+} inside the membrane. Furthermore, it must be added that our transient kinetic analyses of the time-course of P-incorporation suggested that during the initial phase not only a Ca^{2+} gradient but also a Mg^{2+} gradient are important factors which determine the amount of phosphorylation.

The reciprocal of the amount of P incorporated was linearly dependent on the reciprocal of the concentration of P_i. Thus, the dependences of [EP] in the steady-state on the Ca^{2+} gradient and the concentrations of Mg^{2+} ions and P_i in the phosphorylation medium were given by

$$\frac{[EP]}{\varepsilon} = \frac{1}{1+\phi\left(\frac{[Ca^\circ]}{[Ca^i]}\right)^2 \frac{1}{[Mg][P_i]}}.$$

Under our experimental conditions the value of ϕ was calculated to be about 2×10^5 M^2. Thus, the Ca^{2+} gradient from the inside to the outside of the membrane necessary for half-saturation of the phosphorylation was about 4×10^4 in the presence of 5mM P_i and 20mM $MgCl_2$. Under these conditions the free energy obtained by transporting 2 moles of Ca^{2+} ions from the inside to the outside of the membrane was estimated to be about 12 kcal.

Figure 9 shows the pH-stability curves of phosphorylated intermediates, which had been denatured by adding perchloric acid. Thus, at various pH values the stability of the intermediate formed by the reaction with P_i, coupled with the Ca^{2+} gradient, was equal to that of the intermediate formed by the reaction with ATP. The decomposition of acid-denatured EP formed with the reaction of P_i was accelerated by hydroxylamine in the same way as that of EP formed by reacting with ATP. These results suggest that EP is an intermediate of an acyl-phosphate type. Furthermore, the phosphorylation of the Ca^{2+}-loaded SR with $^{32}P_i$ was competitively inhibited by ATP.

The distribution of ^{32}P was measured on the SDS-gel electrophoretogram of EP. The peak of ^{32}P of the intermediate formed by the reaction with $^{32}P_i$, coupled with a Ca^{2+} gradient, coincided with that formed by the reaction with AT^{32}P. The peak of radioactivity was equal to the major protein component. The main component accounted for 65% of the total protein, and its molecular weight was estimated to be 10.5×10^4, which was equal to the value reported for the ATPase protein (cf. chapter 11,

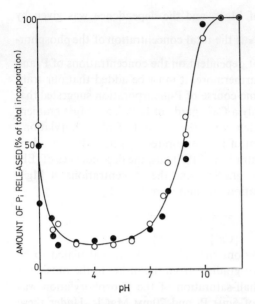

Fig. 9. pH-dependence of stability of phosphorylated intermediate after denaturation with perchloric acid. Phosphorylated intermediate was prepared by SR reacting with AT^{32}P (●) or by Ca^{2+}-loaded SR reacting with ^{32}P$_i$ under standard conditions (○). Liberation of ^{32}P$_i$ from the intermediate, denatured by 4% perchloric acid was measured in solution containing 25mM pH buffer after incubation at 20°C for 1 hr.

section 8). Thus, it is evident that the ATPase is phosphorylated with P$_i$ at the same site as that of phosphorylation with ATP.

When the Ca^{2+}-loaded SR, which was prepared by incubating SR with 20mM CaCl$_2$ for 1.5 hr, was added into solution containing 1mM CaCl$_2$, 10mM EGTA, 20mM MgCl$_2$ and 5mM ^{32}P$_i$, incorporation of P into the SR occurred rapidly and amounted to 1.7 moles/10^{6}g protein, as shown in Fig. 6. When 0.3mM ADP was added 30 sec after starting the phosphorylation experiment, the amount of P incorporated was reduced rapidly to 0.45 mole/10^{6}g protein. At appropriate intervals after the addition of 0.3mM ADP, the amount of AT^{32}P formed was measured. As shown in Fig. 6, 17.5 moles of ATP per 10^{6}g protein was formed within 15 sec after adding ADP. The amount of ATP synthesised was exactly 1/2 of the amount of Ca^{2+} ions loaded during the preincubation, 34 moles/10^{6}g (cf. Fig. 5), as was already reported by Makinose and Hasselbach[42] and

Panet and Selinger[43a]. On the other hand, when the Ca^{2+}-unloaded SR was used, only about 0.5 mole of ATP per 10^6g was formed.

These results established the formation of a phosphorylated intermediate of an energy-rich type coupled with cation translocation, and indicated the following mechanism of the phosphorylation reaction, on the basis of the molecular mechanism of the ATPase reaction described in chapter 11. When Mg^{2+} ions and P_i are added externally to Ca^{2+}-loaded SR, EP is formed coupled with translocation of Mg^{2+} ions from the outside to the inside of the SR membrane. The Mg^{2+} ions on EP are replaced by two Ca^{2+} ions inside the membrane, and the EP is stabilised when the Ca^{2+} gradient across the membrane is high. The EP is further stabilised by its binding with Mg^{2+} ions. Then, when ADP is added externally to EP, 1 mole of ATP is formed coupled with translocation of two Ca^{2+} ions on cation sites of the enzyme from the inside to the outside of the membrane. The formation of ATP continues until ion gradients decrease to critical levels.

It should be added that a carboxyl group of the SR is phosphorylated by ATP, as described above and also in chapter 11, section 4, and that the Ca^{2+}-Mg^{2+}-dependent ATPase of the SR is inhibited when specific groups are blocked by SH reagents[46]. These results might suggest that an acyl-thiol bond is formed as an intermediate for formation of a phosphorylated intermediate of an energy-rich type. However, during the SR-ATPase reaction ^{18}O was not incorporated from $H_2^{18}O$ into the phosphorylated intermediate of an acyl-phosphate type, thus excluding the possibility of the above mechanism[47]. However, the time-course of P-incorporation showed a lag phase when the reaction was started by adding the Ca^{2+}-loaded SR to solution containing EGTA, Mg^{2+} and P_i, while it showed no lag phase, when the reaction was started by adding P_i to the Ca^{2+}-loaded SR which had been incubated with EGTA and Mg^{2+}. This suggests that the phosphorylation reaction occurs via a non-phosphorylated energised state, which is formed in the SR coupled with a Ca^{2+} gradient across the membranes.

Another important problem is the physiological function of Ca^{2+}-release by the reversal of the ATPase reaction:

$$E^o + Mg + P_i + 2Ca^i \rightleftharpoons EP^i \cdot Ca_2^i + Mg,$$

$$EP^i \cdot Ca_2^i \rightleftharpoons EP^o \cdot Ca_2^o \underset{\underset{EP^o + 2Ca^o \xrightleftharpoons{Mg^o} EP^o \cdot Mg^o + 2Ca^o}{}}{\overset{ADP^o}{\rightleftharpoons}} \begin{matrix} E^o \cdot ATP^o \cdot Ca_2^o \longrightarrow E^o + ATP^o + 2Ca^o \\ \end{matrix}$$

where superscripts o and i indicate the outside and the inside of the membrane of the SR, respectively. We[45] showed that, when ADP is added to a Ca^{2+}- loaded SR which has been phosphorylated with P_i, the amount of EP decreases very rapidly, *i.e.*, release of Ca^{2+} ions occurs very rapidly. Furthermore, Ca^{2+}-release by the reversal of the ATPase reaction cannot occur in the presence of high concentrations of external Ca^{2+} ions, since it is inhibited by the binding of Ca° with E°. On the other hand, its rate is also markedly retarded in the complete absence of external Ca^{2+} ions, since the rate of steps, $EP^\circ \cdot Ca_2^\circ \rightarrow EP^\circ + 2Ca^\circ \rightarrow EP^\circ \cdot Mg^\circ + 2Ca^\circ$, becomes higher than that of steps, $EP^\circ \cdot Ca_2^\circ \dashrightarrow E^\circ + ATP + 2Ca^\circ$. In other words, Ca^{2+}-release by this mechanism is triggered by low concentrations of external Ca^{2+} ions, as is observed on Ca^{2+}-release from the SR in muscle fibres under the physiological conditions (see chapter 11, section 8). Thus, Ca^{2+}-release by the reversal of the ATPase reaction has properties as those actually occurring under the physiological conditions. Furthermore, the formation of ATP (accordingly the formation of CrP) accompanying with Ca^{2+}-release from the SR might be one of the causes for the apparently high enthalpy change of CrP-hydrolysis *in vivo*, calculated from the relation between the enthalpy change and $-\Delta$ CrP during contraction (see chapter 1, section 2).

4. Oxidative Phosphorylation

The three main pathways of biological phosphorylation are glycolysis, oxidative phosphorylation and photosynthetic phosphorylation. The mechanism of phosphorylation in glycolysis is well known, and that of photosynthetic phosphorylation is very similar to oxidative phosphorylation. For example, Melandri *et al.*[47a] have recently presented evidence for the exchangeability of coupling factors in photosynthetic and respiratory energy conversion, using a nonsulphur purple photosynthetic bacterium which can obtain energy for growth from either anaerobic photosynthetic phosphorylation or dark aerobic phosphorylation. Therefore, in this section we will briefly discuss the molecular mechanism of oxidative phosphorylation,* emphasising its relationship to the mechanisms of the contractile and transport ATPases.

Mitchell *et al.*[48,52-54] have proposed that the ATPase present on the membranes of mitochondria and chloroplasts has vectorial properties, and that the energy source which it uses to produce ATP from ADP and P_i is derived from the asymmetrical distribution of cations—particularly protons—inside and outside the membrane, caused by the electron flow.

* Refs. *48-51a* give general reviews of the mechanism of oxidative phosphorylation.

The most dramatic evidence for the Mitchell's hypothesis was obtained by Jagendorf and Uribe[55], who demonstrated ATP-formation caused by acid-base transition of chloroplasts. Quite recently, Hinkle et al.[55a] have reported that vesicles formed from cytochrome oxidase and phospholipids catalyse the oxidation of reduced cytochrome c in a manner similar to that of intact mitochondria and a respiration-dependent release of protons and uptake of K^+ ions occur in the presence of valinomycin and a reductant for cytochrome c. Furthermore, the facts that oligomycin inhibits both oxidative phosphorylation and the transport ATPase (chapter 12, section 1), that the Na^+- and Ca^{2+}-pumps working backwards can synthesise ATP (section 3), that mitochondria can use energy from the hydrolysis of ATP to accumulate cations[56], and that mitochondria can synthesise ATP via a downhill movement of cations[57–59], make the relationship between the ATPase in mitochondria and chloroplasts and the transport ATPase extremely interesting.

However, we do not know how energy conserved in the form of an ion gradient can be linked to anhydro-bond formation. There is a remarkable resemblance between the mitochondrial membrane and actomyosin-ATPase[60]. Both systems not only engage in mechano-chemical coupling, but also share a number of characteristic biochemical properties. The most striking are the effect of 2,4-dinitrophenol, the bimodal sulphhydryl dependence (chapter 4, section 3), and the rather specific ^{18}O-exchange reaction between the aqueous medium and inorganic phosphate (chapter 3, section 8) which has not been observed with other phosphate-transferring enzyme systems. Boyer[61] has suggested that the energy coupling in oxi-

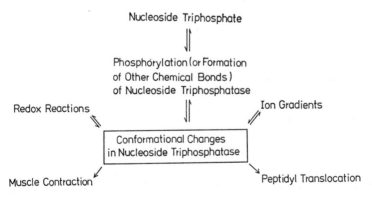

Fig. 10. A model of biological energy transduction.

dative and photosynthetic phosphorylation may be mediated by conformational changes of proteins. Hackenbrock[62] and Green and co-workers [49,63,64] have found that changes in the ultrastructure of the inner mitochondrial membrane are influenced by the metabolic state, although the changes have recently been shown to be induced by a minute amount of ADP even when oxidative phosphorylation is inhibited by oligomycin[64a].

On the other hand, Azzi et al.[64b] suggested a conformational change in mitochondria coupled with energy transduction, using a hydrophobic probe. Quite recently, Ryrie and Jagendorf[65] demonstrated the incorporation of a small amount of tritium from a labelled aqueous medium into the coupling factor (ATPase) of chloroplasts, which indicates a conformational change in the protein since the bound tritium is released only after denaturation of the protein. The tritiation of this protein does not occur in the absence of light, and is markedly inhibited by uncouplers, suggesting that ATP-formation in chloroplasts and mitochondria may be linked to a conformational change in the phosphorylating enzyme. The tritiation also occurred during acid-base transition and the ATPase reaction, and was decreased by inclusion of ADP and P_i[65a]. These results further strengthened the above suggestion. McCarty et al.[65b] reported that light markedly enhances the inhibition of photophosphorylation of chloroplasts by NEM. The properties of the inhibition were consistent with the concept that light causes a conformational change in the coupling factor (cf. Ref. 66). Cross and Boyer[67,68] have recently obtained results which suggest the formation of a protein bound phosphate intermediate of an acyl-phosphate type in the oxidative phosphorylation of mitochondria, as in the case of ATP-formation in the SR membrane coupled with cation gradients (section 3), although there remains a possibility that '^{32}P bound to protein' is not due to a phosphorylated intermediate but is derived from AT^{32}P molecules, newly synthesised by oxidative phosphorylation and bound tightly to proteins in mitochondria[69]. Therefore, the identification of the reaction intermediate in oxidative phosphorylation remains to be solved, and it seems probable that the phosphorylated intermediate in the oxidative phosphorylation occurs transiently and is TCA-labile, as in the case of myosin-ATPase, even if it exists. On the other hand, Horio and his co-workers[70] recently showed the existence of a phosphorylated intermediate in the reaction of the ATP-ADP exchange enzyme purified from chromatophores of *Rhodospirillum rubrum*.

It appears that a conformational change in the ATPase coupled with phosphorylation of a carboxyl group and its dephosphorylation, which is suggested in this monograph to play a vital role in muscle contraction, cation transport and peptidyl translocation, may also be the key process in oxidative and photosynthetic phosphorylation in mitochondria and chloroplasts (Fig. 10). We have already emphasised the importance of phosphorylation and transconformation of proteins in various physiological functions[71], and we hope that in the near future studies of muscle contraction and cation transport at the molecular level will enable us to elucidate the molecular basis, and hence establish the basic principles, of biological energy transduction.

REFERENCES

1 T. W. Conway and F. Lipmann, *Proc. Natl. Acad. Sci. U.S.*, **52**, 1462 (1964).
2 J. D. Watson, *Bull. Soc. Chim. Biol.*, **46**, 1399 (1964).
3 R. W. Erbe, M. M. Nau and P. Leder, *J. Mol. Biol.*, **39**, 441 (1969).
4 Y. Nishizuka and F. Lipmann, *Arch. Biochem. Biophys.*, **116**, 344 (1966).
5 F. Lipmann, *in* "The Role of Nucleotides for the Function and Conformation of Enzymes (A. Benson Symposium I)," ed. by H. M. Kalckar, H. Klenow, M. Ottesen, A. Munch-Petersen and J. H. Thaysen, Scandinavian Univ. Books, p. 292 (1969).
6 S. S. Thach and R. E. Thach, *Proc. Natl. Acad. Sci. U.S.*, **68**, 1791 (1971).
7 R. Heintz, H. McAllister, R. Arlinghaus and R. Schweet, *Cold Spring Harbor Symp. Quant. Biol.*, **31**, 633 (1966).
8 A. S. Spirin, *Cold Spring Harbor Symp. Quant. Biol.*, **34**, 197 (1969).
9 D. Chuang and M. V. Simpson, *Proc. Natl. Acad. Sci. U.S.*, **68**, 1474 (1971).
10 M. H. Schreier and H. Noll, *Proc. Natl. Acad. Sci. U.S.*, **68**, 805 (1971).
11 J. E. Allende, R. Monro and F. Lipmann, *Proc. Natl. Acad. Sci. U.S.*, **51**, 1211 (1964).
12 Y. Nishizuka and F. Lipmann, *Proc. Natl. Acad. Sci. U.S.*, **55**, 212 (1966).
13 Y. Kaziro and N. Inoue, *J. Biochem.*, **64**, 423 (1968).
14 A. Parmeggiani, *Biochem. Biophys. Res. Commun.*, **30**, 613 (1968).
15 Y. Kaziro, N. Inoue, Y. Kuriki, K. Mizumoto, M. Tanaka and M. Kawakita, *Cold Spring Harbor Symp. Quant. Biol.*, **34**, 385 (1969).
16 Y. Kuriki, N. Inoue and Y. Kaziro, *Biochim. Biophys. Acta*, **224**, 487 (1970).
17 A. Parmeggiani and E. M. Gottschalk, *Biochem. Biophys. Res. Commun.*, **35**, 861 (1969).
18 N. Brot, C. Spears and H. Weissbach, *Biochem. Biophys. Res. Commun.*, **34**, 843 (1969).
19 J. W. Bodley, F. J. Zieve and L. Lin, *J. Biol. Chem.*, **45**, 5662 (1970).
20 W. O. Godtfredsen, S. Jahnsen, H. Lorck, K. Roholt and L. Tybring, *Nature*, **193**, 987 (1962).

21 N. Tanaka, T. Kinoshita and H. Masukawa, *Biochem. Biophys. Res. Commun.*, **30**, 278 (1968).
22 J. W. Bodley, F. J. Zieve, L. Lin and S. T. Zieve, *Biochem. Biophys. Res. Commun.*, **37**, 437 (1969).
23 J. W. Bodley, F. J. Zieve, L. Lin and S. T. Zieve, *J. Biol. Chem.*, **245**, 5656 (1970).
24 A. Okura, T. Kinoshita and N. Tanaka, *Biochem. Biophys. Res. Commun.*, **41**, 1545 (1970).
25 T. Yamamoto, Y. Kuriki and Y. Tonomura, *J. Biochem.*, **72**, 1327 (1972).
26 L. Skogerson and K. Moldave, *J. Biol. Chem.*, **243**, 5354 (1968).
27 H. Gutfreund, *Disc. Faraday Soc.*, **20**, 167 (1955).
28 L. Ouellet and K. J. Laidler, *Can. J. Chem.*, **34**, 146 (1956).
29 L. Ouellet and J. A. Stewart, *Can. J. Chem.*, **37**, 737 (1959).
30 Y. Kaziro, presented at a seminar at the Institute for Protein Research, Osaka Univ., Osaka (1971).
31 M. Nomura, *Bact. Rev.*, **34**, 228 (1970).
32 K. Kischa, W. Möller and G. Stöffler, *Nature New Biology*, **233**, 62 (1971).
33 M. Meselson, M. Nomura, S. Brenner, C. Davern and D. Schlessinger, *J. Mol. Biol.*, **9**, 696 (1964).
34 W. Möller, H. Castleman and C. P. Terhorst, *FEBS Letters*, **8**, 192 (1970).
35 I. Suzuka, *J. Jap. Biochem. Soc.*, **43**, 722 (1971) (in Japanese).
36 E. Hamel and T. Nakamoto, *Federation Proc.*, **30**, 1203 Abs (1971).
37 A. A. Infante, D. Klaupiks and R. E. Davies, *Science*, **144**, 1577 (1964).
38 P. J. Garrahan and I. M. Glynn, *Nature*, **211**, 1414 (1966).
39 P. J. Garrahan and I. M. Glynn, *J. Physiol.*, **192**, 237 (1967).
40 V. L. Lew, I. M. Glynn and J. C. Ellory, *Nature*, **225**, 865 (1970).
41 I. M. Glynn and V. L. Lew, *J. Physiol.*, **207**, 393 (1970).
42 M. Makinose and W. Hasselbach, *FEBS Letters*, **12**, 271 (1971).
43 B. Borlogie, W. Hasselbach and M. Makinose, *FEBS Letters*, **12**, 267 (1971).
43a R. Panet and Z. Selinger, *Biochim. Biophys. Acta*, **255**, 34 (1972).
44 S. Yamada and Y. Tonomura, *J. Biochem.*, **71**, 1101 (1972).
45 S. Yamada, M. Sumida and Y. Tonomura, *J. Biochem.*, **72**, 1537 (1972).
46 W. Hasselbach and K. Seraydarian, *Biochem. Z.*, **345**, 159 (1966).
47 P. D. Boyer, personal communication.
47a B. A. Melandri, A. Baccarini-Melandri, A. San Pietro and H. Gest, *Science*, **174**, 514 (1971).
48 G. D. Greville, *Curr. Topics Bioenergetics*, **3**, 1 (1969).
49 D. E. Green and H. Baum, "Energy and the Mitochondrion," Academic Press, New York (1970).
50 E. Racker, "Membranes of Mitochondria and Chloroplasts," Van Nostrand (1970).
51 Y. Kagawa, *in* "Mitochondria," ed. by B. Hagihara, Asakura Shoten, Tokyo, p. 105 (1971) (in Japanese).

51a Y. Kagawa, *Biochim. Biophys. Acta*, **265**, 297 (1972).
52 P. Mitchell, *in* "The Molecular Basis of Membrane Function," ed. by D. C. Tosteson, Prentice-Hall, Englewood Cliffs, N.J., p. 483 (1969).
53 P. Mitchell, *in* "Membranes and Ion Transport," ed. by E. E. Bittar, Wiley-Interscience, London, Vol. 1, p. 192 (1970).
54 R. N. Robertson, "Protons, Electrons, Phosphorylation and Active Transport," Cambridge Monographs in Experimental Biology, Vol. 15, Cambridge Univ. Press, Cambridge (1968).
55 A. T. Jagendorf and E. Uribe, *Proc. Natl. Acad. Sci. U.S.*, **55**, 170 (1966).
55a P. C. Hinkle, J. J. Kim and E. Racker, *J. Biol. Chem.*, **247**, 1338 (1972).
56 A. L. Lehninger, E. Carafoli and C. S. Rossi, *Advan. Enzymol.*, **29**, 259 (1967).
57 R. S. Cockrell, E. J. Harris and B. C. Pressman, *Nature*, **215**, 1487 (1967).
58 R. A. Reid, J. Moyle and P. Mitchell, *Nature*, **212**, 257 (1966).
59 E. Rossi and G. F. Azzone, *Eur. J. Biochem.*, **12**, 319 (1970).
60 A. L. Lehninger, *J. Biol. Chem.*, **234**, 2187 (1959).
61 P. D. Boyer, *in* "Oxidases and Related Redox Systems," ed. by T. E. King, H. S. Mason and M. Morrison, John Wiley and Sons, New York, Vol. 2. p. 994 (1965).
62 C. R. Hackenbrock, *J. Cell Biol.*, **37**, 345 (1968).
63 D. E. Green, *Proc. Natl. Acad. Sci. U.S.*, **67**, 544 (1970).
64 J. H. Young, G. A. Blondin, G. Vanderkooi and D. E. Green, *Proc. Natl. Acad. Sci. U.S.*, **67**, 550 (1970).
64a N. E. Weber and P. V. Blair, *Biochem. Biophys. Res. Commun.*, **41**, 821 (1970).
64b A. Azzi, B. Chance, G. K. Radda and C. P. Lee, *Proc. Natl. Acad. Sci. U.S.*, **62**, 612 (1969).
65 I. J. Ryrie and A. T. Jagendorf, *J. Biol. Chem.*, **246**, 3771 (1971).
65a I. J. Ryrie and A. T. Jagendorf, *J. Biol. Chem.*, **247**, 4453 (1972).
65b R. E. McCarty, P. R. Pittman and Y. Tsuchiya, *J. Biol. Chem.*, **247**, 3048 (1972).
66 A. Bennun, *Nature New Biology*, **233**, 5 (1971).
67 R. L. Cross and P. D. Boyer, *Federation Proc.*, **30**, 1245 Abs (1971).
68 P. D. Boyer, R. L. Cross, O. Chude, A. S. Dahms and T. Kanazawa, *in* "Biochemistry and Biophysics of Mitochondrial Membranes," ed. by F. Fox, Academic Press, New York, p. 343 (1972).
69 M. Klingenberg, *in* "Mitochondrial Structure and Compartmentation," ed. by E. Quagliariello, S. Papa, E. C. Slater and J. M. Tager, Adriatica Editrise, Bari, p. 320 (1967).
70 N. Yamamoto, Y. Horiuti, K. Nishikawa and T. Horio, *J. Biochem.*, **72**, 599 (1972).
71 Y. Tonomura, T. Kanazawa and K. Sekiya, *in* "Molecular Biology; Problems and Perspectives," ed. by A. E. Braunstein, Acad. Nauk U.S.S.R., Moscow, p. 213 (1964).